Introduction to
SOFT MATTER PHYSICS

Introduction to SOFT MATTER PHYSICS

Luwei ZHOU

Fudan University, China

NEW JERSEY • LONDON • SINGAPORE • BEIJING • SHANGHAI • HONG KONG • TAIPEI • CHENNAI • TOKYO

Published by

World Scientific Publishing Co. Pte. Ltd.
5 Toh Tuck Link, Singapore 596224
USA office: 27 Warren Street, Suite 401-402, Hackensack, NJ 07601
UK office: 57 Shelton Street, Covent Garden, London WC2H 9HE

Library of Congress Cataloging-in-Publication Data
Names: Zhou, Luwei, author.
Title: Introduction to soft matter physics / Luwei Zhou (Fudan University, China).
Description: New Jersey : World Scientific, 2019. | Includes bibliographical references and index.
Identifiers: LCCN 2018035674| ISBN 9789813275096 (hardcover : alk. paper) | ISBN 981327509X (hardcover : alk. paper)
Subjects: LCSH: Soft condensed matter. | Matter--Properties.
Classification: LCC QC173.458.S62 Z46 2019 | DDC 530.4/13--dc23
LC record available at https://lccn.loc.gov/2018035674

British Library Cataloguing-in-Publication Data
A catalogue record for this book is available from the British Library.

For any available supplementary material, please visit
https://www.worldscientific.com/worldscibooks/10.1142/11124#t=suppl

Typeset by Stallion Press
Email: enquiries@stallionpress.com

Printed in Singapore

This is dedicated to my mother.

Preface

Soft matters, or complex fluids, are condensed matters between hard matters and simple fluids. Soft matter physics is rapidly developing in recent decades, not only because of its numerous research output in each branch of the subject, but also for its new branches that are emerging one after another. This book introduces the basic knowledge, some well-known research methods and techniques of the field, and it can be used as an introductory textbook for undergraduate and graduate students.

In Chapter 1, the overview of soft matter physics is introduced. The interactions between the fundamental units, such as colloidal particles, molecules or atoms, are described in Chapter 2. This textbook puts emphasis on the experimental process, as well as the development of experimental ideas and experimental techniques. Most soft matters consist of light elements, such as hydrogen, carbon or oxygen, and their neutron scattering cross-sections are much larger than their x-ray ones. Small angle neutron scattering is especially useful for detecting the structures of soft matters, so Chapter 3 is devoted to the topic of neutron scattering.

Since the interaction of soft matters is between that of hard matters and gases, the characteristics of complex systems, such as chaos and fractals, usually appear more easily in soft matters. Viewing soft matters as complex systems would help in better understanding their behaviors. These are discussed in Chapter 4.

As an example of interactions of matters and fields in soft matters, also as an example of suspensions and colloids, Chapter 5 describes the knowledge of electrorheological fluids. Not only on the physical properties of the new materials, nor only on the development of the related models, this chapter emphasizes on the important theoretical and experimental methods of solving the interactions of colloidal systems under external fields. The dynamic characteristics of such systems under external fields is a difficult point in their study; Chapter 6 introduces the work of Ping Sheng's group on solving such problems using the Onsager principle of least energy dissipation. Knowing the related theoretical and experimental methods would help readers to deal with the dynamic properties of similar systems of soft matters.

The study of granular materials is the most fascinating part of soft matter physics. Chapter 7 describes the properties of granular materials as granular liquid, solid and gas; however, readers should put their attention on the theoretical and experimental methods, as well as related physical problems.

The first edition of this textbook was published by Fudan University Press in 2011. In this new edition, the physical background of some interesting applications has been added. Those applications include liquid body armors (Chapter 2), possible relations between granular systems and earthquakes and also between creep and sand soilization (both in Chapter 7). The issue of reducing particle sedimentation in liquid is important in the fabrication of magnetorheological fluids. We will review with readers the classical derivation of Stokes' formula in Chapter 5.

The path of the development of soft matter physics shows interested readers how a strongly interdisciplinary subject is growing with great difficulty step by step. These contents draw special attention in this textbook in the hope that readers understand the basic contents and methods of soft matter physics and use them in other subjects as well.

I would like to thank Profs. Ruibao Tao and Rongjia Tao for encouraging me to enter the research field of soft matter physics, Profs. Ping Sheng, Kunquan Lu, Jixing Liu, Meiying Hou, Penger Tong, Weijia Wen, Hongru Ma, Yuqiang Yu, Jiping Huang, Wei Chen, Jianwei Zhang, Peng Tan and many others for numerous discussions, Prof. Xun Wang for reading part of the manuscript and suggesting modifications. I have enjoyed lecturing on the course of Soft Matter Physics for both undergraduate and graduate students for more than a decade, during which time the students and teaching assistants have truly enriched the textbook — they are most appreciated.

Luwei ZHOU, June 2018

Contents

Dedication v
Preface vii
List of Tables xvii
List of Figures xix

Chapter 1 Characteristics of Soft Matters 1
1.1 Why Soft Matters 1
1.1.1 Why should we study soft matter physics 1
1.1.2 The interests of soft matter physics 4
1.2 Classifications of Soft Matters 7
1.2.1 Complex fluids 7
1.2.2 Basic concepts of non-Newtonian fluids 8
1.2.3 Major characteristics of non-Newtonian fluids . . 13
1.3 Self-organization of Soft Matters 16
1.3.1 Scale invariance 17
1.3.2 Entropy driven self-organization 20
1.3.3 Measurements of depletion effect 21
1.3.4 Calculations of depletion effect 23
1.4 Modern Methods used in the Study of Complex Systems 30
References 33

Chapter 2 Basic Interactions in Soft Matters 35

2.1 Intramolecular Interactions 36
2.1.1 Ionic bonds 36
2.1.2 Covalent bonds 37
2.1.3 Metallic bonds 37
2.1.4 Hydrogen bonds 38
2.2 Intermolecular Interactions 38
2.2.1 Double-layer forces 38
2.2.2 Electric dipole interaction 41
2.2.3 Induced dipoles, polarizability 43
2.2.4 Repulsive forces 45
2.2.5 The origin of van der Waals interaction 46
2.3 Structural Forces 53
2.3.1 Wettability of colloidal particles 53
2.3.2 Lyophilic repulsive force 55
2.3.3 Slip length change on nanostuctured surface . . . 56
2.4 Hydrodynamic Interactions 59
2.4.1 Shear thickening effect 59
2.4.2 Essential role of friction in DST 60
2.4.3 Hydroclustering and jamming for an infinite system 63
References 68

Chapter 3 Structure Determination of Soft Matters 71

3.1 Why Neutrons 71
3.1.1 Advantages of neutron scattering 71
3.1.2 Discovery of neutrons 74
3.1.3 Neutron imaging 76
3.2 Neutron Diffraction 78
3.2.1 Diffraction of radiation 78
3.2.2 Wave properties of neutrons 79
3.2.3 Neutron elastic scattering 80
3.2.4 Neutron inelastic scattering 82
3.3 Structure Determination of Soft Matters 84
3.3.1 Neutron scattering of light elements 84
3.3.2 The neutron scattering of liquid 84

3.3.3 Radial distribution function g(r) of liquid 85
3.3.4 Form factor and structure factor of neutron scattering spectrum 86
3.3.5 Small angle neutron scattering 89
3.4 Optical Microscopy and Light Scattering 93
3.4.1 Structure determination with optical microscopy 93
3.4.2 Static and dynamic light scattering 94
3.4.3 Diffusing-wave spectroscopy 96
3.4.4 Applications of DWS 101
References 102

Chapter 4 Complexity of Soft Matters 105

4.1 Examples of Chaos in Soft Matters 105
4.1.1 Rheochaos 106
4.1.2 Chaos in ECG 106
4.1.3 Neural system 108
4.1.4 Self-similarity 109
4.1.5 Fractal dimension 110
4.1.6 Measurements of fractal dimension 112
4.2 Physical Mechanism of Fractals 115
4.2.1 Butterfly effect 115
4.2.2 Necessary and sufficient conditions for fractal structures 118
4.3 Quantitative Analysis of Chaos 119
4.3.1 The broad band power spectrum 119
4.3.2 The positive maximum Lyapunov exponents . . . 120
4.3.3 Conditions for deterministic chaos of time series 123
4.4 Complexity Helps Better Understanding of Soft Matters 128
4.4.1 Fractal growth in colloidal aggregation 128
4.4.2 Settling of fractal aggregates in water 128
4.4.3 Chaos helps mix microfluids 130
4.4.4 Life system is a dissipative structure 132
References 134

Chapter 5 Static Electrorheological Effect **137**

5.1 Electrorheological Effect 137
5.1.1 Basic phenomena 137
5.1.2 Static particle structure of ER fluid 139
5.1.3 Colloidal electrorheological effect 142
5.1.4 Polarization types and electric double layer . . . 143
5.2 Suspension ER Models 146
5.2.1 Dielectric ER models 146
5.2.2 Conduction ER models 166
5.3 Colloidal ER Models 175
5.3.1 Giant ER effect 175
5.3.2 Polar molecule ER effect 178
5.4 Sedimentation of Particles 191
5.4.1 Importance of sedimentation study 191
5.4.2 Derivation of Stokes' formula 192
5.4.3 Derivation of Stokes' law 196
References 200

Chapter 6 Dynamic Electrorheological Effects **201**

6.1 Dynamic Behaviors of ER Fluids 201
6.1.1 Dynamic phenomena 201
6.1.2 Lorentz local field 203
6.1.3 Shear stress under static shear flow
and transient electric field 209
6.2 Lamellar Structure 212
6.2.1 Lamellar structure stability under shearing 212
6.2.2 Criterion of ER activity 213
6.3 Two-Fluid Model of Continuous Phase 222
6.3.1 Two-fluid model of continuous phase 222
6.3.2 Electric field to a quiescent suspension 225
6.3.3 Electric field to a flowing suspension 227
6.4 Onsager Principle of Least Energy Dissipation 229
6.4.1 Derivation of the Onsager principle 229
6.4.2 Establishment of the Navier–Stokes
equations 232
6.4.3 Numerical calculation 234

6.5 Shear Banding 237
6.5.1 Experimental phenomena of shear banding 237
6.5.2 Constitutive models of shear banding 239
References 243

Chapter 7 Granular Systems 245

7.1 Introduction 245
7.2 Granular Fluid-Pattern Formation 248
7.2.1 Vibration convection 248
7.2.2 2D pattern formation 251
7.2.3 3D pattern formation 259
7.3 Granular Flow 262
7.3.1 Jamming of granular flow 262
7.3.2 Self organization criticality 266
7.4 Grain Segregation 271
7.4.1 Granular liquids — stratification 272
7.4.2 Rotation drum 273
7.4.3 Segregation by vertical vibration — Brazil nut problem 279
7.5 Granular Solid 281
7.5.1 Counterintuitive phenomenon: construction history 281
7.5.2 Thermodynamics of sands 286
7.6 Granular Gas 291
7.6.1 Experiment of sand as Maxwell's demon 291
7.6.2 Model of flux function 292
7.7 Applications on Granular Systems 296
7.7.1 Earthquake and granular systems 296
7.7.2 Creep and sand soilization 305
7.7.3 Final words: granular system is a complex fluid 314
References 316

List of Tables

2.1 Typical molecular interactions in vacuum. 47

3.1 A comparison of characteristics of x-ray and neutron diffraction. 75

4.1 Three basic models of fractal growth. 129

7.1 Particle size classification. 246

List of Figures

1.1 Two types of typical shear flow: a plane Couette flow (left) and a tube Poiseuille flow (right). 10
1.2 The shear stress τ– shear rate $\dot{\gamma}$ relations. Curve 1: Bingham plastic; Curve 2: Dilatant; Curve 3: Newtonian; Curve 4: pseudoplastic; Curve 5: cytological fluid. 12
1.3 Newtonian (a, c) vs. non-Newtonian (b, d) fluid. 13
1.4 Newtonian (a, c) vs. non-Newtonian (b, d) fluid. 14
1.5 Newtonian (a, c) vs. non-Newtonian (b, d) fluid. 15
1.6 Newtonian (a, c) vs. non-Newtonian (b, d) fluid. 15
1.7 Newtonian (a, c) vs. non-Newtonian (b, d) fluid. 16
1.8 Particles in a ferrofluid between two pieces of glass form their structure under a magnetic field. 17
1.9 Shock wave after a nuclear explosion gives an example of swelling symmetry. 18
1.10 Various animals attempting to follow a scaling law drawn by P.G. de Gennes. 18
1.11 The entropy-driven ordering. It is interesting to see that the natural ordering process leads to a more uneven, rather than a more even configuration when entropy is increased. 21
1.12 (a) The trace of a large sphere on a focal plane is observed under a microscope. (b) If there are no small spheres, the large sphere freely diffuses all available space inside a vesicle. The lighter the color of a spot is (or the larger N, according to the vertical sidebar), the more times the

large sphere passes that spot. (c)'If the vesicle also contains small spheres, the large sphere is attracted to the wall. 22
1.13 The pair interaction energy $F(r)$, which measures the depletion potential of two large spheres in the system as functions of large particle separation r, when small sphere volume fraction changes from 0 to 0.42. 24
1.14 Free energy difference of a system consisting of two large spheres and many small ones when the gap between the two large spheres is D, and the radius ratio r/R of the small spheres over larger ones is 0.2, $L_x = L_y = 18r, L_z = 36r$. (a) $\eta = 0.341, N = 867$; (b) $\eta = 0.229, N = 580$ and (c) $\eta = 0.116, N = 294$. See text for the definitions. 28
1.15 Depletion effect of a hard (but flexible) tube of radius t in a solution of smaller hard spheres of radius r_s leads the tube to form a helical conformation. 29

2.1 An HCl molecule is formed with a covalent bond between Cl^- and H^+ ions. 41
2.2 Ion hydration is caused by an ion-dipole interaction when an ion is surrounded by water molecules. q is the angle between the direction of dipole moment and the line joining the dipole center and the ion. 42
2.3 (a) Original electric dipole; (b) formation of an induced dipole when there is an electric field E either applied or caused by a neighboring molecule. 43
2.4 L-J function incorporates many-body effects. 51
2.5 The expressions of the non-retarded van der Waals potentials for the micro- and macroscopic bodies. The Hamaker constant A is defined as $A = \pi^2 \rho_1 \rho_2$ where ρ_1 and ρ_2 are the number of atoms per unit volume in the two bodies, D is the gap between the two bodies, and C is the coefficient in the atom-atom pair potential. 52
2.6 Schematic representation of a colloidal system (a) and a wetting film (b). 53

2.7 Fluid speed profiles. At the fluid-solid boundary, slip length λ is defined as the distance between the point where the fluid speed is zero and the boundary. The slip length is zero for no-slip case, λ for partial slip case and infinitive for the perfect slip case. 56
2.8 Wenzel state A and Cassie state B with air padding. . . 57
2.9 (a), (b) Shear rate and stress dependence of the relative viscosity η_r, respectively. The open and filled symbols indicate the results of $n = 512$ and 2048, respectively. The friction coefficient is $\mu = 1$ except for the dashed and dotted curves, for which $\mu = 0.1$ and 0, respectively. (c) DST (solid thick curve) and CST (dashed thick curve) are shown in the phase diagram. 62
2.10 Particle contacts are visualized as bonds at the two shear rates $\dot{\Gamma} = 0.05$ and 0.1 exhibiting low and high viscosity states, respectively ($\mu = 1$, $\phi = 0.56$, $n = 2048$). 63
2.11 Schematic picture of structures in the flow. Gradient is perpendicular to the flow direction indicated by two long arrows in opposite direction. 64
2.12 Plot of $E(\phi_v)\eta_0$ against ϕ_v obtained from the fit to the stress-strain data. The units are chosen so that $a = \eta_0 = 1 = \dot{\gamma}$, thus the physical stress s is in units of $\eta_0\dot{\gamma}$. The circles for 700 particles and the diamond for 1400 particles show good agreement. The solid line for comparison is a theoretical prediction, Eq. (2.57). (b) Same for $\sigma(0)\dot{\gamma}\eta_0$; the theoretical prediction is from Eq. (2.58). 68

3.1 Phase change in a scattering. 79
3.2 Different particle sizes as compared to the wavelength of the probe wave lead to different 2θ, where the scattering wave is out of phase. 90
3.3 Intensity of the scattering wave versus 2θ for small (curve 1) and large (curve 2) particle sizes as compared to use of the wavelength. Curve 3 is for the intermediate particle size. 90

3.4 Projections of neutron scattering density of a derivant of B_{12}. 92
3.5 N scatterings in a photon path. 98

4.1 (a-c) Stress relaxation dynamics for a shear-thinning wormlike micellar system with certain amount of NaCl for different shear rate; and (d) the broad band Fourier power spectrum of (a) shows the chaotic characteristics of the system. 107
4.2 The EEGs of normal people (upper) and a patient of epileptic disease (lower). 108
4.3 A self-similar pattern grown from the branch center using a diffusion-limited monomer-aggregation. 109
4.4 The Sierpinski triangle. 110
4.5 An example of measuring fractal dimension of a pattern shown. 113
4.6 Measuring fractal dimension of a pattern shown with box-counting. 114
4.7 E. Lorenz's curve of calculation of atmospheric convection: A little error would bring tremendous disaster. . . . 116
4.8 Lorenz system solution projected onto X-Z and Y-Z planes. 117
4.9 The orbits and attractors of steady solutions. The left diagram shows the orbits of a damped pendulum. Four x's denote four different initial positions of the pendulum, and all the orbits come to an attractor of a steady point A (at the origin). The right diagram shows the orbits of a clock pendulum, and the attractor is a steady loop A. 117
4.10 A schematic representation of the evolution and replacement procedure used to estimate Lyapunov exponents from experimental data. The largest Lyapunov exponent is computed from the growth of length elements. 123
4.11 Decorrelation time is the first zero of an autocorrelation function. 125

4.12 The trajectories of the dynamics plotted in phase space $x(t-\tau)$ vs $x(\tau)$ for (a) Lorenz chaos and (b) randomness. 126
4.13 Λ is the indicator to determine whether a time series is deterministically chaotic or randomized. A system is deterministic chaotic when the weighted average of V_j over all the occupied boxes $\Lambda = 1$, and random when $\Lambda = 0$. 127
4.14 Mean Lyapunov exponent distribution in the $A_p \sim \omega$(Hz) parameter space for the mixed microfluids. 131

5.1 Particles are distributed randomly under zero electric field (a), when the electric field increases, ER particle polarization increases, particles aggregate into "chains" (b, c) and "columns" (d). The black blocks at the right and left boundaries of each panel are electrodes. 140
5.2 Laser diffraction pattern of glass spheres. 141
5.3 The body-centered structure of glass spheres. 142
5.4 Diagram of electric double layer. 145
5.5 Interaction between two spherical particles due to electric field polarization. 149
5.6 Two closed dielectric spheres A and B in spherical coordinate systems (r,θ) and (R,Θ), respectively. 152
5.7 Comparison between the experimental results and theoretical calculations based on the different modeling. 157
5.8 Particle structure is sheared by shear angle θ. The maximum stress at a certain strain θ is defined as yield stress. 164
5.9 β_{eff}^2 as a function of $\omega_e t_{\text{MW}}$ for (a) $\beta_d = 10\beta_f$ and (b) $\beta_d = 0.1\beta_f$. 170
5.10 Left: A particle configuration in calculation of electric potential between particles; Right: The electric potential between particles when the particles are treated as equipotential spheres. 171
5.11 Left: Electric field distribution when oil conductivity is considered; Right: Forces at different particle distances counted for different models. 172

5.12 (a) The two parallel plates are used to explain that when the distance between the two electrodes t_L is very small, there is a limitation in the field increment between particles; (b) spherical particle arrangement, where t_L is the particle distance, and t_L changes with position x. When $x < \delta$, the maximum value of electric field is E_m. If E_0 is the average external field ($E_0 = V_0/2R$), then the field enhancement factor E_m/E_0 is given by the ratio of the contact resistance $R_c(x < \delta)$ and surface resistance $R_s(x > \delta)$. 174

5.13 The static yield stress curves for two volume fractions, measured under DC electric fields. The corresponding current densities are shown in the inset. The current density J is below $4\,\mu\,\mathrm{A\,cm^{-2}}$ at $E < 2\,\mathrm{kV\,mm^{-1}}$ for the 30% sample. 176

5.14 Size effect of GER fluids. 177

5.15 The yield stress of some polar molecule electrorheological (PM-ER) fluids. Triangles indicate nano-TiO2 particles adsorbed with $(NH_2)_2CO$ molecules; circles are for nano-Ca-Ti-O with C=O, O-H groups; squares are sintered Ca-Ti-O particles. 178

5.16 The polar molecules aligned in the gap of the particles under zero field (left) and under an electric field (right). 179

5.17 A simplification of a water bridge between two plates. Gap thickness is D and the radius of the water bridge is W_{neck}. 182

5.18 (a) Experimental result for yield stresses y and $y_0 \cdot y$: TiO_2 modified by polar molecules (with liquid-solid ratio 0.45 ml/g or mol ratio 1.5); y_0: bare TiO_2. (b) Theoretical fitting experimental results. The inset shows the number n of polar molecules (dipoles) within the gap between the two particles as a linear function of external electric fields. 186

5.19 The difference between wetting force $|F_w|$ and van der Waals force $|F_v|$. 190

6.1 (a) Arrangement of two metallic spheres and $LiNbO_3$ (LN) crystal. (Not to scale.) (b) The experimental data (symbols) and the theoretical fitting (lines) of the LLF E_L at the center P. (c) Relaxation times τ vs. rotating speed, also in the inset in a logarithmic scale. 204

6.2 The diameter of the above circular diagrams is 35 mm. (a)–(f) are the top views at 1.0, 2.0, 3.0, 4.0, 5.0 and 9.0 s after an electric field of 1.6 kV/mm is applied in the experiment. The dark area is the polarized particle aggregation area. 213

6.3 The lamellar structure of an ER fluid at equilibrium state formed between two relatively rotating discs at different electric and shear fields. 214

6.4 A possible phase diagram of the ER fluid on the plane of electric field and shear rate. 214

6.5 Simulation of models for flow of ER fluids simultaneously under electric and shear fields for a couette and parallel disc flow geometrics. (a) and (b) show how a chain breaking model might appear; (c) and (d) show how the alternative model presented here would appear. 216

6.6 Dry $BaTiO_3$ ER fluid (a, b) and wet $BaTiO_3$ ER fluid (c, d). Particle structure formation when only electric field is applied (a, c) or when both electric and shear fields are applied (b, d). A thick column may be a sign of active ER effect in an static case provided that it also shows lamellar structure under both electric and shear fields. 217

6.7 The formation of particle structure of polystyrene ER fluid (a, b) and sulfonated polystyrene ER fluid (c, d) when only an electric field is applied (a, c) or both electric and shear fields are simultaneously applied (b, d). . . 218

6.8 Schematic construction of two-phase shear banded flow. The units of shear stress and shear rate are k_BT/a^3 and $\sqrt{k_BT/ma^2}$, respectively. 218

6.9 A flowing-hexagonal layer (FHL) flow state on the x-z plane at left and x-y plane at right. The applied electric field is along the z-axis and the flow is along x. 219

6.10 The bottom view of the lamellar structure of the PMER fluid at different applied electric field when rotational speed of the upper electrode is 300 rpm. 220

6.11 The shear stress of polar-molecules dominant ER fluids at different fields when the rotating speeds of one disc electrode is 300 rpm and the other, zero. Part of its lamellar structures at certain electric fields are shown in Fig. 6.10. 220

6.12 Particle concentration and linear speed v of particle rings in the bottom image of lamellar structure of the PMER fluids at different fields while the rotational speed is kept at 330 rpm. 221

6.13 (a) Initial structure and (b) structure after the application of an electric field to a quiescent suspension ($tM/L^2 = 0.05$). Dark and light parts represent positive and negative concentration fluctuations, respectively. . . 226

6.14 Contour plots of sL^2/M as a function of k_xL and k_yL for (a) $\dot{\gamma}L^2/M = 10^3$ and (b) $\dot{\gamma}L^2/M = 10^4$. Stable and unstable regions are denoted by S and U, respectively. . 228

6.15 (a) Initial structure and (b) structure after the application of an electric field to a flowing suspension ($\dot{\gamma}L^2/M = 10^4$, $tM/L^2 = 0.05$). Dark and light parts represent positive and negative concentration fluctuations, respectively. 228

6.16 Unstable fluctuations parallel to the applied electric field (along z-axis) would rotate by simple shear flow and become stable fluctuations perpendicular to the electric field. 229

6.17 The shear stress of an ER fluid changed with shear rate numerically calculated on the basis of the Onsager principle of least energy dissipation. 235

6.18 The result of numerical calculation depicts the lower and upper lamellar structure of a colloidal ER fluid when applied electric field is 1000 V/mm and rotating speed of one electrode is 300 rpm. The gap between the electrodes is 1 mm, and the radius of the electrodes is 17.5 mm (the unit of the vertical axes is m). 237

6.19 The basic phenomenon of gradient shear banding. Homogeneous flow is unstable to the formation of macroscopic coexisting bands of unequal shear rates $\gamma\mathrm{dot}_1$, $\gamma\mathrm{dot}_2$. For a comparison of gradient banding and vorticity banding, see Fig. 6.20. 238

6.20 (a) and (b) [(c) and (d)] are the candidate constitutive curves (*thin lines*) of average total stress $\bar{T}_{xy}$ vs. average shear rate $\bar{\dot{\gamma}} \leftarrow$ for shear thinning and shear thickening flow, respectively, in case of gradient [vorticity] shear banded flows. The *thick lines* could correspond to shear banding under either gradient (in (a) and (b)) or vorticity (in (c) and (d)) shear banding conditions. In all cases, homogeneous states could be observed for imposed conditions corresponding to the *thin solid lines*, while inhomogeneous shear-banding states could be observed for imposed conditions corresponding to the *thick solid lines*, while the *dashed lines* are putative unstable steady flows that must be resolved into inhomogeneous shear banding states. The upper (lower) left diagram shows schematically the arrangement of the gradient (vorticity) shear bands in curved Couette flow. 238

6.21 (a) Homogeneous constitutive curve and steady state flow curve of the non local Johnson–Segalman model. (b) Corresponding steady state shear banded profile predicted by one-dimensional calculations. Shown are the shear rate and the various components of the viscoelastic stress. All quantities here are reduced ones and are dimensionless, and y is a normalized distance between the two plates (see Fig. 6.19 (b)). 240

6.22 Sketch map of Eq. (6.70) with different value of η. a^2 is set to be 0.1, and other parameters are set to 1. 242

7.1 The apparent mass vs. filling mass, the straight line indicates ρgh. 247
7.2 The relation of amplitude and frequency shows a power law. 249
7.3 Vibration is effected by the air pressure P in the container. 249
7.4 (a) Patterns formed from vibrations with different frequency, acceleration and particle size; (b) Oscillons can be formed under certain condition. 252
7.5 Evolution of patterns in time. (a)–(d): The real images of the free surface of granular layers for $\Gamma_i = 2.4$ and $\Gamma_f = 2.8$ at t = 2 (a), $t = 10$ (b), $t = 200$ (c), and $t = 1000$ periods after quench (d). The bright parts correspond to the crests of the free surface, and the dark parts correspond to the troughs of the free surface. (e)–(h): The images of the numerical results of the 2D Swift–Hohenberg equation in time for $\varepsilon = 0.2$. 253
7.6 2D linear pattern of granules under vertical vibration when $N = 1$. 257
7.7 Two orthogonal waves are superimposed to result in a square pattern. Red means the maximum value and blue, the minimum. 258
7.8 Clearly, hexagons or triangles, depending on the relative phases of the amplitude A, can be found in the pattern. . 259
7.9 Quasi-periodic pattern when $N = 4$. 260
7.10 The result of the computer simulation of the pattern formation under vertical vibration. 261
7.11 Phase diagram obtained in the experiments. The parameter values for the comparison in Fig. 7.10 of patterns from the simulations and experiments are indicated by (a) through (h). The transitions from a flat layer to square patterns are hysteretic: solid lines denote the transition for increasing Γ, while dotted lines denote it for decreasing Γ. Shaded areas show transitional regions between stripes and squares. 262

7.12 Dynamic phenomena of flow granular materials in a hopper. 263

7.13 The left side is a ketch of the apparatus. The right side shows flow rates Q_{A2} and Q_B vs. V. Q_B is essentially constant; Q_{A2} decreases monotonically with increasing V. Regime B disappears at a voltage $Vc = 2.0\,kV$. The two curves meet at V_0. 264

7.14 The change of the average flux of flowing mass as function of an external electric field. 266

7.15 Sketch diagram of a 1-dimentional sand pile. 268

7.16 A process of local avalanches after a sand grain drops onto the center of this 5×5 2D "sand pile". 269

7.17 Self-organized critical state of minimally stable clusters for a 100×100 array. 270

7.18 Distribution of cluster sizes at criticality in a 2D system of 50×50 array, averaged over 200 samples. 271

7.19 Stratification. Courtesy of. 272

7.20 Explanation of repose angle for different cases. 273

7.21 (a) Dependence of the repose angle on the two types of rolling grains on the concentration of the surface of large grains ϕ_2. An essential ingredient to obtain stratification is that $\theta_{22} > \theta_{11}$. For the numerical integration, we use the linear interpolation between $\phi_2 = 0$ and $\phi_2 = 1$ as plotted here. (b) Picture with the different quantities appearing in the text. The dashed circle is the kink. 273

7.22 A cross section of segregation of a mixture of sugar (white) and iron power (black) in a horizontal drum with 25 cm diameter. The top two photos show a radial segregation when the drum is under a slow rotation of 0.54 rpm when stratification can be observed. The bottom two photos show how stratification disappears at a fast rotation (>3 rpm) while radial segregation remains. 275

7.23 An axial segregation in a drum rotating horizontally around its axis. 276

7.24 10% suspension (a) before and (b) after shearing, with the inner cylinder rotating at 9 rpm when the Couette is filled up to 95% of the available gap volume. 277
7.25 A 15% suspension at the 95% fill level is sheared at (a) 2.5 rpm (b) 9 rpm. 277
7.26 The mixture of 15% suspension (a) 50% fill level (b) 95% fill level, sheared at 9 rpm. 278
7.27 Time-dependent band formation: 15% suspension, 95% fill level sheared at 9 rpm after (a) 1 min.; (b) 20 min; (c) 3 h. 278
7.28 The axial segregation. 279
7.29 While the upper packing shows an energy difference, the lower sphere packing shows that, as far as energy is concerned, neither configuration is more favorable than the other. 280
7.30 The triangular symmetry of the upper part is greatly disturbed. 281
7.31 (a) Sandpiles prepared by point source have a "dip" in the center of the bottom (b) Sandpiles prepared by extended source have no "dip". 282
7.32 The force chains in a 2D (left) and a 3D sand-pile (right). Right: 3 mm diameter Pyrex glass beads in $70 \times 70 \times 40\,\mathrm{mm}^3$ box saturated by glycerol and water mixture under 200 N compression are viewed using the photoelastic method. 283
7.33 An experiment that shows a force chain. 284
7.34 (a) The stress profile at different depths for ordered packing. Near the bottom, there are two diffusive peaks; (b) Stress profile at different depths for disordered packing; the inset shows the linear dependence of peak width on depth. 284
7.35 (a) The 6-fold preferred contact angle direction. (b) The stress profile (point source) with the distance from bottom and a "dip" at the bottom center. 285

7.36 (a) The randomly distributed contact angle direction. (b) The stress profile (extended source) with the distance from bottom and no "dip" at the bottom center. 285
7.37 (a) A schematic drawing of the apparatus. (b) and (c) The packing fraction ρ from capacitors 4 and 2, respectively, vs. tap time that is offset by one. The curves are parameterized by vibration strength Γ. 287
7.38 The dependence of ρ on the vibration history. 288
7.39 Fluctuations in the volume density $\delta\rho(t) = \rho(t) - \rho_{ss}$ after the system has had sufficient time to relax to a steady-state density ρ_{ss}. 289
7.40 The average variance of the experimental volume fluctuations (open symbols) as a function of the steady-state volume. 289
7.41 Sand as Maxwell's demon. 292
7.42 Bifurcation diagram for the 2-box system ($k = 1, 2$). The solid line represents stable and the dashed line unstable equilibria of the flux model. The dots are experimental measurements. In both cases, the transition to the clustering state is seen to take place via a pitchfork bifurcation. 295
7.43 (a) Stress-strain relations for free and confined compressions, and τ_y and τ_y' denote the yield stresses of a material under free and confined compressions, respectively; (b) the differential stress $\sigma_1 - \sigma_3$ under a confined compression; (c) a scratch of confined shear action, where σ_s, σ_n and σ_r are external stress, normal pressure and resistive blocking stress, respectively. 299
7.44 Sketch of various forces applied on part of the crust. Arrows starting from the dash curve represent tectonic forces; M and N respectively are the frontier and posterior of the rock block. 303
7.45 Illustration of the primary, secondary and tertiary phases of the creep. 306

7.46 Kelvin and Voigt rheological model with demonstration of the stress σ and strain ε behavior. 307
7.47 Basic scheme of the oedometer. 311
7.48 Sketch of direct shear apparatus. 312
7.49 Contact constraint is exactly satisfied. 313
7.50 Left: Creep strains of soilized sand in its ODI state with different water content. Right: Typical curve of the Kelvin model in viscoelasticity theory. 314

Chapter 1

Characteristics of Soft Matters

1.1 Why Soft Matters

1.1.1 *Why should we study soft matter physics*

French scientist Pierre-Gilles de Gennes won the Nobel Prize in Physics in 1991 "for discovering that methods developed for studying order phenomena in simple systems can be generalized to more complex forms of matter, in particular to liquid crystals and polymers" as described by the Nobel Prize committee. His Nobel Lecture was entitled Soft Matter [1.1]. The French version of soft matter, *matière molle,* was invented as a joke by M. Veyssié in Orsay around 1970 [1.2]. The term *matière molle* has a double meaning in French: both soft matter and useless or weak matter. "Soft matter" was introduced to describe something that goes plastic with soap bubbles, from gels, elastomers, liquid crystals, cosmetic creams, mud, ceramic paste, etc. Soft matters are usually called complex fluids in North America.

Soft matters refer to the soft condensed matters — the materials other than those in gas and solid states, but usually not including simple fluids. From the point of view of materials, soft matter physics is concerned with physical principles governing the behaviors of foams, liquid crystals, polymers, colloidal dispersions, micro emulsion, micelle and various types of biological liquids, suspensions, and even granular materials, because of their wide applications.

What are the major differences between "soft" and "hard" matters?

A common feature that all soft matters or complex fluids have is that they respond a lot even under a small action which is not necessarily mechanical. Soft matters are usually very sensitive to external actions, while "hard" matters are usually not sensitive to them. P.G. de Gennes likes using the example of the Indian boot to explain the essence of soft matters. Indians take the sap from the Hevea tree and smear their feet with it. After 20 minutes, under the affect of oxygen, the liquid sap, or latex, coagulates and becomes a solid boot. Oxygen was later substituted with sulfur to make a stable boot. It has been noted that only 1 in 200 carbon atoms in the latex react with a sulfur atom, and yet the sulfuration is an extremely weak chemical reaction. Nevertheless, the liquid matter becomes solid matter. A drop of friction reducer made of a kind of polymer would greatly reduce the friction of water flow in a water hose, and greatly increase the height of a water column out of a water hose. Small causes create large effects: soft matter can be transformed by weak external actions. P. G. de Gennes says "this is the central and fundamental definition of soft matter." [1.3]

The reason why soft matters respond strongly to weak actions is that entropy plays an important role in soft matters, while in hard matters, inner energy does so. The basic characteristics of soft matters include nonlinear responses to external forces, self-organization and dilation symmetry, which can be seen in the later part of this chapter.

As pointed out by the Royal Swedish Academy of Sciences in the press release about the Nobel Prize in Physics in 1991, "P. G. de Gennes has by some judges been called 'the Isaac Newton of our time'. The reason for this highly appreciative epithet is probably that de Gennes has succeeded in perceiving common features in order phenomena in very widely differing physical systems, and has been able to formulate rules for how such systems move from order to disorder. Some of the systems de Gennes has treated have been so complicated that few physicists had earlier thought it possible to incorporate them at all in a general physical description. Physicists often take pride in dealing with systems that are as simple and 'pure' as possible, but de Gennes' work has shown that even 'untidy' physical systems can successfully be described in general terms. In this way he has opened new fields in physics and stimulated a great deal of theoretical and experimental work in these fields. While this is pure research, it has also meant the laying of a more solid foundation for the technical exploitation of the materials mentioned here: liquid crystals and polymers." I. Lindgren of the Royal Swedish Academy of Sciences said in his Presentation Speech (December 10, 1991) that "the major progress in science is often made by transferring knowledge from one discipline to another. Only few people have sufficiently deep insight and sufficient overview to carry out this process. De Gennes is definitely one of them."

Facing such honor, the reply of P.G. de Gennes was very calm: in his Nobel Lecture in 1991 he quoted "a poem from an experiment on soft matter" by Boudin from the following drawing: Have fun on sea and land/Unhappy it is to become famous/Riches, honors, false glitters of this world/All is but soap bubbles [1.1].

The aim of condensed matter physics is to understand the collective properties of large assemblies of atoms and molecules in terms of interactions between their component parts. Soft condensed matter physics concerns the study of the structure, physical properties and electronic states of soft matters. It includes also the critical behaviors of liquid, such as liquid-solid, liquid-gas and metal-nonmetal transition of liquid. Soft matter physics studies surface, interface and wetting of liquid, transportation and properties of liquid in porous media.

Why should we study soft matters? As pointed out by Prof. Feng Duan, the frontiers of physics in the 21st century are the interaction of strongly correlated electrons, microstructure of nanometer-sized materials and soft condensed matters. The wide potential applications of soft matters and the exploration of new science have become two major reasons for increasing interest in the study of soft condensed matter physics.

P. G. de Gennes pointed out [1.1] that "granular matter is a new type of condensed matter, as fundamental as a liquid or a solid and showing in fact two states: one fluid-like, one solid-like. But there is as yet no consensus on the description of these two states! Granular matter, in 1998, is at the level of solid-state physics in 1930". It has been found that soft matter physics is a brand new area of physics, where one may find new phenomena that cannot be explained with current knowledge of physics. For instance, in Chapter 7, on granular systems, we will mention that axial segregation of granular material of different sizes in a cylinder remains unexplained.

Soft matters are complex systems, which also show many types of nonlinear effects. This is also a reason why people are interested in the study of soft matters.

1.1.2 *The interests of soft matter physics*

The amazing aspect of any science is that it is full of unknowns. In a young discipline such as soft condensed matter physics, unknowns as well as unsolved problems occupy the majority of the subject.

(1) Nature of glasses and the glass transition

Glass is an amorphous solid that exhibits a glass transition [1.4]. P.W. Anderson says that the deepest and most interesting unsolved problem in solid state theory is probably the theory of the nature of glass and the glass transition [1.5]. Glass state is strongly dependent on the processing history and it has complicated aging effects. Analogous freezing arises in colloids and magnetic materials. It is not completely known what the physical principles governing these various frozen fluids are. In some cases, a flowing state can be restored by exceeding a threshold of external force. In others, such a force seemingly creates a glass where none was before ("jamming").

Glass transitions are also full of mysteries. It is still not clear when, and why, stressed glasses melt and freeze. An argument claims that the behavior of our familiar glasses is produced by partially realized phase transitions [1.6, 1.7]. There are two such transitions. Higher temperature transitions drop the fluid into a glassy state in which the system permits jumping back and forth between some sorts of configurations but greatly inhibits jumps to other configurations. As the system ages, it slowly finds more favorable configurations. However, it cannot directly return to the previous, less favorable, ones. If the temperature were lowered still further, the system would undergo another phase transition and fall into a disordered, low temperature state. However, this lower temperature transition is almost always preempted by a drop into a solid configuration. This conceptual framework was used to describe how structures form within the glasses, how these structures get jammed, how glasses age, and how the material retains memory of its past.

The dynamics of glasses is an interesting area of research. Mode coupling theory [1.6] starts from a nonlinear feedback for supercooled liquid, and it is successful in describing glass transition in colloids. It predicts an ideal glass transition temperature, which has not been found in experiments.

The coupled dynamic molecular string model [1.7] assumes that moving units in glass are dynamically correlated molecular strings (super-molecules). This model is based on the observation of dynamical heterogeneity, or so-called "cooperatively rearranging regions" in glass systems. There is structural inhomogeneity in glass. Starting from a reduced Hamiltonian, a process of renormalization turns a strongly correlated super-cooled liquid into a normal liquid of super-molecules that are weakly correlated. The string model describes the relaxation dynamics of super-cooled liquid state well.

Researchers try to quantitatively compare all existing models with experiments to develop a general model that can describe all major behaviors of glasses and the glass transition.

(2) What principles govern the organization of matter away from equilibrium? [1.8]

Physics these days is mostly describing matter in equilibrium where the rule of minimum of free energy, from principles of thermodynamics and statistical mechanics, governs the static or slow-moving world. However, in dynamic systems where matters move fast and systems are in a non-equilibrium state, different rules apply to describe the state and motion of matters. In a state not too far from equilibrium, the Onsager principle, which requires a minimum Onsager action functional, a summation of time derivative of free energy dissipation rate, will dominate the behavior of the system.

If a system is far from an equilibrium state, the world becomes more complicated, and we will see phenomena such as turbulent fluid flow, waves at the sea-shore, electricity driven through an atom, and even life itself. It is still not certain whether it is possible to make a theoretical model to describe the statistics of a turbulent flow (in particular, its internal structures). Also, under what conditions do smooth solutions to the Navier–Stokes equations exist? This is probably the last unsolved problem in Classical or Newtonian Physics [1.8].

(3) What principles govern the flow of granular materials? [1.8]

Granular materials include sand in dunes, snow (or rocks) in avalanches, sediments on the sea floor, and powders used commercially e.g. to make pharmaceutical pills. As the grains collide with each other they dissipate energy, causing new and unpredictable behavior. Granular materials can sustain ripples and waves, including shock waves; can evolve spontaneous avalanches; and can un-mix (rather than mix, as fluids would) when stirred together. Even at rest, the grains can carry subtly imprinted memories of earlier flows, through the particular set of positions and contacts that remain. A deeper knowledge of the physical principles of granular flow will help us understand how the earth's surface came to be as it is.

(4) Can statistical mechanics be applied to a system as complex as the living cell? [1.8]

Statistical mechanics describes the emergent properties of large collections of atoms and molecules caused by thermal excitation. Such excitation is certainly present in living cells, but there is much else happening, too: the processes of life drive the system far from a state of thermal equilibrium. The growing field of non-equilibrium statistical mechanics addresses these problems, but which aspects of cellular life can it help us to understand? Are some of these aspects just too complicated for the "physicist's view" — that simplification is the first step towards understanding — to be useful?

(5) What are the physical principles of biological self-organization? [1.8]

We only partially know how DNA encodes the structure of proteins; while the "standard model" works for bacteria and viruses, we are now realizing that, for multi-cellular organisms the transcription from DNA to mRNA is a highly "edited" process. What is the mechanism that controls this editing process? How does DNA encode the software and the large-scale organization of an organism? At a higher level, how do 10^9 bits of information in the human genome encode the information for the 10^{11} neurons and the 1015 synapses of the human brain?

There are many more other unsolved problems in soft matter physics. An example would be the confirmation and the cause of sonoluminescence, which is a phenomenon of an emission of short bursts of light from imploding bubbles in a liquid when excited by sound [1.9].

Another example is the mechanism of formation of singularities in collective matter and in space-time. One basic behavior of matter, especially tenuous matter, is its characteristic of spontaneously creating sharp, singular spatial structures. The vortex in a draining sink and a crumpled sheet of paper are two examples. Scientists want to find out what the range of this singularity-forming behavior is; to what degree does it condense the energy of a system into the singularity, and what analogous singularities might occur in space-time? [1.8]

1.2 Classifications of Soft Matters

1.2.1 *Complex fluids*

Simple fluid is a fluid in which the elements of the fluid act independently. This is not true of ionic fluids such as soap and ketchup. It is generally a good simplification for polymer melts and motor oil. A kind of simple fluid is a solution where there are small molecules, a fraction of a nanometer in size, that are close to each other with the distance of a fraction of a nanometer. The interaction is close to thermal energy — about the order of a fraction of an eV at room temperature ($1\,\text{eV} = 1.60207 \times 10^{-19}$ J; the interaction of gas $\ll$ thermal energy, and that of solid $\gg$ thermal energy.) Fluid flows easily under shearing: the time for a molecule to move to its neighbor (= collision time) is about 10^{-12} s — if a molecule has a mass of 20 amu, the distance between two molecules is 0.2 nm.

We may take a closer look at the inside of complex fluid to find out why there are so many differences between simple and complex fluids.

In some fluid, the molecules and composites are as large as several nanometers, for instance, non-Newtonian fluid model airplane glue is a complex fluid, and it is also called structured fluid. In North America, it is often called soft matter. In soft matter, chain-like long molecules melt in a small molecule solution, and the mass of long molecules are much larger than that of solution molecules. This type of solution of macro molecules is a complex fluid. The existence of these long molecules is also the reason for tubeless siphonage.

Soft matter can be classified by size of fluid components. A *colloid* is a fluid that consists of solid particles, liquid drops or gas bubbles of 1–100 nm in liquid. Some examples of colloids with solid particles are paint (TiO_2 particles), cellulose (pulp), ice cream and ferrofluid. Colloids with liquid particles are milk (fat in water), egg yolk salad sauce, dish detergent and so on. An example of colloids with gas particles is foam. *Suspension* is a fluid that consists of solid particles larger than 1 μm in liquid (electrorheological fluids or magnetorheological fluids, for instance). *Microemulsion* is a fluid that consists of liquid drops from a fraction of a nanometer to more than a micrometer in liquid. *Micelle* and *reversed micelle* are fluids with long molecules with a polarized head and hydrophobic chain. A micelle is head-outward, for example, oil drops in water, while a reversed micelle is chain-outward, water drops in oil.

1.2.2 *Basic concepts of non-Newtonian fluids*

(1) Newtonian law

According to the viscosity, all fluids can be characterized as Newtonian fluids or non-Newtonian fluids. A Newtonian fluid is the fluid that obeys the following Newtonian law of viscosity when it is in a laminar flow

$$\tau = \mu\frac{\partial u}{\partial y}. \tag{1.1}$$

Equation (1.1) implies that the shear stress, τ, between layers is proportional to the velocity (u) gradient, $\partial u/\partial y$, in the direction perpendicular to the layers. The proportional constant μ (or η in some literatures and also used in other chapters of this book) is called viscosity or dynamic viscosity. $\partial u/\partial y$ is also known as a rate of strain, a shear rate, or a velocity field which is known as plane Couette flow, simple shear flow. The unit for shear rate is 1/s. Viscosity has a unit of Pa · s in SI units, and g/cm · s or Poise in cgs units. 1 Poise = 0.1 Pa · s. Poise is a short form for Poiseuille, and it is associated with a pipe flow. The dynamic viscosities of some fluids are as follows: μ(air at 15° C) = 17.9 mPa · s, μ(water at 20° C) = 1.002 mPa · s, μ (blood at 37°C) $\approx 3 \sim 4$ mPa · s. The viscosity of Newtonian fluids do not change with velocity, and its velocity distribution shows a parabolic curve. Most fluids, such as water, alcohol, gasoline, bean oil, glycerin, air and so on belong to this type of fluid. In a stress-strain plot, Newtonian fluids behave as $\tau = \mu\dot{\gamma}$, as curve 3 shown in Fig. 1.1. A science studying stress-strain relationship is rheology.

Kinematic viscosity ν is related to the density and the transport of momentum. The relation between kinematic and dynamic viscosities is $\nu = \mu/\rho$, where ρ is the density of the liquid. The unit of kinematic viscosity is m^2/s (no special name) in SI units, and cm^2/s or St (Stoke) in cgs units. 1 $m^2/s = 10^4$ St $= 10^6$ cSt (centistoke). The kinematic viscosity of air ν(air at 15°C) $= 1.47 \times 10^{-5}$ m^2/s, ν (water at 20°C) $= 1.004 \times 10^{-6}$ m^2/s. These numbers determine the time to approach a steady state of corresponding fluids. Kinematic viscosity ν can be analogous to the thermal diffusivity $\alpha = L^2/t$, where t is the time for an L long metal poker to get thermal equilibrium while α (metal) $= 1.1 \times 10^{-5}$ m^2/s.

(2) Couette flow and Poiseuille flow

Under the Couette flow condition, if the pressure gradient is neglected and the shear rate is homogeneous, one would have the following form of the Navier–Stokes equations:

$$\frac{d^2u}{dy^2} = 0, \tag{1.2}$$

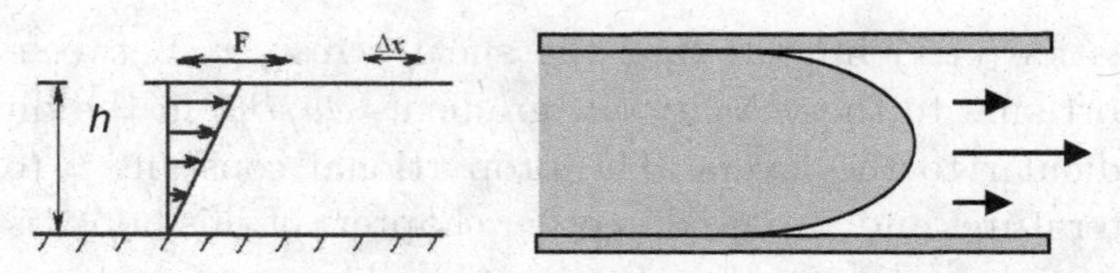

Fig. 1.1 Two types of typical shear flow: a plane Couette flow (left) and a tube Poiseuille flow (right).

where y is a spatial coordinate normal to the plates and $u(y)$ is the velocity distribution. If $y = 0$ at the lower plate, the boundary conditions are $u(0) = 0$ and $u(h) = u_0$, and the exact solution of the previous equation is

$$u(y) = u_0 y/h. \tag{1.3}$$

A more general Couette flow situation arises when a pressure gradient is imposed in a direction parallel to the plates. The Navier–Stokes equations, in this case, simplify to

$$\frac{d^2u}{dy^2} = \frac{1}{\mu}\frac{dp}{dx}. \tag{1.4}$$

Integrating the above equation twice and applying the same boundary conditions as in the case of Couette flow without a pressure gradient yields the following exact solution:

$$u(y) = u_0\frac{y}{h} + \frac{1}{2\mu}\left(\frac{dp}{dx}\right)(y^2 - hy). \tag{1.5}$$

A flow of a viscous fluid in a channel is the Poiseuille flow. The channel has a width a in the y-direction, a length l_z in the z-direction, and a length l_x in the x-direction, the direction of flow. There is a pressure drop Δp along the length of the channel, so that the constant pressure gradient is $\Delta p/l_x$. Assuming that the flow is steady, $dv_x/dt = 0$, and the flow is symmetric, the no-slip boundary condition at the top and bottom edges of the channel reads $v_x(y = a/2) = v_x(y = -a/2) = 0$. The Navier–Stokes equation then becomes

$$\eta \frac{\partial^2 v_x}{\partial y^2} + \frac{\Delta p}{l_x} = 0. \tag{1.6}$$

Integrating twice under above boundary conditions, one has

$$v_x(y) = \frac{1}{2\eta} \frac{\Delta p}{l_x} [(a/2)^2 - y^2]. \tag{1.7}$$

(3) Shear stress – shear rate relations

Non-Newtonian fluids are the fluids that do not obey Newtonian law of viscosity, such as airplane glue, egg white, syrup, honey, vaseline, tomato sauce, paint, tooth paste, dish detergent, cement paste, etc. The stress-strain relations for Newtonian and non-Newtonian fluids are different. The shear stress τ is proportional to shear rate $\dot{\gamma}$ for Newtonian fluids as shown by line 3 in Fig. 1.2. For non-Newtonian fluids, the τ–$\dot{\gamma}$ relation is not linear, or not a line through the origin as shown by curves 1, 2 and 4 in Fig. 1.2. Curve 1 represents the behavior of a kind of Bingham plastic fluid such as frozen orange juice, mayo, etc. Their τ–$\dot{\gamma}$ relation is linear, but there is a minimum yield stress needed to make a liquid move:

$$\tau = \tau_0 + \mu_p \dot{\gamma}, \tag{1.8}$$

where μ_p is the plastic viscosity and τ_0 is the initial stress. Curve 2 corresponds to dilatant plastic fluids, such as clay suspensions, whose μ increases with $\dot{\gamma}$:

$$\tau = k\dot{\gamma}^n, \quad n > 1. \tag{1.9}$$

Curve 4 is very common in polymer melts. This kind of soft matter is called pseudoplastic fluid:

$$\tau = k\dot{\gamma}^n, \quad n < 1, \tag{1.10}$$

where k, the thickness factor, and n, the rheological index, are determined by experiments.

A fluid may exhibit a combination of the phenomena displayed in Fig. 1.2. For example, liquid chocolate exhibits both yield stress and shearing thinning. This behavior is important in fabricating chocolate

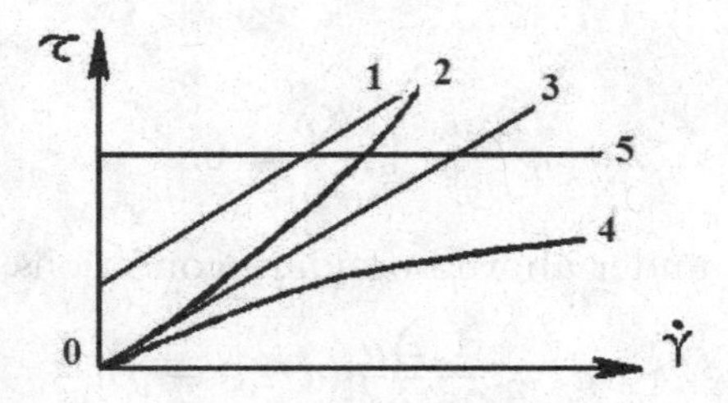

Fig. 1.2 The shear stress $\tau-$ shear rate $\dot{\gamma}$ relations. Curve 1: Bingham plastic; Curve 2: Dilatant; Curve 3: Newtonian; Curve 4: pseudoplastic; Curve 5: cytological fluid.

figurines. Another example is cytological fluid, shown in Fig. 1.2 as Curve 5, where shear stress is shear-rate independent.

(4) Viscoelasticity

In an ideally (Hookean) elastic material, stress σ is proportional to strain ε, and one has Hooke's law, $\sigma = E\varepsilon$, where E is Young's modulus. In an ideally viscous material, stress σ is proportional not to strain ε itself, but to strain rate, $\dot{\varepsilon} = d\varepsilon/dt$. So the following relation is satisfied in ideal viscous materials: $\sigma = F\dot{\varepsilon}$, where F is viscosity. If the above equation is expressed with shear stress, $\sigma_\mathrm{s} \equiv \tau$, and shear rate, $\varepsilon_s \equiv \dot{\gamma}$, one has $\tau = \mu\dot{\gamma}$.

Generally speaking, a common material is not ideally elastic, nor ideally viscous, but somewhere between the two extremes, namely, it is viscoelastic. A viscoelastic material is one in which hysteresis is seen in the stress-strain curve; stress relaxation occurs when step constant strain causes decreasing stress; and creep occurs when step constant stress causes increasing strain.

If a material can be equivalently combined by one ideally elastic and one ideally viscous material in series, the material is called Maxwell fluid. It satisfies $\dot{\varepsilon} = \dot{\sigma}/E + \sigma/F$.

If a material can be equivalently combined by one ideally elastic and one ideally viscous material in parallel, the material is called Kelvin solid. It satisfies $\sigma = E\varepsilon + F\dot{\varepsilon}$.

In shearing cases, one has shear modulus G defined as

$$G = \frac{\tau}{\gamma} = \frac{\sigma_s}{\varepsilon_s}. \tag{1.11}$$

If the stress is time dependent, for example $\tau(t) = \tau_0 \, e^{i\omega t}$, the strain will change with the stress as $\gamma(t) = \gamma_0 \, e^{i\omega t + \phi}$, where ϕ is a phase difference. The shear modulus is then a complex quantity $G^* = G' + iG''$, and the imaginary part G'' is called loss modulus.

Almost all polymers exhibit viscoelastic behavior. Polymers (and other viscoelastic materials) behave more like solids at low temperatures and fast deformation speeds. They are more like liquids at high temperatures and slow deformation speeds.

Linear viscoelasticity is usually applicable only for small deformations; nonlinear viscoelasticity is used when the function is not separable. It usually happens when the deformations are large or the material changes its properties under deformations.

1.2.3 *Major characteristics of non-Newtonian fluids*

We can do some experiments to identify non-Newtonian fluids from Newtonian ones. The left sides of the following ten diagrams are Newtonian fluids, while the right sides are non-Newtonian fluids. How do we know Newton's law of viscosity is inadequate for polymeric liquids? There are many fascinating experiments [1.11–1.14] which show that the flow of polymeric fluids is qualitatively different from that of Newtonian fluids. Figures 1.3–1.7 show some of these experiments, the first three of which involve rotation.

Figures 1.3 (a) and (b) give the comparison of the behaviors of Newtonian and polymeric fluids near a rotating rod. The surface of the Newtonian fluid is depressed near the rod, whereas the

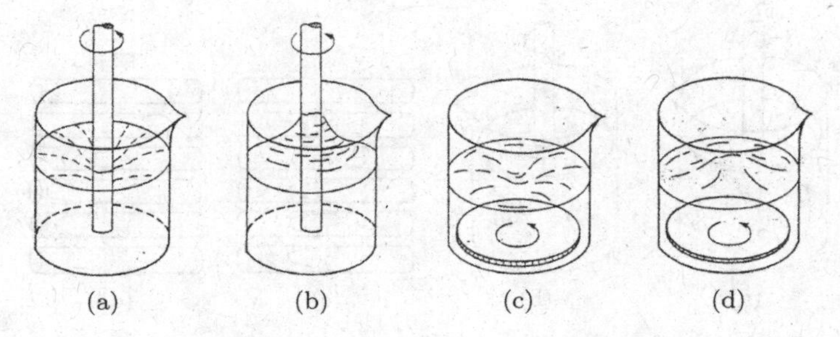

Fig. 1.3 Newtonian (a, c) vs. non-Newtonian (b, d) fluid [1.10].

polymeric liquid tries to climb the rod. This climbing is known as the "Weissenberg effect."

Figures 1.3 (c) and (d) show the behaviors closely related to that of Figs. 1.1 (a) and (b). Here a rotating disk at the bottom of the beaker causes a depression in the surface of the Newtonian fluid, but a rise in the surface of the polymeric liquid.

In Figs. 1.4 (a) and (b), a rotating disk placed at the surface of either fluid causes a primary flow in the tangential direction, but superposed on this primary flow is a secondary flow. Newtonian fluids are shoved outward by the rotating disk, move downward near the beaker wall, and then move upward near the axis of the beaker. Polymeric liquids also have a secondary flow — but in the opposite direction!

In Figs. 1.4 (c) and (d) we see how fluids behave as they are pumped down a circular tube. We follow the motion by watching a streak of dye that is inserted before the motion starts. Six successive snapshots of the streak are shown. When the pump is turned off at the fourth snapshot, the Newtonian fluid comes to rest, but the polymeric liquid "recoils" as shown in the fifth and sixth snapshots. This illustrates the "memory" of polymeric fluids. Because they do not return all the way to their initial configuration (as a rubber band would after being stretched), we say that these fluids have "fading memory".

Figures 1.5 (a) and (b) show how a polymeric liquid swells when it emerges from a tube of silt. The cross-sectional area can increase by as much as a factor of five.

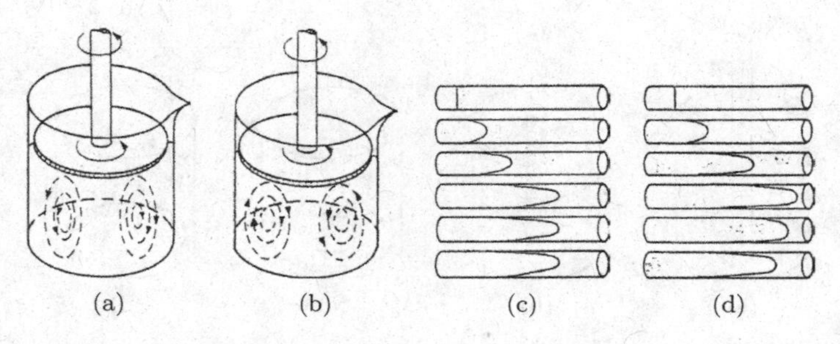

Fig. 1.4 Newtonian (a, c) vs. non-Newtonian (b, d) fluid [1.10].

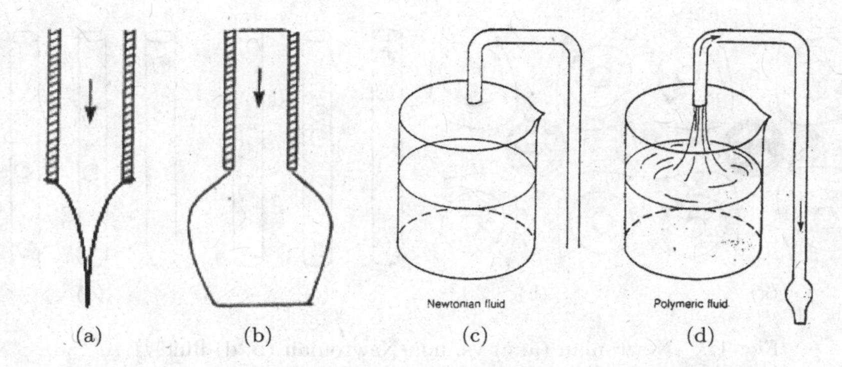

Fig. 1.5 Newtonian (a, c) vs. non-Newtonian (b, d) fluid [1.10].

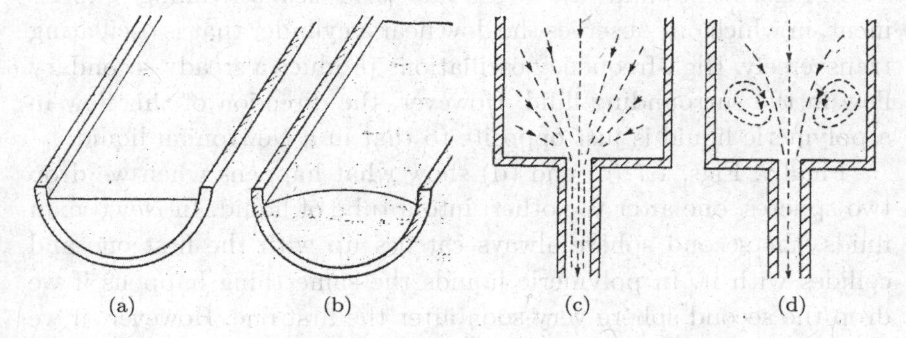

Fig. 1.6 Newtonian (a, c) vs. non-Newtonian (b, d) fluid [1.10].

Figures 1.5 (c) and (d) show a siphon experiment. For Newtonian fluids, siphons work only as long as the upstream end of the tube is beneath the surface of the liquid. One can siphon polymeric fluids even if there is a gap of several centimeters between the surface of the liquid and the end of the tube!

In Figs. 1.6 (a) and (b) we see a cross section of what happens when a liquid flows down a tilted, semi-circular trough. The flow is laminar in each case. The surface of the Newtonian liquid is flat except for meniscus effects, whereas the surface of the polymeric liquid is slightly convex. This is a small effect, but it is reproducible.

In Figs. 1.6 (c) and (d) we see how fluids flow from a large-diameter tube into a small-diameter tube in slow flow. In polymeric liquids, a vortex forms upstream. Fluid particles trapped in this vortex do not move on into the small-diameter pipe.

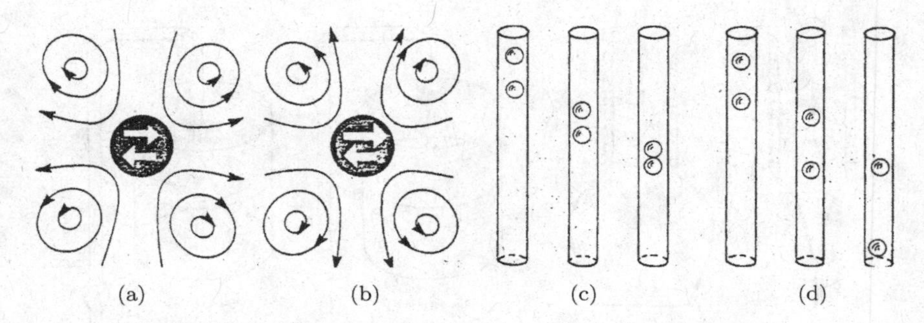

Fig. 1.7 Newtonian (a, c) vs. non-Newtonian (b, d) fluid [1.10].

In Figs. 1.7 (a) and (b), we see the "acoustical streaming" experiment, in which one observes the flow near a cylinder that is oscillating transversely. High-frequency oscillations produce a steady secondary flow in the surrounding fluid. However, the direction of this flow in a polymeric liquid is just opposite to that in a Newtonian liquid.

Finally, Figs. 1.7 (c) and (d) show what happens when we drop two spheres, one after the other, into a tube of liquid. In Newtonian fluids, the second sphere always catches up with the first one and collides with it. In polymeric liquids the same thing happens if we drop the second sphere very soon after the first one. However, if we wait longer than a critical time interval, then the spheres tend to move apart while falling.

In the experiments just described, the response of the polymeric liquid is *qualitatively* different from that of the Newtonian liquid. We are thus not dealing with minor variations on an old theme. Rather, we are faced with striking differences that can be explained only by rejecting Newton's law of viscosity and replacing it with some new and more general expression that can account for fading memory, recoil the reversal of secondary flows, and other bizarre behavior. That is a challenge!

1.3 Self-organization of Soft Matters

The major character of soft matters is the self-organization of particle structure. As shown in Fig. 1.8, particles in a ferrofluid between two pieces of glass form their structure under a magnetic field.

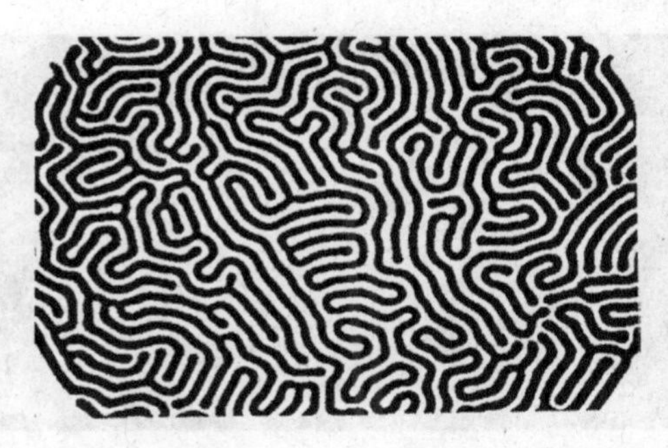

Fig. 1.8 Particles in a ferrofluid between two pieces of glass form their structure under a magnetic field [1.15].

1.3.1 *Scale invariance*

The first character of self-organization in soft matters (polymers, for example) is swelling order (or symmetry). An example of swelling symmetry is that an expanded small-sized sock would completely match the large-sized one. Just as time translational symmetry leads to energy conservation, space translational symmetry leads to linear momentum conservation, and rotational symmetry leads to angular momentum conservation, dilation symmetry would lead to a scale invariance.

Scaling structures come from swelling symmetry. As shown in the picture of the shock wave of a nuclear explosion, the envelopes of shock waves of nuclear explosions with different energies form a series of semi-spheres. This shows a swelling symmetry, and gives a scaling law as follows: G. I. Taylor's scaling law for the shock-wave radius r_f after a nuclear explosion can be formulated as

$$r_f = \left(\frac{Et^2}{\rho_0}\right)^{1/5}, \tag{1.12}$$

where E is the explosion energy, t is the time after explosion, and ρ_0 is the air density. The picture drawn by P.G. de Gennes shows various animals exhibiting a swelling symmetry. One may find a scaling for the breathing rate R of animals as follows $R = AW^n$, where W is the body mass of animals.

This is because the complex and continuous single processes are independent of subject size, event location, and time: they are

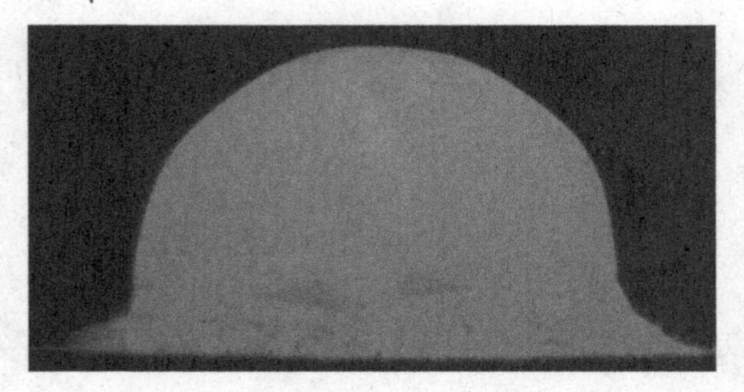

Fig. 1.9 Shock wave after a nuclear explosion gives an example of swelling symmetry [1.16].

Fig. 1.10 Various animals attempting to follow a scaling law drawn by P.G. de Gennes [1.17].

universal scaling transformations, and they exist in self-similarities between large and small scaled objects. The amplifying ratio of similarities from small to large objects or events is the universal Feigenbaum constant (see Chapter 4 for details). To a physicist, a constant ratio means a scaling law, which means that the physical characteristics reappear in the smaller and smaller scales.

In a polymer solution, each segment randomly expands relative to its previous one, forming a molecular random walk. A scale-dependent density reflects a space order, and this is independent of rotation and translation, but dependent of dilation. For dilation symmetry, one may find a scaling index. A commonly seen scaling in soft matter is from Flory. If R_G is a radius of a gyration of a polymer coil and N is the degree of polymerization, $R_G = CN^v$, where C is a constant, the scaling index $v = 3/(\mathrm{D}+2)$, and D is the dimension. In three-dimensional space, $\mathrm{D} = 3$ and $v = 0.6$.

Another example of scale-invariant functions is the monomial $f(x) = x^n$, for which one has $\Delta = n$, in that clearly $f(\lambda x) = (\lambda x)^n = \lambda^n f(x)$. In mathematics, one can consider the scaling properties of a function $f(x)$ under re-scaling of the variable x. That is, one is interested in the shape of $f(\lambda x)$ for some scale factor λ, which can be taken to be a length or size rescaling. The requirement for $f(x)$ to be invariant under all re-scaling is usually taken to be $f(x) = \lambda^{-\Delta} f(\lambda x)$ for some choice of exponent Δ, and for all dilations λ [1.18].

Correlation functions are used to quantitatively study dilation symmetry [1.19]. The measurement of correlation functions will be covered in the chapter on neutron scattering. The correlation function of two points in a space is the probability of occupation of one point by a particle while the other point is also occupied by a particle of the same type.

Let us take a polymer chain as an example. The polymer chain is formed with k segments (monomers), and the chain passes through the origin. $\rho(r)$ is the random density. One way to characterize the variations of ρ over the images is by means of the correlation functions $\langle\rho(r_1)\rho(r_2)\ldots\rho(r_k)\rangle_0$. The subscript 0 indicates that the sample pictures are selected to have the origin occupied. The points $r_1,\ldots,r_k$ are specific points on each image. For example, r_1 could be one centimeter inward from the upper left corner. Any of these correlation functions can be measured to arbitrary precision using a large enough sample of images. If the picture is magnified by a factor λ^{-1}, the scale of the picture is enlarged by a factor λ^{-1}. Due to the dilation symmetry, which leads to an invariance of the polymer picture under a scale transformation, the magnified picture will not be identified from the original ones. In the magnified picture, all length scales are magnified by a factor λ^{-1}, including the size of the polymer. This implies that certain length r in the original length would be identified as λr in the magnified picture. The correlation function in the picture magnified by a factor λ^{-1} becomes $\langle\rho(\lambda r_1)\rho(\lambda r_2)\ldots\rho(\lambda r_k)\rangle_0$. Dilation invariance implies that the two statistics are the same:

$$\langle\rho(r_1)\rho(r_2)\ldots\rho(r_k)\rangle_0 = \mu_k(\lambda)\langle\rho(\lambda r_1)\rho(\lambda r_2)\ldots\rho(\lambda r_k)\rangle_0. \quad (1.13)$$

The $\mu_k(\lambda)$ prefactor accounts for a possible overall change in "contrast" upon dilation.

1.3.2 *Entropy driven self-organization*

The self-organization of soft matters is often dynamics-driven. The colloidal particles aggregate due to the attractive interaction between particles (often by van der Waals potential). In a fluid consisting of small particles, the attraction between small particles may be balanced by the collisions between the small particles and the surrounding solvent molecules to form stable dispersions. On the other hand, in a fluid consisting of larger particles, the Brownian motion cannot balance the particle attraction and would destroy the stability of colloids; irreversible flocculation and sedimentation would occur.

A major character of self-organization of soft matters differing from hard matters is the entropy-driven self-organization.

Thermodynamic equilibrium is obtained from the minimization of the free energy

$$F = U - TS \tag{1.14}$$

where entropy $S = k_B \ln W$ (the Boltzmann equation), and W is the number of microscopic states in a macroscopic state.

In hard matter systems, the contribution of inner energy U (mainly interaction) to the free energy is much larger than that of entropy. It is the inner energy (interaction) that decides the restoring force when a hard matter system deviates from a free energy minimum state and the thermal fluctuation functions as a perturbation only.

On the other hand, in soft matter systems the molecular interaction is weak, and when a configuration changes, there is almost no change in the inner energy. A minor perturbation may cause a complicated deformation and flow of a soft matter system. It is the thermal fluctuation that has a decisive influence on the structures and behaviors of the soft matter systems.

The entropy force in soft matter systems is a restoring force when the system deviates from the state of entropy maximum

$$f = \partial F/\partial l = -\partial S/\partial l. \tag{1.15}$$

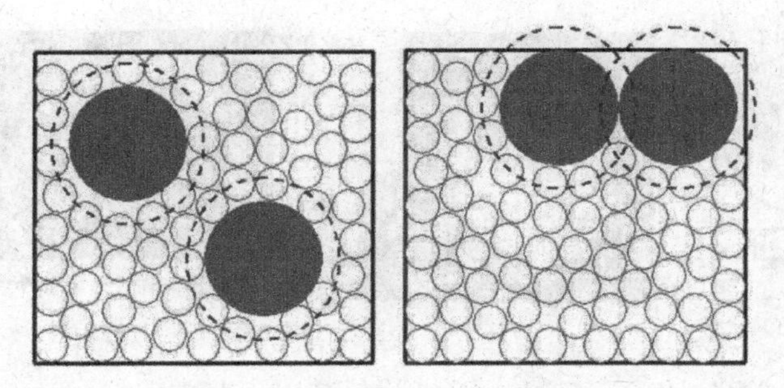

Fig. 1.11 The entropy-driven ordering. It is interesting to see that the natural ordering process leads to a more uneven, rather than a more even configuration when entropy is increased [1.20].

It is the same as the effect of the force produced by a potential gradient.

The following is an example of the entropy-driven ordering. When one mixes a system consisting of two types of spheres, if the total volume of large spheres is equal to that of small ones, then the number of large spheres must be much larger than that of small ones. This means that the number of the states of the configurations of the small ones must be much larger than that of large ones. In other words, the entropy of configuration of the small spheres dominates. There is a trend in this system that large spheres would get close to each other, aggregating near the wall or surface, or even to a corner allowing more free space or more configuration states for small spheres. It seems that there is an attraction force between large spheres — this is usually called entropy force. The physical picture is clear, however the quantitative theoretical study is difficult, and physicists use molecular dynamics simulation to calculate the entropy-driven self-organization.

1.3.3 *Measurements of depletion effect*

To observe a depletion effect, physicists from the University of Pennsylvania [1.21] observe the trace of a single 0.237-μm-radius polystyrene sphere on a focal plane under a microscope (see Fig. 1.12). If there are no small spheres in the vesicle, the single

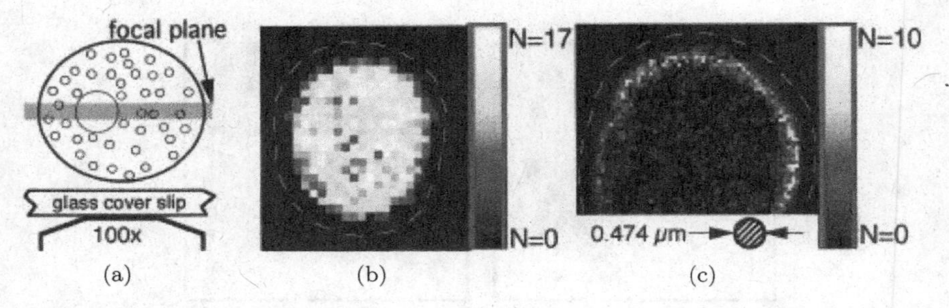

Fig. 1.12 (a) The trace of a large sphere on a focal plane is observed under a microscope. (b) If there are no small spheres, the large sphere freely diffuses all available space inside a vesicle. The lighter the color of a spot is (or the larger N, according to the vertical sidebar), the more times the large sphere passes that spot. (c) If the vesicle also contains small spheres, the large sphere is attracted to the wall [1.21].

large sphere freely diffuses all available space inside a vesicle as indicated by the 2000 events shown in Fig. 1.12 (b). The lighter the color of a spot is (or the larger N, according to the vertical sidebars in Figs. 1.12 (b) and (c), the more times the large sphere passes that spot. Figure 1.12 (c) shows the case when the vesicle also contains small spheres. The volume fraction of small spheres $\phi_S = 0.30$, the radius of small spheres $R_S = 0.042\,\mu$m, namely, the ratio of large to small radii $(R_L/R_S) = \alpha = 5.7$. In 2,300 events, one can see clearly that, in this case, the large sphere is attracted to the wall.

The reason why the single large sphere prefers the curved wall is because once it is there, small spheres can find more space to stay, and then the entropy of small spheres increases. Since the number of small spheres is much larger than that of large ones, the entropy of small spheres S_S is dominant in the total entropy $S(= S_L + S_S)$. Eventually the free energy $F(= U - TS)$ of the system is reduced.

The depletion potentials are quantitatively measured in many different methods, such as scattering [1.22] or optical tweezers [1.22–1.26]

To measure the depletion potential quantitatively, Crocker *et al.* [1.26] built A line optical tweezers. In the apparatus, a galvanometer mirror scans an IR laser beam focused along a line for about $10\,\mu$m back and forth at a frequency of 180 Hz. As a result, two large PMMA spheres can only float within this narrow region as if they are undergoing a nearly one-dimensional Brownian motion. At the same time, the small colloidal spheres are free to diffuse in one dimension, along the scan direction, while being strongly confined to the two perpendicular directions. A camera attached to the microscope videotaped the trace of the large sphere for 1 hour, gathering some 2×10^5 separation measurements. These video clips give an equilibrium probability $P(r)$ of finding the two large spheres with separation r. From this, $F(r)$, the Helmholtz free energy, or the pair interaction energy, of the system as a function of large sphere separation r can be obtained according to the Boltzmann equation $P(r) \sim \exp[-F(r)/k_B T]$.

The sample system consists of large spheres with volume fraction less than 10^{-7}, and small spheres with volume fraction ϕ_s ranging from 0 to 0.42. They found that, when ϕ_s is about a few percent (see Figs. 1.13 (b) and (c), the potential is monotonically attractive — which agrees with the Asakura–Oosawa model. The rest of the curves in Fig. 1.13 show that when ϕ_s gets larger and larger, repulsive barriers appear, and when ϕ_s is larger than 0.34, the pair interaction energy becomes fully oscillatory.

The authors of the above work explained the oscillatory depletion potential by assuming that the small spheres form shells in the region between the two large spheres.

1.3.4 *Calculations of depletion effect*

L. Onsager first discovered the depletion effect [1.27]. In the 1950s, S. Asakura and F. Oosawa [1.28, 1.29] used Onsager's idea in analyzing the interaction between two plates or two spheres suspended in a solution of macromolecules, and their method is called the Asakura–Oosawa model in literature. P. J. Flory [1.30, 1.31] developed statistical thermodynamics of particle interaction in the solution of chain molecules or rod-like particles.

There are three methods commonly used in calculation of depletion force [1.32], such as the integral equation method [1.33] which usually starts with OZ integral equation [1.34], density functional theory [1.35, 1.36] and numerical simulation [1.37].

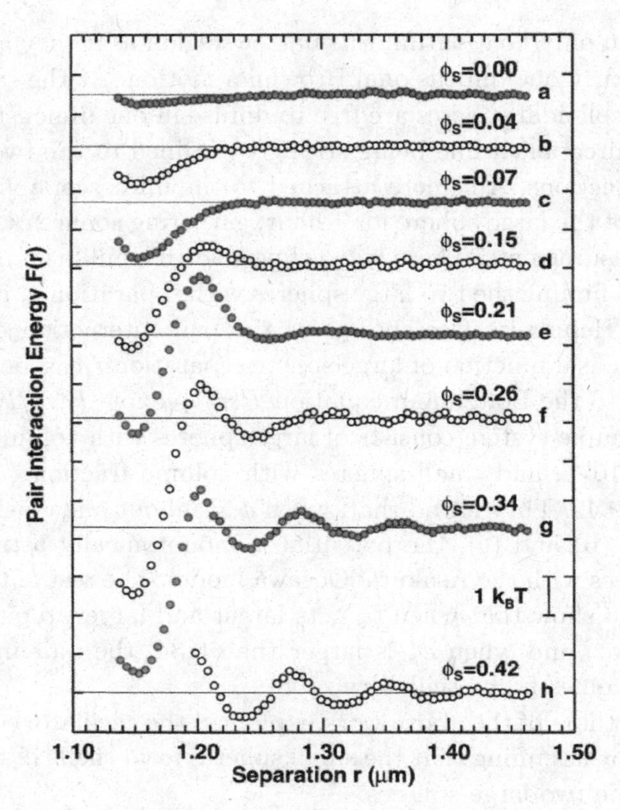

Fig. 1.13 The pair interaction energy $F(r)$, which measures the depletion potential of two large spheres in the system as functions of large particle separation r, when small sphere volume fraction changes from 0 to 0.42 [1.26].

(1) Asakura–Oosawa Model

S. Asakura and F. Oosawa [1.29] propose first a mechanical picture. One may consider two large particles immersed in a solution of rigid spherical macromolecules. When the distance between the inner surfaces of these two large particles is smaller than the diameter of solute macromolecules, none of these molecules can enter the space between the large particles and this space becomes a phase of the pure solvent. Therefore, a force equivalent to the osmotic pressure of the solution of macromolecules acts on the outer surfaces of these large particles.

The interaction of the two large particles can also be considered in a depletion picture. This "battle" comes from an overlapped volume. The more volume is overlapped in the space, the more microscopic states one can find for a component of the system, and the larger the entropy the component can have, the larger the entropy the whole system can achieve. This would lead to a decrease in the overall free energy.

When the concentration of macromolecules is sufficiently low, the force F between two particles suspended in the solution is given by [1.28]

$$F = kTN\partial \ln V_{\text{free}}/\partial a, \tag{1.16}$$

$$V_{\text{free}} = \int_V \exp[-w(x.a)/kT]dx, \tag{1.17}$$

where $\boldsymbol{N}$ denotes the total number of solute macromolecules, V, the total volume of the solution, and a, the distance between the centers of two particles. The quantity $w(x, a)$ is the potential energy of a macromolecule situated at x generated by its interaction with the suspended particles or, more rigorously, w is the average free energy of a macromolecule at x. The negative value of F corresponds to the attraction between the two large particles.

If macromolecules are rigid spheres with no interaction between them and suspended particles, then V_{free} is equal to the volume of space in which each macromolecule can move freely. In other words, we can obtain V_{free} by subtracting from the total volume of the solution the volume into which the centers of macromolecules cannot enter owing to hindrance by the particles. When particles are also spherical, we have

$$V_{\text{free}} = \begin{cases} V - \pi(4/3)[((D+d)/2)^3 + (D+d)^2 a - (1/12)a^3] \\ \qquad D \leq a \leq D+d \\ V - \pi(8/3)[(D+d)/2]^3 \\ \qquad D+d \leq a \end{cases} \tag{1.18}$$

where D and d are the diameters of suspended particles and solute macromolecules, respectively. Consequently, we obtain $F = -p_0 S$, where

$$\begin{aligned} S &= (\pi/4)D'^2 = (\pi/4)[(D+d)^2 - a^2] \quad D \leq a \leq D+d \\ &= 0 \qquad\qquad\qquad\qquad\qquad\qquad\quad D+d \leq a \end{aligned} \tag{1.19}$$

and $p_0 = kTN/V$.

When two particles approach each other, two spherical volumes of diameter $D + d$, which exclude macromolecules, begin to overlap at mutual distance a equal to $D + d$. With further decrease of a, the volume of the overlapping region increases and therefore the attraction acts between particles. The quantity S in Eq. (1.19) is an area of the circular cross section of the overlapping region in the middle of a, and p_0 is the osmotic pressure of the solution of macromolecules. The interaction between the macromolecules was neglected in deriving the above result.

(2) The phase separation of colloidal systems

To discuss the experimental result in Section 1.4.3, the free energy difference of the system similar to the one described in that experiment is calculated with an acceptance ratio method [1.38]. When the canonical ensemble is applied, the free energy

$$F = k_B T \ln Q, \tag{1.20}$$

where Q is the canonical partition function which encodes how the probabilities are partitioned among the different microstates based on their individual energies.

The sphere-sphere and wall-sphere models [1.32,1.38] are adopted in finding out the free energy difference of a system consisting of two large hard spheres (radius $= R$) and many small hard ones (radius $= r$) in a box of $L_x L_y L_z$. The number of small spheres N is determined by the given volume fraction $\eta = N v_s/(V - 2v_b)$ in the sphere-sphere model, where v_s and v_b are the volumes of a small and a big sphere, respectively. Suppose the gaps between the two big spheres are D_0 and D_1 for systems 0 and 1 with corresponding external potentials U_0 and U_1, partition functions Q_0 and Q_1, respectively [1.39].

$$Q_0 = \frac{\Lambda_0^{-3N}}{N!} \int dr \exp\{-\beta U_0\} = \frac{\Lambda_0^{-3N}}{N!} Z_0, \tag{1.21}$$

$$Q_0 = \frac{\Lambda_1^{-3N}}{N!} \int dr \exp\{-\beta U_1\} = \frac{\Lambda_1^{-3N}}{N!} Z_1, \tag{1.22}$$

where $\beta = 1/(k_B T)$. The free energy difference is

$$\beta \Delta F = \beta F_1 - \beta F_0 = -\ln\left(\frac{Q_1}{Q_0}\right) = -\ln\left(\frac{Z_1}{Z_0}\right) - 3N \ln\left(\frac{\Lambda_0}{\Lambda_1}\right). \tag{1.23}$$

If a proper weight function $W(r)$ is chosen as

$$W(r) \propto \left(\frac{Z_0}{n_0} \exp[-\beta U_1] + \frac{Z_1}{n_1} \exp[-\beta U_0]\right)^{-1}, \tag{1.24}$$

one can get

$$\begin{aligned} \frac{Z_1}{Z_0} &= \frac{Z_1 \int dr W(r) \exp[-\beta(U_0 + U_1)]}{Z_0 \int dr W(r) \exp[-\beta(U_0 + U_1)]} \\ &= \frac{\int dr W(r) \exp[-\beta U_1)] \exp(-\beta U_0)/Z_0}{\int dr W(r) \exp[-\beta U_0)] \exp(-\beta U_1)/Z_1} \\ &= \frac{\langle W(r) \exp[-\beta(U_1)]\rangle_0}{\langle W(r) \exp[-\beta(U_0)]\rangle_1} \\ &= \frac{n_0 Z_1 \langle (1 + \frac{Z_1 n_0}{Z_0 n_1} \exp[-\beta(U_0 - U_1)])^{-1}\rangle_0}{n_1 Z_0 \langle (1 + \frac{Z_0 n_1}{Z_1 n_0} \exp[-\beta(U_1 - U_0)])^{-1}\rangle_1}, \end{aligned} \tag{1.25}$$

where the ensemble average $\langle \ldots \rangle_i$ is taken in the reference system U_i $(i = 0, 1)$.

Using the Fermi distribution function, $f(x) = 1/[1 + \exp(x)]$ further simplifies the above ratio:

$$\ln\left(\frac{Z_1}{Z_0}\right) = \ln\left(\frac{\langle f(\beta(U_1 - U_0) + C)\rangle_0}{\langle f(-\beta(U_1 - U_0) - C)\rangle_1}\right) + C \tag{1.26}$$

where $C = \ln(Z_1 n_0 / Z_0 n_1)$. Since n_0 and n_1 are arbitrary constants, C is also an arbitrary constant, and one may choose $C = 0$. The above becomes

$$\ln\left(\frac{Z_1}{Z_0}\right) = \ln\left(\frac{\langle f(\beta(U_1 - U_0))\rangle_0}{\langle f(-\beta(U_1 - U_0))\rangle_1}\right). \tag{1.27}$$

Note that in sampling according to system 0, $f[\beta(U_1 - U_0)] = 1/2$ if $U_1 = 0$, and $f[\beta(U_1 - U_0)] = 0$ if $U_1 = \infty$; while in sampling according to system 1, $f[-\beta(U_1 - U_0)] = 1/2$ if $U_0 = 0$, and $f[-\beta(U_1 - U_0)] = 0$ if $U_0 = \infty$.

When the Monte Carlo method is applied, a sample is taken at every few (say 5) steps, and the total number of sampling times is M. Suppose the number of accepted ratio when $D = D_0$, namely the external potential upon small particles, is U_0, while $U_1 = 0$ is M_{10}/M, and that when $D = D_1$ with external potential U_1 while $U_0 = 0$ is M_{01}/M. One then gets $\beta\Delta F = -\ln(Z_1/Z_0) = -\ln(M_{10}/M_{01})$.

This equation leads to curves of free energy shown in Fig. 1.14 for different gaps between the two large spheres.

The oscillation form implies that small spheres form shells between the two large ones, and this is quite consistent with the experimental results shown in Fig. 1.13 [1.26].

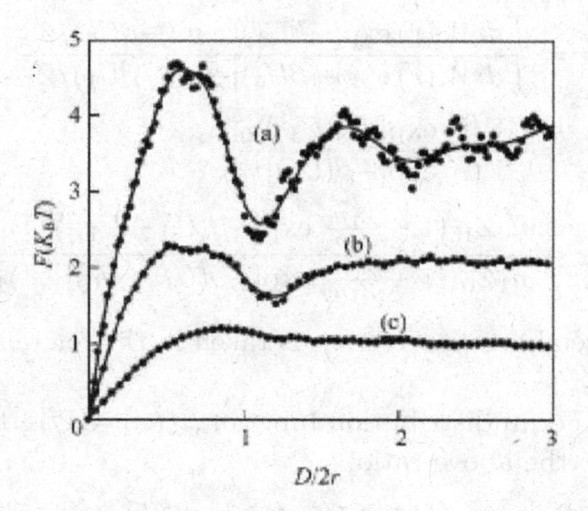

Fig. 1.14 Free energy difference of a system consisting of two large spheres and many small ones when the gap between the two large spheres is D, and the radius ratio r/R of the small spheres over larger ones is 0.2, $L_x = L_y = 18r, L_z = 36r$. (a) $\eta = 0.341, N = 867$; (b) $\eta = 0.229, N = 580$ and (c) $\eta = 0.116, N = 294$. See text for the definitions [1.38].

(3) Depletion force may lead a hard tube into a helical form

Physicists at the University of Pennsylvania [1.40] studied the depletion effect of a hard (but flexible) tube of radius t in a solution of smaller hard spheres of radius r_s (see Fig. 1.15). In analogy to the halo around the large spheres there is a halo of excluded volume (light green in Fig. 1.15) around straight and bent tubes for all points within r_s where the spheres cannot be. In the example of the large spheres there is an effective attraction, while in this situation the tube has an effective self-attraction, forcing it to bend. When the tube bends into the helical conformation there is overlap of the excluded volume between the layers of the helix and in the central core region of the helix. The excluded volume decreases by the overlap volume when the tube forms the helix. When the tube bends, it is able to overlap the excluded volume from the two tube segments, leading to a decrease in the overall free energy. Thus, one may seek the shape of the tube that minimizes the free energy of the small spheres. The physicists at the University of Pennsylvania found helices of specific pitch to radius ratio 2.512 to be optimally compact.

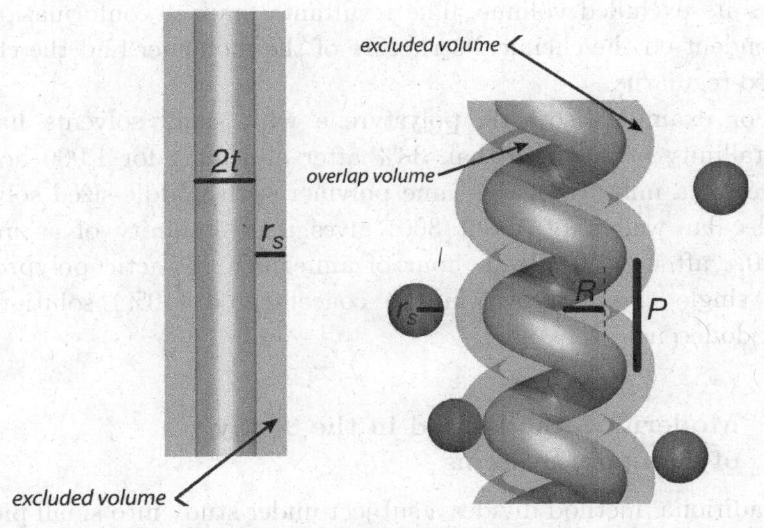

Fig. 1.15 Depletion effect of a hard (but flexible) tube of radius t in a solution of smaller hard spheres of radius r_s leads the tube to form a helical conformation [1.40].

This is the same geometry that minimizes the global curvature of the curve defining the tube.

(4) Depletion potential can be used to make new materials

Chemists at Nanjing University studied the depletion effect in binary polymer mixtures, which are traditionally referred to as polymer solutions consisting of polymers and common small solvents or as polymer blends consisting of two kinds of polymers with comparable molecular weight [1.41]. They studied phenomena in cases where the molecular size of the other component was larger than the common small solvents but much smaller than that of the polymer; more specifically, where the molecular size of the solvent was a few (2–6) times larger than that of the repeating unit of the polymer (they refer to it as middle-sized solvent). The authors found that the depletion interaction also plays a role in the solution of polymer/middle-sized solvent. That is, there is also entropy competition between the polymer conformational entropy and the solvent translational entropy. In appropriate conditions, such competition will drive the solution into instability and drive polymer chains to a conformation that minimizes its excluded volume. The resultant compact conformation is dependent on the chemical structure of the monomer and the chain stereo-regularity.

For example, isotactic polystyrene with small solvents has a crystallinity of no more than 33% after annealing for 1,000 hours. However, a mixture of the same polymer with middle-sized solvent (molecular weight of about 300) gives a crystallinity of as much as 70% after only half an hour of annealing. Isotactic polypropylene single crystal grows in the concentrated (10%) solution of cyclododecene.

1.4 Modern Methods used in the Study of Complex Systems

A traditional method divides a subject under study into small pieces from its surroundings, and then combines the results from the small ones to form a general picture. Modern methods, on the other hand, pay great attention to the scaling structures between the large subjects and small ones. Ilya Prigogine pointed out that Western sciences emphasize the entity (such as atoms, molecules, elemental particles, biological molecules, etc.), while the Chinese view of nature, on the other hand, is based on the "relation", so it is based on the view of an organized physical world. He thinks that there has been a combination of Western tradition that emphasizes experiments and quantitative descriptions with the Chinese tradition centered on the perspective of a "spontaneous and self-organized world".

Modern methods are especially useful in studying complex systems. The scaling theory in polymer science, renormalization group theory, and mathematical tools developed in the research of nonlinear systems such as chaos, fractals, dissipation structures, stochastic process, catastrophe theory, synergetics, negative entropy, cellular automata, etc., are important in studying complex systems such as soft condensed matter.

Catastrophe theory was originated by the French mathematician René Thom (1923–2002) in the 1960s, and is a special branch of dynamical systems theory. It studies and classifies phenomena characterized by sudden shifts in behavior arising from small changes in circumstances. Catastrophes are bifurcations between different equilibria, or fixed point attractors. Due to their restricted nature, catastrophes can be classified based on how many control parameters are being simultaneously varied. For example, if there are two controls, then one finds the most common type, called a "cusp" catastrophe. If, however, there are move than five controls, there is no classification.

Synergetics was created by H. Haken, a German scientist, in the 1970s, and it deals with systems composed of many subsystems which may each be of a very different nature. In particular, synergetics treats systems in which cooperation among subsystems creates organized structure on macroscopic scales (Haken 1993). Systems such as an atom, a molecule, a cell, an organ, or a man, consist of large number of subsystems. Examples of problems treated by synergetics include bifurcations, phase transitions in physics, convective instabilities, coherent oscillations in lasers, nonlinear oscillations in electrical circuits, population dynamics, etc. Synergetics comes from Greek, meaning "a science of cooperation". Synergetics is a self-organization theory that studies the common characteristics existing between these totally different types of systems, and studies how these systems cooperate through subsystems to form structured, ordered states.

Cellular automata are the simplest models of spatially distributed processes. They consist of an array of cells, each of which is allowed to be in one of a few states. At the same time, each cell looks to its neighbors to see what states they are in. Using this information, each cell applies a simple rule to determine what state it should change to. This basic step is repeated over the whole array, again and again. Some of the patterns produced by several simple cellular automata are shown on this page. Cellular automata were invented in the 1940s by the mathematicians John von Neuman and Stanislaw Ulam, while they were working at the Los Alamos National Laboratory in northern central New Mexico. The most famous cellular automaton is the "Game of Life" invented by mathematician John Conway, in the 1960s. Despite the simplicity of the rules governing the changes of state as the automaton moves from one generation to the next, the evolution of such a system is indeed complex.

Modern methods used in the study of soft matters include the functional method, the black box method and the information method. The functional method is based on the similarity of functions and behaviors of complex systems and uses models to simulate the functions and behaviors of the systems. For example, models for intellectual robots are used to simulate the human brain. The black box method treats an unknown world as a black box. It examines the relation between the input and output of a system to study the functions of a system and find a mathematical form of an equivalent regulation. The information method treats a system as an information system, and finds the regulations governing the events or objects via analyzing the information flow of the system. It first sets required information for a system, then gets real information from a system, compares the above two types of information, and finally issues an order to control the system.

References

[1.1] P. G. de Gennes, Soft Matter, *Rev. Mod. Phys.* **64** (1992), 645; P.G. de Gennes, Soft matter: more than words, *Soft Matter* **1** (2005), 16.

[1.2] O. Jacob, La matière molle, Chap. 7, ed. Domain, *La Physique*, SBN 2-7381-1458-X, (2004).

[1.3] P. G. de Gennes and J. Badoz, *Fragile Objects, Soft Matter, Hard Science, and the Thrill Discovery*, Copernicus, Springer-Verlag New York, (1994).

[1.4] S. R. Elliott, *Physics of Amorphous Materials*, 2nd ed. Longman Scientific & Technical, New York, (1990).

[1.5] P. W. Anderson, H. Weintraub, M. Ashburner, P. N. Goodfellow, *et al.*, Through the glass lightly, *Science* **267** (1995), 1609.

[1.6] U. Bengtzelius, W. Götze and A. Sjolander, Dynamics of super- cooled liquids and the glass transition, *J. Phys. C: Solid State Phys.* **17** (1984), 5915; W. Götze, and L. Sjögren, Relaxation processes in supercooled liquids, *Rep. Prog. Phys.* **55** (1992), 241; W. Götze, Wolfgang Götze, *Complex dynamics of glass-forming liquids: a mode-coupling theory*, Oxford University Press, (2009), 641.

[1.7] Y. N. Huang, Q. Xue, H. J. Zhou, in *10000 Selected Problems in Sciences, Physics*, Science Publishing House, Beijing, 2009, p. 416 (in Chinese); Y. N. Huang, C. J. Wang, and E. Riande, Superdipole liquid scenario for the dielectric primary relaxation in supercooled polar liquids, *J. Chem. Phys.* **122** (2005), 144502.

[1.8] *Eleven big questions about the origins of collective behavior in Matter*, http://frontiers.physics.rutgers.edu (2004).

[1.9] D. Knapp, *Sonoluminescence: an Introduction*, http://www-phys.llnl.gov/N_Div/sonolum.

[1.10] R. B. Bird and C. F. Curtiss, *Physics Today* Jan. (1984), 36–43.

[1.11] K. Q. Lu, *Wuli* (*Physics*) **23** (1994), 257 (in Chinese).

[1.12] K. Q. Lu, Lecture on Soft condensed matter physics, in the Institute of Physics, the Chinese Academy of Sciences, Beijing, (1998).

[1.13] J. P. Gollub, J. S. Langer, Pattern formation in nonequilibrium physics, *Rev. Mod. Phys.* **71** (1999), S396.

[1.14] P. Chaudhari, M. S. Dresselhaus, *Rev. Mod. Phys.* **71**(1999), S3331.

[1.15] T. A. Witten, *Physics Today* (July 1990), 21–28.

[1.16] G. I. Barenblatt, *Scaling Phenomena in Fluid Mechanics*, Cambridge University Press, (1994).

[1.17] P. G. Gennes, *Scaling Concepts in Polymer Physics*, Connell University Press, (1979).

[1.18] www.wikipedia.com.

[1.19] Witten T.A., in Lu Qunquan and Liu Jixing, ed. *Introduction of Soft Matter Physics*, Peaking University Press, 2006. (in Chinese) 136–138.

[1.20] H. R. Ma, K. Q. Lu, *Wuli* (*Physics*) **29** (2000), 516.

[1.21] A. D. Dinsmore, D. T. Wong, P. Nelson, and A. G. Yodh, *Phys. Rev. Lett.* **80** (1998), 409.

[1.22] X. Ye, T. Narayanan, P. Tong, J S. Huang, M. Y. Lin, B. L. Carvalho and L. J. Fetters, *Phys. Rev. E* **54** (1996), 6500.
[1.23] P. D. Kaplan, L. P. Faucheux and A. J. Libchaber, *Phys. Rev. Lett.* **73** (1994), 2793.
[1.24] A. D. Dinsmore, A. G. Yodh and D. J. Pine, *Nature* **383** (1996), 239.
[1.25] Y. N. Ohshima, H. Sakagami, K. Okumoto, A. Tokoyoda, T. Igarashi, K. B. Shintaku, S. Toride, H. Sekino, K. Kabuto and I. Nishio, *Phys. Rev. Lett.* **78** (1997), 3963.
[1.26] J. C. Crocker, J. A. Matteo, A. D. Dinsmore, and A. G. Yodh, *Phys. Rev Lett.* **82** (1999), 4352.
[1.27] L. Onsager, *Annals New York Academy of Sciences* **51** (1948), 627. (This and the next 4 references are cited from Liu, Jixing, *Onsager and proposition of the concept of entropy-driven ordering*, Lecture of Entropy-driven Ordering in Guizhou Univ., June 2007.)
[1.28] Sho Asakura and Fumio Oosawa, *J. Chem. Phys.* **22** (1954), 1255.
[1.29] Sho Asakura and Fumio Oosawa, *J. Polymer Sci.* **33** (1958), 183.
[1.30] P. J. Flory, Statistical Thermodynamics of Semi-Flexible Chain Molecules, *Proceedings of the Royal Society of London. Series A, Math. Phys. Sci.*, **234** (1956), 60.
[1.31] P. J. Flory, Phase equilibria in solutions of rod-like particles, *Proceedings of the Royal Society of London. Series A, Math. Phys. Sci.* **234** (1956), 73.
[1.32] H. R. Ma, Lecture in Guizhou University, Guizhou, 2007.
[1.33] Shiqi Zhou, Xiaoqi Zhang, Xianwei Xiang, Hong Xiang, *Chin. J. Chem. Phys.* **17** (2004), 38.
[1.34] J. P. Hansen and D. R. McDonald, *Theory of Simple Liquids*, 2nd Ed., Academic, London, 1986.
[1.35] K. Shundyak and R. van Roij, *Phys. Rev. E.* **69** (2004), 041703.
[1.36] K. Shundyak, *Interfacial phenomena in hard-rod fluids*, Doctoral Dissertation, Utrecht University, The Netherlands, 2004, http://igitur-archive.library.uu.nl/dissertations/2004-0604-113023/title.pdf.
[1.37] D. M. Zhu, W. H. Li and H. R. Ma, *J. Phys.: Condens. Matter* **15** (2003), 8281.
[1.38] Weihua Li, Song Xue, and H. R. Ma, *J. Shanghai Jiao Tong Univ.* **E-6** (2001), 126.
[1.39] M. P. Allen and D. J. Tildesley, *Computer Simulation of Liquids*, Clarendon Press, Oxford, (1994), Chap. 7, Section 7.2.
[1.40] Yehuda Snir and Randall D. Kamien, *Phys. Rev. E* **75** (2007), 051114.
[1.41] Dongsheng Zhou, *Entropy Competition and Local Order in Polymer Solutions*, Doctoral Theses of Nanjing University, 2003 (in Chinese).

Chapter 2

Basic Interactions in Soft Matters

Various unique properties that different soft matters have are based on the fundamental interactions between the component parts that soft matters consist of. The interactions are the building blocks of the research of soft matters. In many cases, if some wonderful properties of a soft matter are beyond understanding, it is often due to the lack of knowledge of the basic interactions involved. To study the properties of the complex systems (one extreme case is structures and functions of humans), physicists concentrate on the basic interactions — the interactions between cells, between proteins, between the base pairs of a DNA double helix, and so forth.

Although there are quite many theoretical models of the various interactions [2.11–2.13], the real interactions are based on the experiments [2.14]. The scientists who study soft matters are still doing experiments using many basic tools such as electronic scales, optical tweezers, and atomic force microscopy to explore the fundamental interactions between two particles under various conditions. One skill that all good experimentalists who study complex systems should have is to build a system as simple as possible and as pure as possible, and try to isolate the system from the complex and "untidy" environment. A long march begins with the first step — the study of a fundamental interaction is the first step of studying a complex system.

The DLVO theory (named after Derjaguin–Landau–Verwey–Overbeek) is dealing with the interaction of colloidal particles in liquid. This theory includes electrostatic, molecular (van der Waals) and structural interactions. The intramolecular interactions will be covered in Sec. 2.1, intermolecular (van der Waals) interactions in Sec. 2.2. The structural forces mean both lyophobic attractive force and lyophilic repulsive force, and the former will be introduced in Sec. 2.3. In fact, all the above interactions are electrostatic in nature. In Sec. 2.4, hydrodynamic interaction will be discussed as we try to understand the physics behind soft-matter-based body armors.

2.1 Intramolecular Interactions

In general, most of the physical and chemical properties of soft matters are determined by the interactions between the constituent atoms in the materials. Particularly important in the context of soft condensed matters is the order of magnitude of the bond energies compared to the thermal energies. If the bond energy is very much larger than the thermal energy k_BT, where k_B is the Boltzmann constant and T is the temperature in K, then the probability of the bond breaking and reforming owing to thermal agitation is vanishingly small; and these bonds can be thought of as permanent or chemical bonds. In contrast, if the bond energy is comparable to, or only a few times bigger than k_BT, then there is a finite probability that the bond may be broken and subsequently reformed by thermal agitation. This kind of bonds can be thought of as temporary or physical bonds.

According to the nature and magnitude of interactions, there are four basic kinds of bonds, which are described in the following.

2.1.1 *Ionic bonds*

In an ionic solid, the transfer of electrons between atoms is essentially complete, and the resulting ions interact via a straightforward Coulomb potential U_C, which for ions carrying charges q_1 and q_2 at separation r is

$$U_C = \frac{q_1 q_2}{4\pi\varepsilon_0 r}, \tag{2.1}$$

where ε_0 is the dielectric constant of vacuum. This interaction is non-directional, and is substantially stronger than the van der Waals interaction. Typically the order of magnitude for an interaction between a pair of ions in a crystal is 100×10^{-20} J. This is about two orders of magnitude larger than the thermal energy k_BT at room temperature. The interaction is, however, strongly modified if the ions are in solution. In such a case, the field due to a given ion is screened, because other ions can move to take up the positions, partially canceling the field of the test ion. This will be discussed in more detail in Sec. 4.3.3.

2.1.2 *Covalent bonds*

The covalent bonds exist in a molecule where the atoms bond together through the sharing of their valence electrons. In a molecule, electrons which originate from the component atoms interact with more than one nucleus, and the effect of these interactions is that the total energy of the molecule is lower than the sum of the energies of all the separated component atoms. This lowering of energy gives rise to an effective bonding between the atoms. Covalent bonding is short-ranged and highly directional, and is not conveniently represented by simple interatomic potentials. Typical covalent bond energies are in the range of 30 to 100×10^{-20} J, very much greater than k_BT at room temperatures.

For instance in methane each H atom contributes one electron to the C atom and on the other hand the C atom donates one electron to each H atom. The covalence of C and H atoms makes all the valence shells of C and H atoms closed to form a stable configuration.

2.1.3 *Metallic bonds*

The metallic bond could be viewed as a particular variant of covalent bond, in that it involves electrons being delocalized so as not to be associated with a single nucleus. However, in a covalent bond the extent of delocalization is limited to a pair, or at most a few nuclei, whereas in metals the electrons are delocalized throughout a macroscopic volume of material. This delocalization, as in the covalent bond, results in a reduction of the energy of the system. Metallic bonds are comparable in the magnitude of their bond energy to covalent bonds, but they are somewhat less directional.

2.1.4 *Hydrogen bonds*

This is a special kind of interaction which occurs when a hydrogen atom is covalently bonded to an electronegative atom such as oxygen or nitrogen. Because of the small size of the hydrogen atom and with only a single electron involved in it, the side of the hydrogen atom which is furthest away from the electronegative atom presents a significant unshielded positive charge, which can interact with another electronegative atom. This results in an attractive energy intermediate in magnitude between a full covalent bond and a van der Waals interaction. The typical values are between 2 and 6×10^{-20} J, corresponding to 25–100 k_BT at room temperature. Hydrogen bonding accounts for many of the peculiar properties of water, and also plays an important role in many biological macromolecules.

2.2 Intermolecular Interactions

2.2.1 *Double-layer forces*

In many cases, the surfaces are charged, so one may expect that the direct electrostatic interactions are important in determining the forces between colloidal objects with charged surfaces. However, when the objects are suspended in water, dissolved ions are always present, and the interaction of the charged bodies with the free ions profoundly modifies the nature of the electrostatic interaction. In particular, the electrostatic interactions are screened by dissolved ions, rather than a direct Coulomb interaction between two charged bodies. Thus, one finds a screened Coulomb interaction, which exponentially decays in strength with distance.

Suppose there is a surface that is ionized because some chemical groups at the surface ionize, or some ions from solution adsorb on it. The overall charge neutrality will be maintained by a layer of counter-ions, which will be attracted to the surface by the electrostatic field. Some of these counter-ions may be tightly bound to the surface (this layer of tightly bound ions is known as the Stern layer), but more will form a diffuse concentration profile away from the surface.

How can we determine the form of this concentration profile? There will be an electrostatic potential $\psi(x)$ at a distance x from the surface, and the density of ions $n(z)$ will be determined by the Boltzmann equation:

$$n(z) = n_0 \exp\left(\frac{-ze\psi(x)}{k_B T}\right), \tag{2.2}$$

where n_0 is the ion density at the surface, and ze is the charge of the ions.

Now the potential $\psi(x)$ is itself determined by the distribution of net charge $p(z)$ from the Poisson equation:

$$p(z) = -\varepsilon\varepsilon_0 \left(\frac{d^2\psi}{dx^2}\right). \tag{2.3}$$

ε and ε_0 are relative and vacuum dielectric constants, respectively.

In the simplest case where the only ions present are the counterions needed to balance the charge of the surface, $\rho = ze$, and we can combine the above two equations to give the Poisson–Boltzmann equation:

$$\frac{d^2\psi}{dx^2} = -\left(\frac{zen_0}{\varepsilon\varepsilon_0}\right) \exp\left(-\frac{ze\psi}{k_B T}\right). \tag{2.4}$$

This is an important equation that is also met in plasma physics and the study of electrons in solid.

In a very common case the surface is in contact with a solution of an electrolyte which is a solution of a univalent salt such as sodium chloride. Now we have the concentrations of both negative and positive ions to consider. Taking these concentrations to be n_+ and n_-, we have

$$n_\pm = n_0 \exp\left(\mp\frac{ze\psi}{k_B T}\right), \tag{2.5}$$

where n_0 is the ionic concentration in bulk solution. The net charge density is given by

$$p = ze(n_+ + n_-), \tag{2.6}$$

so

$$\frac{d^2\psi}{dx^2} = \frac{2zen_0}{\varepsilon\varepsilon_0} \sinh\left(\frac{ze\psi}{k_B T}\right). \tag{2.7}$$

We need to solve this equation subject to the boundary conditions, that is, for an isolated plate in solution both the potential Ψ and its gradient $d\Psi/dx$ approach to zero as x approaches infinity.

If the potential is small we can make the approximation $\sinh(ze\Psi/k_BT) \approx (ze\Psi/k_BT)$; in this limit (known as the Debye–Hückel approximation), then Eq. (2.7) has the solution

$$\psi(x) = \psi_0 \exp(-kx), \tag{2.8}$$

where k has the value of

$$k = \left(\frac{2e^2 n_0 z^2}{\varepsilon\varepsilon_0 k_B T}\right)^{1/2}. \tag{2.9}$$

Thus in an electrolyte, the electric fields are screened. The length that characterizes this screening, k^{-1}, is known as the Debye screening length. At distances much greater than the Debye screening length, which is inversely proportional to the square root of the concentration of salt in the electrolyte, the strength of the direct electrostatic interaction between charged objects falls rapidly to zero.

For monovalent salts the Debye screening length in water is given by $k = 0.304 I^{-1/2}$ nm, where I is the concentration of salt in moles/liter. Thus even for relatively modest salt concentrations the electrostatic effect is strongly screened.

What happens when two equally charged plates are brought together? As one might expect, there is a repulsive force between them, but less obviously the origin of this force is not due to the direct effect of electrostatics. The combination of the charged surface and the attracted counter-ions must overall be charge neutral, so this cannot lead to any repulsive force; instead it is the excess osmotic pressure of the counter-ions in the gap between the plates that lead to a repulsive force. It can be shown that in the limit of large separations and low surface potentials the repulsive force per unit area P between plates at separation D is given by

$$P = 64 k_B T n_0 \tanh^2\left(\frac{zx\psi 0}{4k_B T}\right) \exp(-kD). \tag{2.10}$$

This equation emphasizes that the forces of electrostatic origin in electrolyte solutions are short ranged, with the Debye screening length k^{-1} determining that range. The dependence of k on the concentration of ions means in practice that the colloidal stability is profoundly affected by the salt concentration.

As far as the predominant forces are not purely electrostatic or covalent, the bonding energies between particles are weak and we enter the domain of "soft matter." Predominant interactions are now screened ionic, dipolar, and van der Waals interactions.

2.2.2 *Electric dipole interaction*

Molecules of "weakly" ordered media are frequently electrically dipolar. Asymmetric molecules, bound internally by covalent bonds, often show permanent electric dipoles. The dipoles occur when the covalent bond connects two different atoms. Figure 2.1 shows an example of HCl molecule with a covalent bond formed by a spherically symmetric 1s-orbital of hydrogen and an elongated 3p-orbital of chlorine.

Although the molecule is electrically neutral, the mean positions of positive and negative charges do not coincide. The absolute value of the dipole moment $u = ql$ (charge times length) is measured in Debye units (1D = $3.336.10^{-30}$ C.m). The dipole moment of two elementary charges separated by 1 Å is about 4.8 D. The dipole vector is directed from the negative charge to the positive one (in some chemical literature, an opposite direction is taken).

The ion-dipole interaction plays an important role in the salvation of ions, i.e., in the phenomenon of clustering of solvent molecules-dipoles around the ion. If the solvent is water, then the ion is said to be hydrated. For example, a cation (e.g., Na^+) attracts and orients

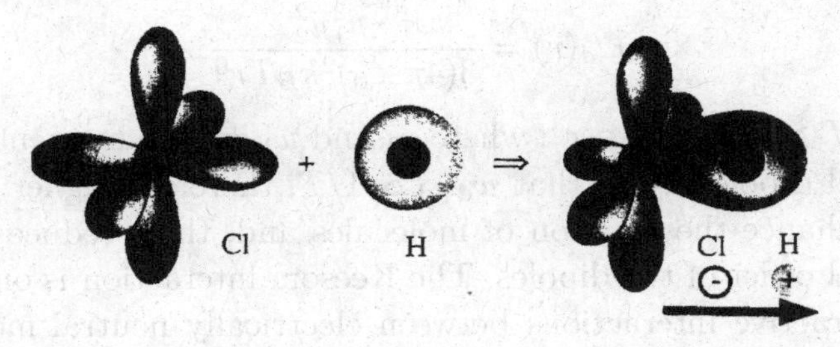

Fig. 2.1 An HCl molecule is formed with a covalent bond between Cl^- and H^+ ions [2.1].

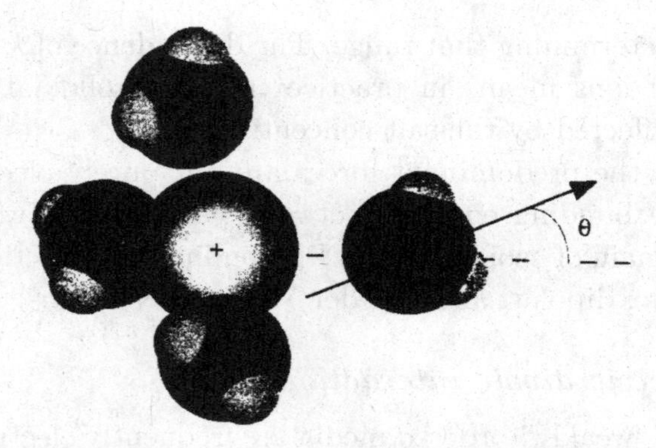

Fig. 2.2 Ion hydration is caused by an ion-dipole interaction when an ion is surrounded by water molecules. q is the angle between the direction of dipole moment and the line joining the dipole center and the ion [2.1].

a neighboring dipole along the line joining the charge and dipole, i.e., along the direction with angle $\theta = 0$ in Fig. 2.2. A hydrated ion might grow to large sizes: this is how the nucleation of rain droplets in clouds is explained. The gain of energy between the dipole oriented at random and the dipole $\theta = 0$ must be equal to or larger than $k_B T$ to stabilize the hydrated complex, which can be very stable if the dipolar ion is small and highly charged.

Attractive dipole-dipole and free dipole-free dipole (or Keesom) interactions are weaker than that of the ion-dipole interaction, and lead to less stable constructions. A free dipole is a dipole capable of taking all directions in space. Therefore, the energy of interaction is calculated by taking an ensemble average over orientations, as in the Langevin model of paramagnetism:

$$w_{d-d}(r) = \frac{-u_1^2 u_2^2}{3(4\pi\varepsilon\varepsilon_0)^2 k_B T r^6}, \tag{2.11}$$

when $\mathrm{k}_B T > u_1 u_2 / 4\pi\varepsilon\varepsilon_0 r^3$, where u_1 and u_2 are the moments of two interacted dipoles. Note that $w_{\mathrm{d-d}} \propto 1/T$, therefore higher temperatures enhance the rotation of molecules and, thus, reduce the orientational order of the dipoles. The Keesom interaction is one of the three attractive interactions between electrically neutral molecules. The other two already mentioned are London interactions between molecules with no permanent dipoles and Debye interactions between permanent and induced dipoles. All these three scale as $1/r^6$ and are generically called van der Waals forces.

2.2.3 *Induced dipoles, polarizability*

An electric field E, either applied or caused by a neighboring molecule, induces an electric dipole $u_{\text{ind}} = \alpha_0 E$ on a neighboring molecule by separating the centers of the positive and negative charges, where α_0 is the polarizability of the molecule.

The model shown in Fig. 2.3 helps to estimate this induced moment u_{ind} in a simple case of an electron rotating around a proton with a radius of R. The electron is subjected to two forces: the Coulomb attractive force with the nuclear charge and the shifting force of the external field E. The change in the Coulomb force caused by the shift is roughly $F_c = (e^2/4\pi\varepsilon_0 R^2)\sin\theta \sim e^2 l/4\pi\varepsilon_0 R^3 = eu_{\text{ind}}/4\pi\varepsilon_0 R^3$, and the field-induced force is $F_E = eE$. At equilibrium, $F_c = F_E$ and

$$u_{\text{ind}} = 4\pi\varepsilon_0 R^3 E = \alpha_0 E. \tag{2.12}$$

Here, we assume that the induced dipole is always oriented along the field; thus, $\alpha_0 E = 4\pi\varepsilon_0 R^3$ is a scalar quantity. With $R = 1$ Å, $\varepsilon_0 = 8.85 \times 10^{-12}$ F/m, one gets an estimate $\alpha_0 = 10^{-40}$ C^{2}m^2/J for

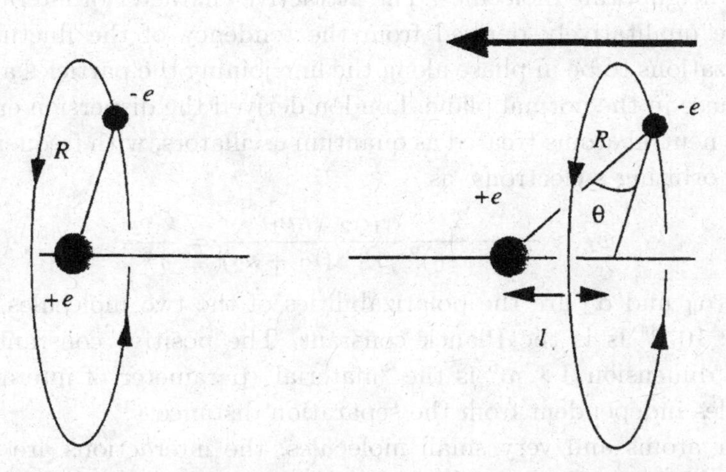

Fig. 2.3 (a) Original electric dipole; (b) formation of an induced dipole when there is an electric field E either applied or caused by a neighboring molecule [2.1].

the polarizablitity of an atom or a small molecule in vacuum. However, larger molecules may have much larger polarizabilities, because $\alpha_0 \propto R^3$. Furthermore, if the freely rotating molecule has a permanent dipole moment u, an external field would restrict free rotation of such a molecule so that the time-average polarization would be different from zero. The resulting orientational polarizability equals $u^2/3k_BT$.

The ion-neutral molecule and dipole-neutral molecule (or Debye) interactions are discussed in terms of field-induced dipoles. The Debye interaction scales as the Keesom interaction but is temperature independent. Finally, the London or dispersion forces are caused by fluctuating interactions between neutral atoms or molecules with no permanent dipoles. Actually, these forces are independent of the particular type of the molecule, because the fluctuation of the charge density is a universal quantum mechanical effect. Thus, the London interactions are present in any condensed matter system. The term "dispersion" originates from the dispersion of light in the visible, UV, and IR parts of the spectrum, and it should not be confused with the term "dispersion" describing colloidal systems: The frequencies at which an electromagnetic field causes a fluctuating dipole in a nonpolar molecule are absorption frequencies.

Fluctuations of charge densities induce mutual dipole moments in the neighboring molecules. The attractive character of interaction can be qualitatively derived from the tendency of the fluctuating polarizations to be in phase along the line joining the particles and in antiphase in the normal plane. London derived the dispersion energy of two neutral atoms treated as quantum oscillators, with frequency ν of the orbiting s-electrons, as

$$w_L \approx -\frac{3}{2(4\pi\varepsilon_0)^2}\frac{\alpha_1\alpha_2}{r^6}\frac{h\nu_1\nu_2}{(\nu_1+\nu_2)} = \frac{C_{12}}{r^6}, \qquad (2.13)$$

where α_1 and α_2 are the polarizabilities of the two molecules, $h = 6.63 \times 10^{-34}$ Js is the Planck constant. The positive constant C_{12} of the dimension J $\times$ m^6 is the "material" parameter of interacting particles independent from the separation distance r.

For atoms and very small molecules, the interactions are weak and unlikely to form stable, ordered phases at room temperature.

Noticing that $I_i = hv_i$ is some characteristic molecular energy that can be approximated by its first ionization potential, which is on the order of 10^{-18} J, and $\alpha_1 = \alpha_2 = 3 \times 10^{-40}\,\mathrm{C^2m^2/J}$ and would correspond to a molecular radius $R_i \approx 1.3$ Å in Eq. (2.12), one estimates the constant $C = 3\alpha^2 I/4(4\pi\varepsilon_0)^2 \approx 3R^6 I/4$ in Eq. (2.13) as $C \approx 0.6 \times 10^{-77}\,\mathrm{Jm^6}$. At a distance $r = 3$ Å, the corresponding attraction energy $|w_L| \approx 8 \times 10^{-21}$ J is larger than k_BT at room temperature. Hence, fairly large nonpolar molecules can be kept in a condensed state exclusively due to the London forces. Even when the molecules are polar, the London forces still make a significant contribution of 20–99% of the total energy. The second strongest interaction of polar molecules are usually Keesom forces.

2.2.4 *Repulsive forces*

Interactions between molecules include the attractive forces considered above and repulsive forces. In most of the cases, the repulsive forces are caused by electrostatic repulsion between particles having charges of the same sign or by short-range steric repulsion. Steric repulsion originates largely from the Pauli exclusion principle, and shows up when the electron clouds of two particles approach each other too closely. These interactions are hard to describe. In the simplest model of hard-core spherical particles of radius r_0, the energy of repulsion is assumed to be infinitely large when the separation between the centers is less than $2r_0$ and to be zero when the separation is larger than $2r_0$. In more realistic approaches, the repulsion potential at $r > 2r_0$ is often represented by an exponential decrease function of r, $w_{\text{rep}} \propto \exp(-r/\text{const})$, where const is positive, or, to simplify algebra in calculations involving both attractive and repulsive forces, as

$$w_{\text{rep}} = \frac{B}{r^{12}}, \tag{2.14}$$

where B is a positive constant.

Both electrostatic and steric repulsions appear in phases of interfaces, for example, in the lamellar phases of membranes. Because of the steric forces, the shape of the molecules has a pronounced effect on the nature of packing. Packing of amphiphilic molecules in monolayers or bilayers and cylindrical and spherical micelles depends in an extremely sensitive way on the molecular shape. Note also that in the lamellar phases, spatial fluctuations of membranes are at the origin of a specific repulsive (Helfrich) potential that stabilizes the lamellar phases against attractive London forces, in the case of nonionic surfactants.

Table 2.1 summarizes the schematic classification of molecular interactions that are assumed to take place in vacuum.

2.2.5 *The origin of van der Waals interaction* [2.4]

The van der Waals interactions (Fig. 2.4) arise between uncharged, weakly interacting atoms and molecules. All such atoms and molecules can be thought of as having a constantly fluctuating random dipole moment, and this dipole moment will induce a corresponding dipole in its neighboring atom or molecule, leading to an attracting force between them. The interaction is not strongly directional and typically the strength of the interaction between two molecules in contact is around the order of magnitude of 10^{-20} J. This is of the same order of magnitude as thermal energy $k_B T$ at room temperature.

The various types of physical forces described so far are fairly easy to understand since they arise from straightforward electrostatic interactions involving charged or dipolar molecules. But there is a further type of force, which — like the gravitational force — acts between *all* atoms and molecules, even totally neutral ones such as helium, carbon dioxide, and hydrocarbons. These forces have been variously known as dispersion forces, London forces, charge-fluctuation forces, electrodynamic forces, and induced-dipole–induced-dipole forces. We shall refer to them as *dispersion forces* since it is by this name that they are most widely known. The origin of this name has to do with their relation to the dispersion of light in the visible and UV regions of the spectrum, as we shall see.

Dispersion forces make up the third and perhaps most important contribution to the total van der Waals force between atoms and molecules, and because they are always present (in contrast to

Table 2.1. Typical molecular interactions in vacuum* [2.1].

Type of interaction	Representative scheme	Potential of interaction	Attributed to	Comments
Ion 1-Ion 2		$\frac{(ze)_1(ze)_2}{4\pi\varepsilon_0 r}$	Coulomb	Strongly directional; sign depends on valences
Ion 1-permanent fixed dipole 2		$\frac{(ze)_1 u_2 \cos\theta}{4\pi\varepsilon_0 r^2}$	Coulomb	Strongly directional; sign depends on z and orientation of the fixed dipole
Ion 1 — Freely rotating permanent dipole 2		$-\frac{(ze)_1^2 u_2^2}{6(4\pi\varepsilon_0)^2 k_B T r^4}$		Always attractive
Permanent fixed dipole 1-permanent fixed dipole 2		$-\frac{u_1 u_2}{4\pi\varepsilon_0 r^3} \times (2\cos\theta_1\cos\theta_2 - \sin\theta_1\sin\theta_2\cos\varphi)$	Coulomb	Sign depends on orientation; $\varphi =$ angle between the planes formed by each dipole and the line joining the dipoles
Permanent freely rotating dipole 1 Permanent freely rotating dipole 2		$-\frac{u_1^2 u_2^2}{3(4\pi\varepsilon_0)^2 k_B T r^6}$	Keesom	Always attractive
Ion 1-Induced dipole 2		$-\frac{(ze)_1^2 \alpha_2}{2(4\pi\varepsilon_0)^2 r^4}$		Always attractive
Permanent fixed dipole 1-Induced dipole 2		$-\frac{u_1^2 \alpha_2}{2(4\pi\varepsilon_0)^2 r^6}(1+3\cos^2\theta)$		Always attractive

(*Continued*)

Table 2.1. (*Continued*)

Type of interaction	Representative scheme	Potential of interaction	Attributed to	Comments
Permanent freely rotating dipole 1-Induced dipole 2		$-\frac{u_1^2\alpha_2}{(4\pi\varepsilon_0)^2 r^6}$	Debye	Always attractive
Induced dipole 1-Induced dipole 2		$-\frac{3h\nu_1\nu_2\alpha_1\alpha_2}{2(v_1+v_2)(4\pi\varepsilon_0)^2 r^6}$	London	Always attractive; h is the Planck constant; ν is the characteristic vibration frequency of electron
Induced dipole 1-Induced dipole 2 (retarded interaction)		$-\frac{23hc\alpha_1\alpha_2}{8\pi^2(4\pi\varepsilon_0)^2 r^7}$	Casimir and Polder	Always attractive; applies when $r > c/v$
Hydrogen bond	X – H . . . Y	Complicated, short-range, roughly $\sim -1/r^2$, where r is the distance between the atoms X and Y.		
Repulsion		$\frac{B}{r^\beta}$		$B > 0$; β in range 9–15, usually, 12

the other types of forces that may or may not be present depending on the properties of the molecules) they play a role in a host of important phenomena such as adhesion, surface tension, physical adsorption, the properties of gases and liquids, the strengths of solids, the flocculation or aggregation of particles in aqueous solutions, and the structures of condensed macromolecules such as proteins and polymers. Their main features may be summarized as follows:

(1) They are long-range forces and, depending on the situation, can be effective from large distance (greater than 10 nm) down to interatomic spacing (about 0.2 nm).
(2) These forces may be repulsive or attractive, and in general the dispersion force between two molecules or large particles does not follow a simple power law.
(3) Dispersion forces not only bring molecules together but also tend to mutually align or orient them, though this orienting effect is usually weak.
(4) The dispersion interaction of two bodies is affected by the presence of other bodies nearby. This is known as the *nonadditivity* of an interaction.

The origin of dispersion forces may be understood intuitively as follows: for a nonpolar atom such as helium, the time average of its dipole moment is zero, yet at any instant there exists a finite dipole moment given by the instantaneous positions of the electrons about the nuclear protons. This instantaneous dipole generates an electric field that polarizes any nearby neutral atom, inducing a dipole moment in it. The resulting interaction between the two nearby dipoles gives rise to an instantaneous attractive force between the two atoms, and the time average of the force is finite.

Now for a simple semi-quantitative understanding of how these forces arise, let us consider the following model based on the interaction between two Bohr atoms. In the Bohr atom an electron is pictured as orbiting around a proton. The smallest distance between the electron and proton is known as the *first Bohr radius* a_0 and is the radius at which the Coulomb energy $e^2/4\pi\varepsilon_0 a_0$ is equal to $2h\nu$, that is,

$$a_0 = e^2/2(4\pi\varepsilon_0)h\nu = 0.053\,\text{nm} \tag{2.15}$$

where h is the Planck constant and ν the orbiting frequency of the electron. For a Bohr atom, $\nu = 3.3 \times 10^{15}\text{s}^{-1}$, so that $h\nu = 2.2 \times 10^{-18}$ J. This is the energy of an electron in the first Bohr radius and is equal to the energy needed to ionize the atom — the *first ionization potential I*.

The Bohr atom has no permanent dipole moment. However, at any instant there exists an instantaneous dipole moment

$$u = a_0 e, \tag{2.16}$$

whose field will polarize a nearby neutral atom giving rise to an attractive interaction that is entirely analogous to the dipole-induced dipole interaction. The energy of this interaction in vacuum can be written as

$$w(r) = -u^2\alpha_0/(4\pi\varepsilon_0)^2 r^6 = -(a_0 e)^2\alpha_0/(4\pi\varepsilon_0)^2 r^6, \tag{2.17}$$

where α_0 is the electronic polarizability of the second Bohr atom, which from

$$\alpha_0 = 4\pi\varepsilon_0 R^3 \tag{2.18}$$

is approximately $4\pi\varepsilon_0 a_0^3$.

Using this expression for a_0 and the quantity of the first Bohr radius for a_0, we immediately find that the above interaction energy can be written approximately as

$$w(r) \approx -\alpha_0^2 hv/(4\pi\varepsilon_0)^2 r^6. \tag{2.19}$$

As one of the important empirical potentials of interactions, Lennard-Jones 6-12 model interaction can be expressed as summation of two terms in the sixth and twelfth powers of r as follows:

$$\Phi(r) = 4\varepsilon\{(\sigma/r)^{12} - (\sigma/r)^6\}. \tag{2.20}$$

Figure 2.4 shows the Lennard-Jones 6-12 potential as a function of r. ε is in the order of few times 10^{-14}erg, and σ is about

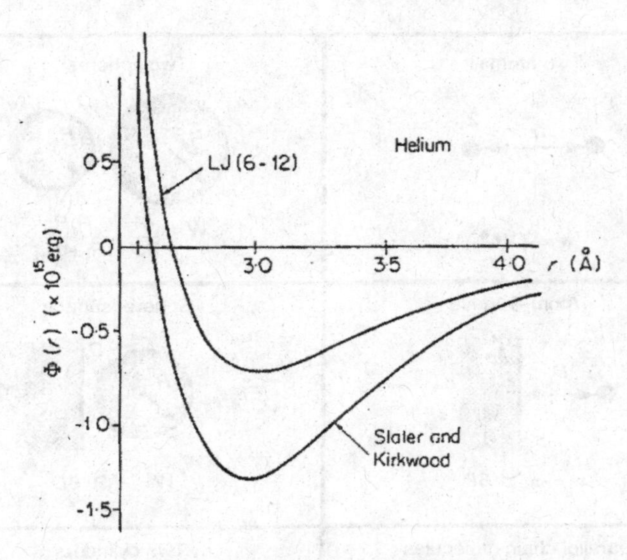

Fig. 2.4 L-J function incorporates many-body effects.

few angstroms. Their values can be determined experimentally to fit the data.

As we know [2.4] that when two atoms are an appreciable distance apart, the time taken for the electric field of the first atom to reach the second and return can become comparable with the period of the fluctuating dipole itself, and the field returns to find that the direction of the instantaneous dipole of the first atom is now different from the original and less favourably disposed to an attractive interaction. This distance d can be estimated as follows: $d = c/\nu$, where $c = 3 \times 10^8\,\mathrm{ms}^{-1}$ is the speed of light and $\nu = 3.3 \times 10^{15}\,\mathrm{s}^{-1}$ is the orbital frequency of the electron, and d is about 100 nm. When the distance between two atoms are much smaller than the above value, this *retardation effect* can be neglected. Figure 2.5 shows the formula to calculate the van der Waals interaction energies in vacuum for pairs of bodies of different geometries when the interaction is *non-retarded* and *additive*. These results of the "two-body" potential are obtained by the summation (or integration) of the energies of all the atoms in one body with all the atoms in the other. This procedure can be carried out for other geometries as well.

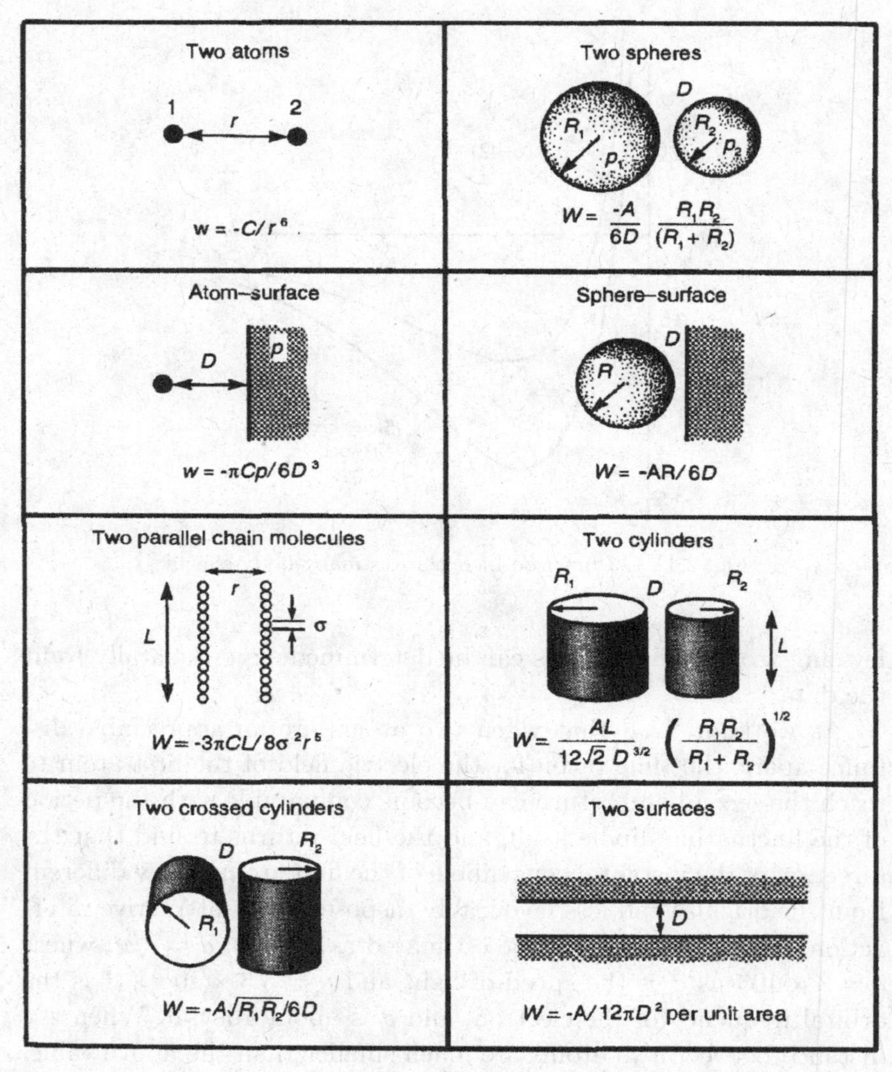

Fig. 2.5 The expressions of the non-retarded van der Waals potentials for the micro- and macroscopic bodies. The Hamaker constant A is defined as $A = \pi^2 \rho_1 \rho_2$ where ρ_1 and ρ_2 are the number of atoms per unit volume in the two bodies, D is the gap between the two bodies, and C is the coefficient in the atom-atom pair potential [2.4].

2.3 Structural Forces

Structural forces of colloidal particles come from their wettability in a liquid. They play a dominant role in the wetting region of contact angle $\theta < 20°$ (lyophilic repulsion force) and partial wetting region $\theta > 40°$ (lyophobic attraction forces). In the intermediate region, the interactions can be explained well by the DLVO theory which is covered in the sections on electrostatic and molecular van der Waals interactions.

2.3.1 *Wettability of colloidal particles*

The nearly contact region of two colloidal particles can be modeled as two interacting parallel plates (Fig. 2.6). A wetting film is formed on the solid surface. P is the pressure acting on the interlayer or film interface, and P_0 is the pressure in a bulk mother liquid [2.12]. In both cases disjoining pressure Π depends on the thickness h and is equal to $\Pi(h) = P - P_0$. In colloidal systems, P_0 is usually kept constant and P varies. A positive disjoining pressure leads to a repulsion of the two plates, while a negative one leads to a attraction of the two plates.

The relation between the contact angle and disjoining pressure can be discussed as follows. One of the fundamental equations related

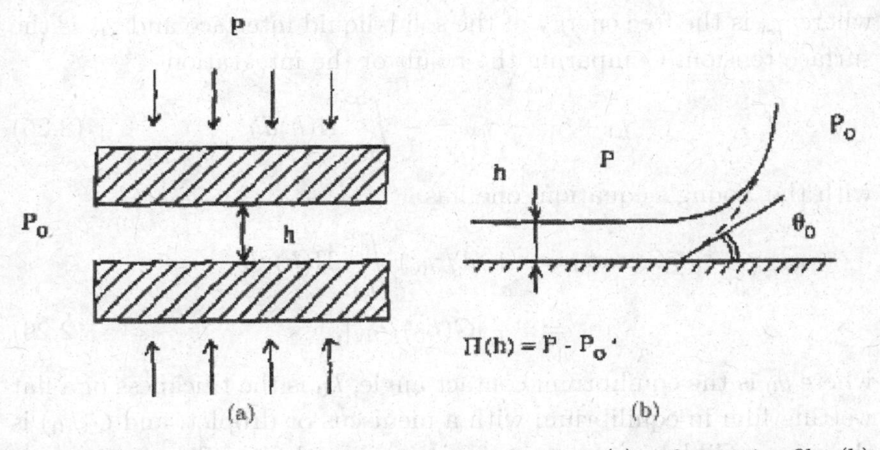

Fig. 2.6 Schematic representation of a colloidal system (a) and a wetting film (b) [2.12].

to wetting phenomena is the Gibbs relation

$$\Gamma = (d\gamma_{\rm sv}/d\mu)_T, \tag{2.21}$$

where Γ is the adsorption (a concentration of vapor or liquid on a surface in dimensions of length^{-1}), $\Gamma = \mathrm{h}/v_{\rm m}$, where $v_{\rm m}$ is the molar volume of the liquid. $\gamma_{\rm sv}$ is the free energy of the solid-vapor interface, and μ is the chemical potential, $\mu = -v_{\rm m}\Pi = RT\ln(p/p_{\rm s})$.

When a solid surface is being wetted, dx, an infinitesimal displacement of contact line (a boundary curve of triple interphases of solid, vapor and liquid), makes an energy change of unit contact line

$$\mathrm{d}E = (\gamma_{\rm SL} - \gamma_{\rm SV})\mathrm{d}x + \gamma_{\rm LV}\cos\theta\mathrm{d}x. \tag{2.22}$$

At the equilibrium condition, minimum energy requirement d$E = 0$ gives the Young's equation

$$\cos\theta = \frac{\gamma_{\rm SV} - \gamma_{\rm SL}}{\gamma_{\rm LV}}. \tag{2.23}$$

When the chemical potential changes from $\mu = 0$ in the bulk to some arbitrary value μ, the interface is changed from $\gamma_{\rm sl} + \gamma_{\rm lv}$ for thick liquid film to $\gamma_{\rm sv}$:

$$\int_{\gamma_{\rm sv}}^{\gamma_{\rm sl}+\gamma_{\rm lv}} d\gamma_{\rm sv} = -\int_{\mu}^{0} \Gamma d\mu = \int_{\Pi}^{0} hd\,\Pi, \tag{2.24}$$

where $\gamma_{\rm sl}$ is the free energy of the solid–liquid interface and $\gamma_{\rm lv}$ is the surface tension. Comparing the result of the integration

$$\gamma_{\rm sl} + \gamma_{\rm lv} - \gamma_{\rm sv} = -\int_{h_0}^{\infty} \Pi(h)dh \tag{2.25}$$

with the Young's equation, one has

$$\begin{aligned}\cos\theta_0 &= 1 + (1/\gamma_{\rm lv})\int_{h_0}^{\infty} \Pi(h)dh \\ &= 1 + [G(h_0)/\gamma_{\rm lv}],\end{aligned} \tag{2.26}$$

where θ_0 is the equilibrium contact angle, k_0 is the thickness of a flat wetting film in equilibrium with a meniscus or droplet, and $G(k_0)$ is the excess Gibbs free energy of wetting liquid film. The word excess implies that $G(k_0)$ is the Gibbs energy of a solution in excess of what it would be if it were ideal; while an ideal solution is a solution of which the enthalpy is zero.

The above equation can be used to calculate contact angle θ_0 on the basis of the $\Pi(h)$ isotherm and surface force theory.

2.3.2 *Lyophilic repulsive force*

The relation between the disjoining pressure and the thickness h is called the isotherm of disjoining pressure [2.13]. In the case of complete wetting, the isotherm can be fitted as

$$\Pi_{\mathrm{s}}(h) = K \exp(-h/\lambda), \tag{2.27}$$

where the parameter K is responsible for the strength of the force and the decay length λ is on the order of the bulk correlation length, λ_0. Since positive K corresponds to positive disjoining pressure, so this corresponds to a hydrophilic repulsion interaction; negative K corresponds to a hydrophobic attraction one.

Using Eqs. (2.26) and (2.27), one may find the disjoining pressure from the measurement of contact angle for the case of complete wetting.

In the case of nonpolar liquids, it is possible to limit the consideration by dispersion forces. Macroscopic theory of dispersion forces gives the following equation for disjoining pressure of thin wetting film of nonpolar liquids.

$$\Pi(h) = \frac{\hbar}{8\pi^2 h^3} \int_0^\infty \frac{(\varepsilon_{\mathrm{s}} - \varepsilon_{\mathrm{l}})(\varepsilon_{\mathrm{l}} - 1)}{(\varepsilon_{\mathrm{s}} + \varepsilon_{\mathrm{l}})(\varepsilon_{\mathrm{l}} + 1)} d\xi = \frac{\mathrm{A}_{\mathrm{slv}}}{6\pi h^3}, \tag{2.28}$$

where $\hbar$ is the Planck constant; $\varepsilon_{\mathrm{s}}(\xi)$ and $\varepsilon_{\mathrm{l}}(\xi)$ are the frequency functions of dielectric permittivity for a solid and liquid, respectively; h is the film thickness, and A_{slv} is the Hamaker constant. Since for condensed matters $\varepsilon > 1$, the sign of dispersion forces depends on the sign of the difference $\varepsilon_{\mathrm{s}}(\xi) - \varepsilon_{\mathrm{l}}(\xi)$.

If the polarity of liquid is larger than that of the solid, $\varepsilon_{\mathrm{l}} > \varepsilon_{\mathrm{s}}$, $\Pi = P - P_0$ is negative, the wetting film becomes unstable, and nonwetting takes place. On the other hand, if the polarity of the liquid is smaller than that of the solid, $\varepsilon_{\mathrm{l}} < \varepsilon_{\mathrm{s}}$, Π is positive, the wetting film is stable, and wetting takes place. This is why we see that nonpolar liquids wet well-cleaned metal.

For non-polar liquid, it is found that [2.13]

$$\cos\Theta_0 = 1 + [A_{\text{slv}}/12\pi h_0^2\gamma_{\text{lv}}]. \tag{2.29}$$

In Chapter 5 we will discuss how the wettability of colloidal particle effects their electrorheological properties.

2.3.3 *Slip length change on nanostuctured surface*

In Chap. 1 when we describe a Couette flow, it was supposed that the fluid speed at the wall is zero, which means that the fluid is dragged by the solid wall and it does not slip on the fluid-solid boundary. However, this is not always the case. When the interaction between the fluid and the wall is less than a certain value, the liquid will slip on the solid wall, that is, the fluid speed on the solid-fluid boundary is not zero. The virtual point where fluid speed is zero would then be found by extending the curve of the speed profile. The distance between the zero-speed point and the solid-liquid boundary is defined as slip length, λ, as seen in Fig. 2.7.

When nanometer-sized cubes are arranged on the solid surface, two cases can be found. One is the Wenzel case (Case A in Fig. 2.8), when the solid is wetted by the fluid and the groves are merged in liquid, namely the air pads disappear; while the other is the Cassie case (Case B in Fig. 2.8), when the solid is dewetted with the liquid, and the groves between the tubes are full of air.

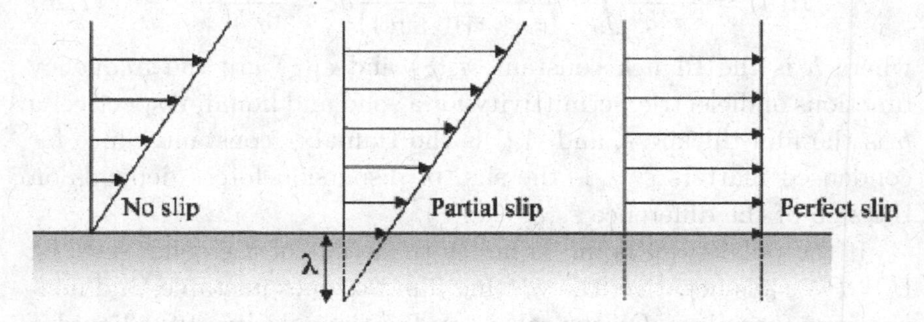

Fig. 2.7 Fluid speed profiles. At the fluid-solid boundary, slip length λ is defined as the distance between the point where the fluid speed is zero and the boundary. The slip length is zero for no-slip case, λ for partial slip case and infinite for the perfect slip case [2.14].

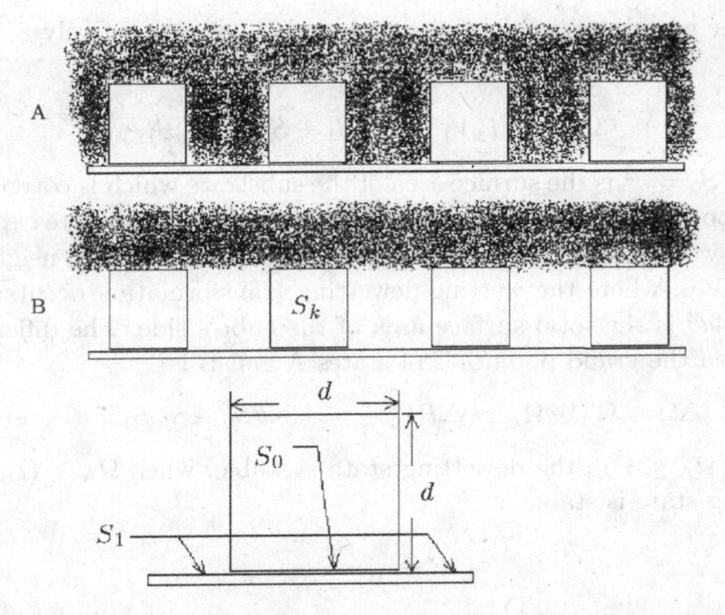

Fig. 2.8 Wenzel state A and Cassie state B with air padding [2.16].

The surface energies of the Wenzel state can be expressed as [2.15]

$$\begin{aligned} \mathrm{d}F &= r(\gamma_{\mathrm{SL}} - \gamma_{\mathrm{SV}})\mathrm{d}x + \gamma \mathrm{d}x \cos\theta^* \\ &= r\gamma\cos\theta_0 \mathrm{d}x + \gamma \mathrm{d}x \cos\theta. \end{aligned} \tag{2.30}$$

Minimizing surface energy of a drop on a structured surface, $\mathrm{d}F = 0$, leads to the following Wenzel equation:

$$\cos\theta = r\cos\theta_0. \tag{2.31}$$

In the Cassie state, displacing the contact line of a quantity $\mathrm{d}x$ parallel to the surface implies a change in surface energy is

$$\mathrm{d}F = \phi_{\mathrm{s}}(\gamma_{\mathrm{SL}} - \gamma_{\mathrm{SV}})\mathrm{d}x + (1 - \phi_{\mathrm{S}})\gamma\mathrm{d}x + \gamma\mathrm{d}x\cos\theta^*. \tag{2.32}$$

At equilibrium, F is minimum ($\mathrm{d}F = 0$), and one has the so-called Cassie equation (using Young's equation)

$$\cos\theta^* = -1 + \phi_{\mathrm{s}}(\cos\theta + 1). \tag{2.33}$$

It is proved that the slip length can be enlarged by locating cube s on the smooth fluid-solid border [2.16].

The grand potentials of states A and B are, respectively,

$$\begin{aligned}\Omega_A &= -P_l V_1 + \gamma_{ls}(S_1 + S_h),\\ \Omega_B &= -P_v V_1 + \gamma_{vs}(S_1 + S_h) + \gamma_{lv} S_1,\end{aligned} \tag{2.34}$$

where $S_0 = d^2$ is the surface area of the substrate which is covered by a nanopattern (cube); S_1 is the residual area of the substrate exposed to the water; $\mathrm{V_1 = S_1 d}$ is the volume of the region between neighboring cubes, where the wetting-dewetting transformation occurs; and $S_1 = 4d^2$ is the total surface area of the cube's side. The difference between the grand potentials of states A and B is

$$\Delta\Omega = \Omega_A - \Omega_B = \Delta P V_1 - \gamma_{lv}\cos\theta S_h - \gamma_{lv} S_1. \tag{2.35}$$

When $\Omega_\mathrm{A} < \Omega_B$, the dewetting state is stable, when $\Omega_\mathrm{A} > \Omega_B$, the wetting state is stable.

$$R \equiv \frac{S_0}{S_0 + S_1} \tag{2.36}$$

is defined as the coverage ratio. It is proved that the dewetting state is stable only when R is larger than a critical value of the coverage ratio, R_c,

$$R_\mathrm{c} = \frac{1 + \cos\theta + d\Delta P/\gamma_{lv}}{1 + d\Delta P/\gamma_{lv} - 3\cos\theta}. \tag{2.37}$$

The theoretical simulation gives proof that the slip length of the Cassie state with the abovementioned cubes and corresponding air pads can be 20 times larger than that on a smooth surface.

The critical condition for the dewetting is the existence of air pads. However, it is proved theoretically that the air pads between the nanometer-sized cubes are quite easily damaged, namely the Cassie state can be transformed to the Wenzel state under some minor pressure [2.17]. If the textures are filled with water, the material loses its water repellent properties. The invasion can occur through the vapor phase (condensation of dew, evaporation of a drop), or by an external pressure (which can be dynamic for impacts, due to the drop curvature for tiny drops, or hydrostatic, for example, a boat with a microtextured coating): invasion occurs quite easily, for a pressure corresponding to that below 20 cm of water.

The lithographically patterned hydrophobic mesa structures can hold the Cassie state only under small liquid pressures (<0.05 atm). On the other hand, a solid surface full of tall and sharp nanoposts with a submicron pitch pressurized (1 atm) flows [2.18]. The posts need to be tall enough so that the meniscus does not touch the bottom surface between posts. The posts also need to be populated densely enough, i.e., the pitch should be small enough, so that the surface tension of the warped meniscus withstands the pressure in the liquid. Nanoturf [2.18] was thus invented and tested to keep the Cassie state even under a pressurized flow. The slip length for water or glycerin on the hydrophobic nanoturf is about 20 or 50 μm, respectively.

2.4 Hydrodynamic Interactions

The phenomena of shear thickening of non-Newtonian fluids are widely known not only in application, but also in the theory of hydrodynamic interactions, such as hydrodynamic clustering and jamming.

2.4.1 *Shear thickening effect*

The shear viscosity in such colloids would decrease (for shear thinning) or increase (for shear thickening fluids) dramatically under certain shear flows. Paints are usually shear thinning colloidal materials. Wall painters like to brush the paint quickly allowing it to be thin enough and distributed evenly on the wall. On the other hand, some shear thickening (or dilatent) colloids are used to make flexible and lightweight "liquid body armors", eventually, to be able to protect arms and legs of soldiers, police officers and others [2.19].

One practical way to distinguish different types of shear thickening is by fitting the Ostwald–de Wale power law $\tau \propto \dot{\gamma}^{\alpha}$ to obtain the exponent α. A Newtonian flow corresponds to $\alpha = 1$. Inertial effects correspond to $\alpha = 2$ in the limit of high shear rates, independent of packing fraction. The discontinuous shear thickening (DST) is characterized by large α which approaches infinity as the critical packing fraction ϕ_c is reached. On the other hand, continuous shear thickening (CST) is typically characterized by α only slightly larger than 1, approaching 1 in the limit of zero packing fraction. While there is not a sharp transition between continuous and discontinuous shear thickening (α increases continuously with packing fraction), the former usually evolves into the latter as the packing fraction is increased [2.20].

Viscosity of shear thickening fluid is closely related with its particle volume fraction ϕ_v. Brady [2.21, 2.22] derived a pair theory from a truncation of the hierarchy of integral equations for the pair distribution function, and he found that a viscosity η diverges close to maximum packing ϕ_m as

$$\eta \propto \eta_0(1 - \phi_v/\phi_m)^{-2}, \tag{2.38}$$

in accordance with the standard phenomenological expression, where η_0 is the viscosity of basic liquid, $\phi_m \approx 0.63$ at low Péclet number (Pe) and $\phi_m \approx 0.71$ for an ordered system at high Pe. Péclet number is defined as the ratio of convection rate over the diffusion rate, and can be written as

$$\mathrm{Pe} = \frac{6\pi\dot{\gamma}\eta_0 a^3}{k_B T} \tag{2.39}$$

where $\dot{\gamma}$ is the rate of strain and a is particle size. The readers may notice that Eq. (2.38) encounters different comments in the literature [2.23].

2.4.2 *Essential role of friction in DST*

To understand the role of particle friction in the discontinuous shear thickening effect, Seto *et al.* [2.24] established a model that deals with the following interparticle interactions: the hydrodynamic force $\boldsymbol{F}_{\mathrm{H}}$, the contact force $\boldsymbol{F}_C$, and a repulsive force $\boldsymbol{F}_{\mathrm{R}}$. Since both fluid and particle inertia are neglected, the dynamics is overdamped and forces (and torques) are balanced for each particle:

$$\boldsymbol{F}_{\mathrm{H}}^{(i)} + \boldsymbol{F}_C^{(i)} + \boldsymbol{F}_{\mathrm{R}}^{(i)} = 0, \quad i = 1, \ldots, N. \tag{2.40}$$

Thus, the hydrodynamic force acting on a particle is approximately given as the sum of the regularized lubrication force

$$\boldsymbol{F}_{\mathrm{lub}}^{(i,j)} = -\alpha(h^{(i,j)})(U^{(i)} - U^{(j)}) \cdot n^{(i,j)} n^{(i,j)} \tag{2.41}$$

and the Stokes drag

$$\boldsymbol{F}^{(i)}_{\text{Stokes}} = -6\pi\eta_0 a\{U^{(i)} - U^{\infty}(r^{(i)})\}, \tag{2.42}$$

where $\alpha(h) = 3\pi\eta_0 a^2/2(h+\delta)$, η_0 is the viscosity of the suspending fluid, a is the particle radius, $h^{(i,j)}$ is the interparticle gap, length $\delta = 10^{-3}a$ is chosen to prevent divergence of α when the interparticle gap $h = 0$, $\boldsymbol{n}^{(i,j)}$ is the unit vector along the line of centers from particle i to particle j, and $\boldsymbol{U}^{(i)}$ and $\boldsymbol{U}^{(j)}$ are the velocities of particles i and j relative to an imposed flow velocity $\boldsymbol{U}^{\infty}(\boldsymbol{r}^{(i)})$ at the position of particle i.

The contact force $\boldsymbol{F}_C$ is activated for the gap $h^{(i,j)} < 0$. A simple spring-and-dashpot contact model is employed to mimic frictional hard spheres; the normal force is proportional to the overlap $-h^{(i,j)}$: $F^{(i,j)}_{\text{C,nor}} = k_{\text{n}} h^{(i,j)} \boldsymbol{n}^{(i,j)}$, where k_{n} is the spring constant. The friction appears as a tangential force and a torque, both proportional to the tangential spring displacement $\boldsymbol{F}^{(i,j)}_{\text{C,tan}} = k_{\text{t}}\xi^{(i,j)}$ and $T^{(i,j)}_C = k_{\text{t}} a n^{(i,j)} \times \xi^{(i,j)}$, where k_{t} is the tangential spring constant. The tangential force is subject to Coulomb's law $\boldsymbol{F}_{\text{C,tan}} < \mu \boldsymbol{F}_{\text{C,nor}}$, where μ is the friction coefficient ranging from 0 (the smoothest) to 1 (the roughest). Even with contact forces, ideal hard-sphere suspensions should be Newtonian, because different $\dot{\gamma}$ result in the same particle trajectories, but with different time ($\sim 1/\dot{\gamma}$) and force ($\sim\dot{\gamma}$) scales. When trying to mimic hard spheres with linear springs, the authors avoid introducing an artificial shear-rate dependence. They choose k_{n} and $k_{\text{t}} \sim \dot{\gamma}$, and tune the dashpot resistance to keep a short contact relaxation time ($=10^{-3}/\dot{\gamma}$).

The repulsive force, which is due to an electrostatic double-layer force,

$$\boldsymbol{F}^{(i,j)}_{\text{R}} = -Cae^{-\kappa h^{(i,j)}} n^{(i,j)} \tag{2.43}$$

is used for $h^{(i,j)} > 0$, with $1/\kappa = 0.05a$. A dimensionless shear rate

$$\dot{\Gamma} \equiv 6\pi\eta_0 a^2 \dot{\gamma}/|\boldsymbol{F}_{\text{R}}(h=0)| \tag{2.44}$$

is introduced and the Lees-Edwards boundary conditions [2.25] are adapted. The result is presented in Fig. 2.9.

Simulations are performed using the Lees–Edwards boundary conditions. The simulation boxes are cubes for $n = 512$ particles

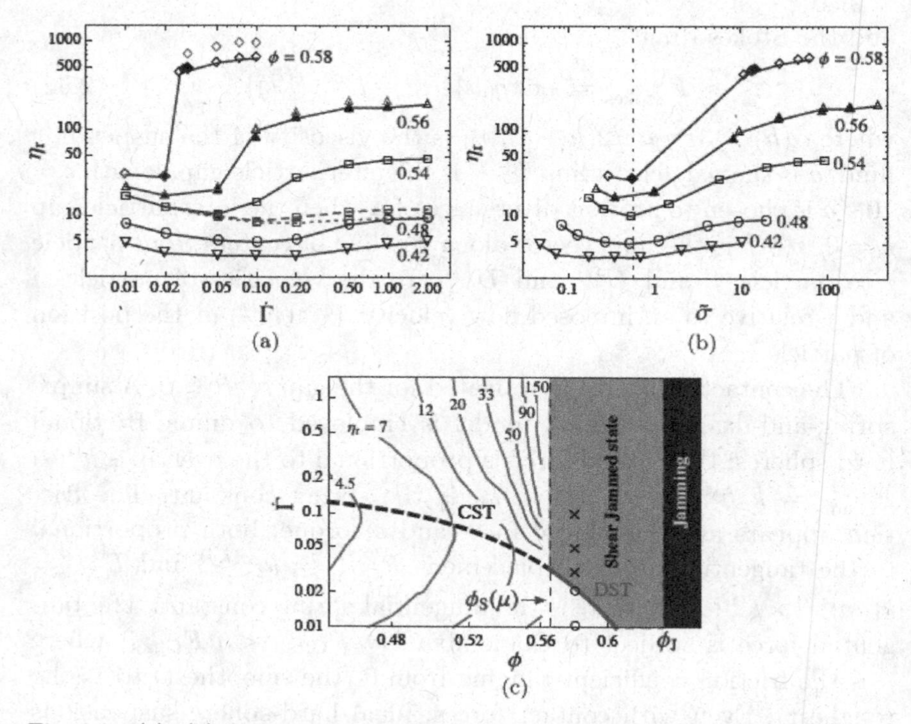

Fig. 2.9 (a), (b) Shear rate and stress dependence of the relative viscosity η_r, respectively. The open and filled symbols indicate the results of $n = 512$ and 2048, respectively. The friction coefficient is $\mu = 1$ except for the dashed and dotted curves, for which $\mu = 0.1$ and 0, respectively. (c) DST (solid thick curve) and CST (dashed thick curve) are shown in the phase diagram [2.24].

and rectangular parallelepipeds (the shear plane is square, and the depth is one half of the other dimensions) for $n = 2048$. They obtain the dependence of the relative viscosity η_r on the dimensionless shear rate $\dot{\Gamma}$ and stress $\tilde{\sigma}(\equiv \eta_r\dot{\Gamma})$ for a range of volume fractions ϕ and friction coefficients η, as shown in Fig. 2.9.

Seto *et al.* [2.24] display in Fig. 2.10 the result on spatial distribution of contact bonds in the system, in both low and high viscosity phases. They found that for frictional spheres, a transition from CST to DST is observed when particle volume fraction ϕ increases. For the roughest particles with friction coefficient $\mu = 1$, the transition occurs at $0.56 < \phi_c < 0.58$. Although the shear rate at the onset of

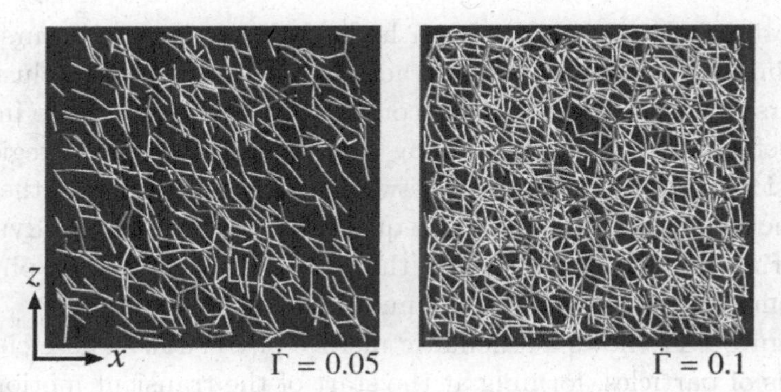

Fig. 2.10 Particle contacts are visualized as bonds at the two shear rates $\dot{\Gamma} =$ 0.05 and 0.1 exhibiting low and high viscosity states, respectively ($\mu = 1$, $\phi = 0.56$, $n = 2048$) [2.24].

thickening decreases with increasing ϕ, the onset stress $\phi_{\rm on}$ is constant, as shown in Fig. 2.9(b). For $\phi > \phi_{\rm c}$, the high viscosity state achieved at high shear rate is strongly dependent on the particle stiffness, while no such dependence is seen for $\phi < \phi_{\rm c}$ or in the low viscosity state.

For friction coefficient $\mu = 0.1$, the thickening is substantially weaker than for $\mu = 1$, and it is completely absent in the frictionless case, even at volume fractions approaching the jamming point $\phi_{\rm J}$. This means that friction is essential for a shear jammed state to exist.

For low shear rate, contact force chains appear, but only as elongated and isolated objects (unique to the combined hydrodynamic-plus-contact force algorithm) along the compression axis (Fig. 2.10, $\dot{\Gamma} = 0.05$), whereas for high shear rate, a network percolates in all directions (Fig. 2.10, $\dot{\Gamma} = 0.1$).

2.4.3 *Hydroclustering and jamming for an infinite system*

In the last section hydroclustering is acquired for systems consisting of some hundreds or thousands of particles. In fact, Farr *et al.* [2.26] constructed a model of the stress carrying "fabric" in a concentrated dispersion of hard spheres at $\mathrm{Pe} = \infty$ based on the Smoluchowski's equation. Their model predicts that above a lower volume fraction

ϕ_l substantially less than 0.63, a hydrodynamic logjam occurs even in an infinite system. This is characterized by an elongated cluster or rod, first forming parallel to the compression axis, growing to infinite size before it tumbles by passing $\theta = 0$ and entering a region of extensional flow. θ is the angle between rods and gradient of the flow. The model quantifies the hitherto qualitative notion of hydrodynamic clustering and jamming, which they believe to be a new physical phenomenon underlying discontinuous shear thickening.

Figure 2.11 shows a schematic picture of the flow. These clusters or rods of particles, forming at the start of the transient motion, are roughly parallel to one another and for simplicity we assume that all the rods which subsequently form and grow by aggregation constitute an approximately parallel population.

The rods contain extremely small interparticle gaps and are defined by this separation of length scales. They take the rod lengths to be incompressible, and for a rod of j particles, each of diameter a, the length will be $L \approx ja$. The bulk average deformation of the

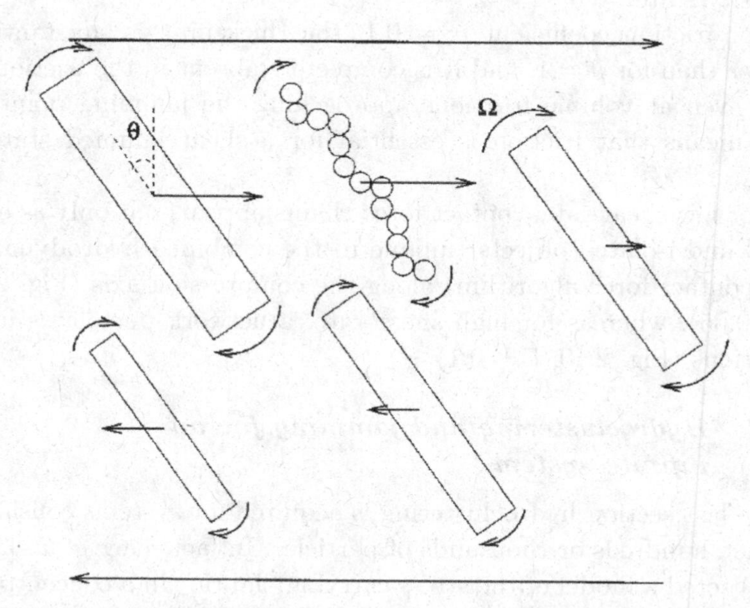

Fig. 2.11 Schematic picture of structures in the flow [2.26]. Gradient is perpendicular to the flow direction indicated by two long arrows in opposite directions.

sample is simple shear; that is, a velocity field given by $\boldsymbol{v} = (0, 0, \dot{\gamma}x)$, where the x, y, and z axes are the "gradient", "flow", and "vorticity" directions, respectively (see Fig. 2.11).

Its center of mass moves with the mean velocity at that point, and the rod rotates at angular velocity

$$\Omega = -\dot{\gamma}\cos^2\theta, \tag{2.45}$$

where $\dot{\gamma}$ is the shear rate, and θ is the angle the rods make with the gradient direction, in the flow-gradient plane shown in Fig. 2.11.

If we consider two rods about to collide, of lengths i and j particle diameters, respectively, we find that in this mean flow approximation, they approach one another with a relative velocity given by

$$V_{\text{rel}} = \frac{\dot{\gamma}}{2}(ai + aj)\cos\theta\sin\theta. \tag{2.46}$$

Thus a more reasonable estimate of the collision cross section Σ is $\Sigma = \omega_1[a\sqrt{\min(i,j)}]^2$, where ω_1 is a dimensionless constant, controlling the width of the walks.

If we now take the rod lengths to be approximately additive on collision, we will obtain a form of Smoluchowski's aggregation equation [2.28]:

$$\frac{dn_k}{dt} = \sum_{i,j}^{\infty} \Sigma V_{\text{rel}} C(\phi_v) n_i n_j (\delta_{i+j\cdot k} - \delta_{i\cdot k} - \delta_{j\cdot k}). \tag{2.47}$$

Here n_k is the number of rods per unit volume with length ka, and $C(\phi_v)$ is a "crowding factor" representing the increase in the collision frequency due to the volume excluded by the particles. Near the random close packed volume fraction $\phi_c \approx 0.63$ at which the dispersion is jammed at the start and cannot flow, one would expect the crowding factor to diverge as $1/[1 - (\phi_v/\phi_c)^{1/3}]$. Note in Eq. (2.47) that the combinatorial factor of 1/2 normally present is canceled by the fact that rods may collide at both ends. We may nondimensionalize in terms of $X_{\text{k}} = n_{\text{k}}/n_0$ where n_0 is the initial concentration of particles, and the scaled time

$$dT = dt\left(\frac{6\phi_v}{2\pi}\right)\dot{\gamma}\cos\theta\sin\theta C(\phi_v)\omega_1, \tag{2.48}$$

where $\phi_{\rm v} = \pi n_0 a^3/6$, to obtain the standard form of Smoluchowski's equation [2.27]

$$\frac{dX_k}{dT} = \frac{1}{2}\sum_{i,j}^{\infty} K_{ij}X_iX_j(\delta_{i+j\cdot k} - \delta_{i\cdot k} - \delta_{j\cdot k}). \tag{2.49}$$

This has the homogeneous kernel

$$K_{ij} = 2(i+j)\min(i,j) = 2ij + 2[\min(i,j)]^2. \tag{2.50}$$

To find the solution of Smoluchowski's equation, the authors of [2.26, 2.27] found the set of Eqs. (2.49) by obtaining the Taylor series solution about the origin, for monomeric initial conditions $[X_1(T=0)=1]$. The first term in the expansion for each cluster size is $X_n = N_nT^{n-1} + O(T^n)$, where the numbers N_n are given by the recurrence relation:

$$(n-1)N_n = \frac{1}{2}\sum_{i+j=n} K_{ij}N_iN_j. \tag{2.51}$$

This is readily generalized, to provide the full Taylor expansion, up to a given order, for

$$X_n(T) = \sum_k X_n^{[k+1]}T^k. \tag{2.52}$$

With $X_1^{[1]} = 1$ and $X_a^{[b]} = 0 \; \forall a > b$, then

$$(k+r-1)X_k^{[k+r]} = \frac{1}{2}\sum_{a=1}^{k-1}\left[K_{a\cdot k-a}\sum_{c=0}^{r} X_{k-a}^{[k+r-a-c]}\right] - \sum_{c=0}^{r-1}\left[X_k^{[k+c]}\sum_{m=1}^{r-c} K_{km}X_m^{[r-c]}\right]. \tag{2.53}$$

This is used by first setting $r=0$, and applying the formula, for $k = 1, 2, \ldots$, then proceeding to $r = 1$, etc. The stress $\sigma_{\rm rod}$ in the fluid due to the rods (which contain at least two particles) is

$$\sigma_{\rm rod} = \frac{E(\phi_v)}{12}\dot{\gamma}(\cos^2\theta)(\sin^2\theta)\left(\frac{6\phi_v}{\pi}\right)\sum_{j=2}^{\infty} j^3X_j. \tag{2.54}$$

At later times, some of the monomers have been incorporated into rods, and so one might expect $\sigma_{\text{mon}} \approx X_1\sigma(0)$ so that the final expression for the stress is

$$\sigma = \sigma_{\text{mon}} + \sigma_{\text{rod}}, \tag{2.55}$$

which involves the three parameters ω_1 , $E(\phi_v)$, and $\sigma(0)$.

Note that in the mean field approximation the passing fluid exerts a frictional force per unit length dF/dx is given by

$$\frac{dF}{dx} = E(\phi_v)\dot{\gamma}(\sin\theta)(\cos\theta)x \tag{2.56}$$

where $E(\phi_v)$ is a parameter for each volume fraction, with the dimensions of viscosity. The authors of [2.26] find also the expression of the sliding force on a length a of the rod due to squeeze lubrication interactions and logarithmic shear lubrication force of sliding particles.

$$E(\phi_v) \approx 7\pi\eta_0 \ln\left(\frac{a}{h_{\text{typ}}}\right). \tag{2.57}$$

The theoretical result of stress $\sigma(0)$ is as follows:

$$\sigma(0) \approx \frac{3\phi_v \nu_{\text{NN}}\dot{\gamma}^2}{a^3\pi}(a + h_{\text{typ}})^2 \left[\frac{\pi\eta_0 a^2}{40h_{\text{typ}}} + \frac{2\pi\eta_0 a}{15}\ln\left(\frac{a}{2h_{\text{typ}}}\right)\right], \tag{2.58}$$

which diverges near random close packing as $1/(1 - \phi_v/\phi_c)$.

Equations (2.57) and (2.58), together with the reasonable guess that $\omega_1^{1/2} \approx 0.5$, or a little less, therefore provide a semiquantitative description of hydrodynamic jamming with no free parameters.

In the context of this theory, the authors [2.26] choose for hc a value dependent upon the expected gap hd in a dimer forming at the start of the flow. This is predicted to collapse like

$$h_d = h_{\text{typ}} \exp\left[\frac{2E(\phi_v)}{3\pi\eta_0}\ln\left(\frac{\cos(\pi/4)}{\cos\theta}\right)\right], \tag{2.59}$$

in which the time dependence enters through η, and we take

$$h_c = \frac{h_d}{\beta}, \tag{2.60}$$

where β is some parameter of order unity.

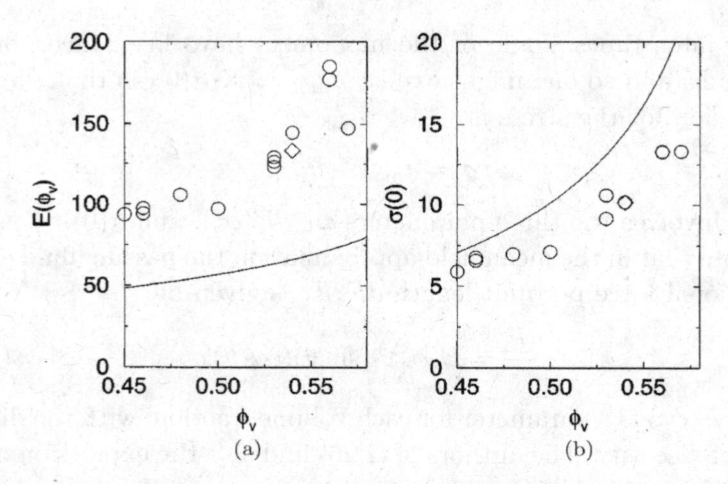

Fig. 2.12 Plot of $E(\phi_v)\eta_0$ against ϕ_v obtained from the fit to the stress-strain data. The units are chosen so that $a = \eta_0 = 1 = \dot{\gamma}$, thus the physical stress s is in units of $\eta_0\dot{\gamma}$. The circles for 700 particles and the diamond for 1400 particles show good agreement. The solid line for comparison is a theoretical prediction, Eq. (2.57). (b) Same for $\sigma(0)\dot{\gamma}\eta_0$; the theoretical prediction is from Eq. (2.58).

A simulation was run at volume fraction $\phi_v = 0.53$, using 700 particles and $E(\phi_v) = 125$, and the configurations partitioned into clusters on the basis of Eq. (2.59).

Figure 2.12 shows the fitted values of $E(\phi_v)$ from the simulation plotted with this theoretical prediction, and shows that Eq. (2.57) underestimates the fitted values, by a factor of about 2. This is quite acceptable for an argument based purely on dimensional grounds, and indeed, one might expect an underestimate on the part of Eq. (2.57), since we have assumed that all gaps surrounding the rods are of size h_{typ}, while in practice, the rod will have to fight through many gaps that are significantly smaller than this.

Thus, the theory of hydrodynamic clustering and jamming applied to an infinite space agrees well with the simulation of limited number of particles.

References

[2.1] R. Jones, *Soft condensed matter*, Oxford, (2002).
[2.2] A. Skjeltorp, *Soft condensed matter: configurations, dynamics, and functionality*, Kluwer Academic Publishers, (2000).

[2.3] F. London, *Trans. Faraday Soc.* **33** (1937), 8.
[2.4] J. N. Israelachvili, *Intermolecular & Surface Forces*, 2nd edition, Academic Press, London, 450 (1992).
[2.5] P. C. Hiemenz, *Principles of Colloid and Surface Chemistry*, 2nd edition, Marcel Dekker, Inc., New York, 790 (1986).
[2.6] D. Tabor, *Solids and Gases*, Cambridge University Press, 19 (1979).
[2.7] H. G. B. Casimir and D. Polder, *Phys. Rev.* **73** (1948), 360.
[2.8] E. M. Lifshitz, *Zh. Eksp. Teor. Fiz.* **29** (1955), 94 [Sov. Phys. JETP **2** (1956), 73].
[2.9] L. D. Landau and E. M. Lifshitz, *Electrodynamics of Continuous Media*, **8**, 2nd edition, Pergamon Press, Oxford, (1984).
[2.10] P. G. de Gennes, *Acad. Sci.* **271** (1970), 469.
[2.11] H. B. G. Casimir, *Proc. K. Ned. Akad. Wet.* **51** (1948), 793.
[2.12] N. V. Churaev, *J. Coll. Interface Sci.* **172** (1995), 479–484.
[2.13] N. V. Churaev, *Adv. Coll. Interface Sci.* **58** (1995), 87–118.
[2.14] Lauga *et al.*, *Handbook of Experimental Fluid Dynamics* (2005).
[2.15] J. Bico, C. Marzolin and D. Quéré, *Europhys. Lett.* **47** (1999), 220–226.
[2.16] Ding Li, Haiping Fang *et al.*, *Chin. Phys. Lett.* **24** (2007), 1021.
[2.17] Lafuma and Quéré, *Nature Mat.* **2** (2003), 457.
[2.18] Chang-Hwan Choi and Chang-Jin Kim, *Phys. Rev. Lett.* **96** (2006), 066001.
[2.19] How Liquid Body Armor Works, HowStuffWorks https://science.howstuffworks.com/liquidbodyarmor.2018/3/10.
[2.20] E. Brown and H. M. Jaeger, *Rep. Prog. Phys.* **77** (2014), 046602.
[2.21] J. F. Brady, *J. Chem. Phys.* **99** (1993), 567.
[2.22] J. F. Brady, *J. Fluid Mech.* **272** (1994), 109.
[2.23] B. Cichocki and B. U. Felderh, *J. Chem. Phys.* **101** (1994), 1757.
[2.24] R. Seto, R. Mari, J. F. Morris, and M. M. Denn, *Phys. Rev. Lett.* **111** (2013), 218301.
[2.25] A. W. Lees and S. F. Edwards, *J. Phys. C*, **5** (1972), 1921.
[2.26] R. S. Farr, J. R. Melrosen and R. C. Ball, *Phys. Rev. E.* **55** (1997), 7203.
[2.27] P. G. J. Van Dongen, Ph.D. thesis, Rijksuniversiteit Utrecht (1983). (Ref. 15 of Ref. [2.26]).
[2.28] M. von Smoluchowski, Versuch einer mathematischen Theorie der Koagulationskinetik kolloider Lösungen (Attempt of a mathematical theory of the coagulation kinetics of colloidal solutions), (Sec. III. Mathematische Theorie der raschen Koagulation (Mathematical theory of rapid coagulation). *Z. Phys. Chem.* **92** (1917), 129–168 (in German).

[illegible] (1963) [illegible]
[illegible]
London, 400 (1929) [illegible]
[illegible] 2nd edition [illegible]
[illegible] (1958) [illegible]
[illegible]
[illegible] Oxford [illegible]
[illegible]
[illegible] (2005) [illegible]
[illegible]
[illegible] 101 [illegible]
[illegible]

Chapter 3

Structure Determination of Soft Matters

To know the structure of soft matters, one has to use electromagnetic waves (x-ray, light), or electrons to determine the atomic positions and/or the positions of the clusters of atoms. Information about the structure of soft matters would be important for one to know the particle-particle interactions (or potential of mean force) and to improve the physical properties of soft matters.

3.1 Why Neutrons

3.1.1 *Advantages of neutron scattering*

What can be determined from neutron scattering? Most information about nature is carried by two kinds of signals: (a) generated by the objects under study, and (b) modified by them [3.1]. This means looking at an object emitting light or at an object reflecting light. The latter method is referred to as scattering. In soft matters, the objects under study can be single atoms, molecules or clusters of them.

Scattering may tell us (a) the size and the shape of the materials, (b) how they are arranged (positions) and (c) how they move. The former two are related to the static problems and the latter is related to the dynamic problems. Problem (b) is about the space correlation among soft matter materials, while problem (c) is about the time correlation among them.

Why are neutrons so interesting in the study of soft matter? Special properties of neutrons make them very useful in the study of soft matters.

(1) Neutrons satisfy two conditions of wavelength and energy.

In order to study the structures of the materials using scattering experiments, the radiation must have a wavelength comparable with atomic distances. In order to study collective excitations, the radiation must have a wavelength comparable with that of the associated quasiparticles such as phonons. The incident beam must carry energies that are comparable with those of molecular or lattice motions so the incident beam can exchange energy with the sample.

The picture described above gives a comparison of the energy-wavelength relations between electromagnetic (photon) and neutron radiations. It can be seen that only neutron radiation satisfies the wavelength and energy conditions simultaneously. The energy-wavelength relation of photons (x-ray) can be expressed as $E_x = hc/\lambda$, where h and c are the Planck constant and the speed of light, respectively. When the wavelength $\lambda = 0.154\,\text{nm}$ (for Cu-K$_\alpha$), $E_x = 8.048\,\text{keV} = 1.289 \times 10^{-15}\,\text{J}$. The energy-wavelength relation of neutrons, on the other hand, is expressed as $E_{\text{N}} = h^2/2m_{\text{N}}\lambda^2$, where m_{N} is neutron mass. For the same wavelength, $E_{\text{N}} = 34.467\,\text{meV} = 278\,\text{cm}^{-1}$, which is 233,500 times smaller than that of photons. The above energy corresponds to the energy of neutrons traveling with the velocity $v = (2E_{\text{N}}/m_{\text{N}})^{1/2} = 2568\,\text{m/s}$, which is in the range of thermal neutrons. The above wavelength value falls within the range of typical molecular vibration modes as seen by infrared and Raman spectroscopy.

The neutron wavelength is similar to the atomic spacing. Neutron scattering may decide the structural information, such as crystal structure and atomic spacing of solid and soft matters, with high resolution in the range of 10^{-15} to 10^{-6} m.

The energy of thermal neutrons is comparable to that of fundamental excitations in both solid and soft matters, and they have similar molecular vibration, crystalline state and atomic kinetics and dynamics.

(2) Neutrons have a strong ability to identify light atoms.

X-rays are scattered by the shell electrons. This means that the probability for a scattering event is directly proportional to the atomic number Z, hence light atoms (hydrogen atoms, for example) are hardly "visible". For light scattering, the momentum transfers are negligibly small and the interaction of electromagnetic radiation with dipole moment is important.

Neutron scattering, on the other hand, is ideal in identifying light atoms, such as hydrogen atoms. A neutron is a chargeless elementary particle with the mass of $m_{\mathrm{N}} = 1.6749286 \times 10^{-27}$ kg and radius of 1.532×10^{-18} m. For neutron scattering, momentum transfer plays an important role. A neutron has a spin of 1/2 and magnetic moment of $\mu_{\mathrm{N}} = -1.04187563 \times 10^{-10}\,\mu_{\mathrm{B}}$ (Bohr magnetons) $= -0.96623707 \times 10^{-26}$ J/T. Neutrons interact with matters via two physical processes: with atomic nuclei through the strong nuclear interaction, and with magnetic moments of any origin (e.g. electronic orbital magnetic moments or nuclear spins).

The ability to identify light atoms is very important in studying the structures and functions of soft matters in which light elements are usually the major components of the building blocks. Neutron scattering can be used to find out how potato roots grow, how a protein is hydrated, how hydrogen atoms are stored in the nanometer-sized carbon tubes, which is the fundamental mechanism of hydrogen storage batteries.

(3) Isotope substitutions

Isotopes of an atom have the same atomic number (same number of protons and electrons surrounding them) but a different number of neutrons. These different masses and magnetic moments give different signals of the isotopes of an atom in neutron scattering, and this opens the whole field of isotope substitution study, which is used to initiate or improve a great number of experiments.

(4) Neutron scattering is a good tool for magnetic measurements.

As mentioned above, neutrons have a finite magnetic moment, so they are a good tool for studying the microscopic magnetic structure and magnetic wave, and developing magnetic materials. Neutrons have a spin, so they may be formed into a polarized neutron beam to study the orientation of atomic nuclei using coherent and incoherent scattering.

(5) Neutrons have a strong ability for penetration.

Neutrons are neutral particles that carry a high momentum and do not bend in magnetic fields, and so have a strong penetrative ability. Neutron scattering can be used for the study of samples in extreme conditions such as low temperature, high pressure, high electric and/or magnetic field with thick and complicated sample chamber walls. In these cases, fast neutrons (0.5–10 MeV) are usually used due to their large penetration depths.

Advanced technology requires stronger, lighter, less expensive and better new materials made in extreme conditions, and x-ray and neutron scattering are used in supervising the manufacturing process of the material at atomic levels.

The penetration depths of thermal neutrons (5–500 meV) in copper (0.7 cm), steel (0.8 mm), hexane (0.9 mm), titanium (1.2 mm), Teflon (2.3 mm), magnesium (4.7 mm), silicon (5.8 mm) and aluminum (6.7 mm) are quite large.

Table 3.1 compares the characteristics of x-ray and neutron diffraction, which clearly summarize the above mentioned advantages of neutron over x-ray scatterings.

3.1.2 *Discovery of neutrons [3.2]*

In 1920, E. Rutherford postulated that there were neutral, massive particles in the nuclei of atoms. This conclusion arose from the disparity between an element's atomic number (protons = electrons) and its atomic mass (usually in excess of the mass of the known protons present).

In 1931, it was discovered by his students, German scientists W. Bothe and H. Becker, that beryllium, when bombarded by α particles, emitted a very energetic stream of radiation. This stream was originally thought to be γ radiation. However, further investigations into the properties of the radiation revealed contradictory results. Like γ rays, these rays were extremely penetrating and since they

Table 3.1. A Comparison of Characteristics of X-ray and Neutron Diffraction.

Properties	X-ray	Neutron
Wavelength	0.154 nm for Cu-K_α characteristic line;	(0.110 ± 0.005) nm;
	Absorption cut-off for long wavelength	Long wavelength advantage
Energy when $\lambda = 0.1\,\text{nm}$	$10^{18}h$	$10^{13}h$, same order of magnitude as a phonon
General characteristics of atomic scattering	Scatterers – Electrons; Form factor depends on $\sin\theta/\lambda$; Angle-dependent polarization coefficient;	Scatterers – Nuclei; Isotropic, angle-independent form factor
	Scattering amplitudes regularly increase with atomic number, and can be calculated with known electric structure;	Do not change with atomic number, nuclear structure dependent, can only be experimentally determined
	No difference among isotopes; Phase difference – 180°	180° for most of nuclei; 0° for H, ^{7}Li, Ti, V, Mr, ^{62}Ni; Abnormal scattering for ^{113}Cd
Magnetic scattering	No additional scattering	Additional scattering for atoms with magnetic moment; (1) Diffusion scattering for paramagnetic materials; (2) Coherent diffraction peak of ferro- and antiferro-magnetic materials;
Absorption coefficient	Very large Absorption > scattering	Small; Absorption < scattering (except for B, Cd and rare earth elements); Change among different isotope
Non-elastic scattering	Wavelength change is negligible	Wavelength change obvious; ω-wave vector relation exist for crystal vibration and magnetic spin wave;
Measuring method	Photo film, Geiger counter; Measurement of absolute intensity is difficult	^{3}He counter; Direct, especially for powder method

were not deflected upon passing through a magnetic field, neutral. However, unlike γ rays, these rays could not cause the photoelectric effect: it would not discharge a positively charged electroscope when the ray shone onto the conductor sphere of the electroscope.

When this news came to France, I. Curie and F. Joliot discovered that when a beam of this radiation hit a substance rich in protons (paraffin, for example), protons were knocked loose, which could be easily detected by a Geiger counter. However, all the experimental facts proved that the product could not be γ ray — if it were so, the energy of that "γ ray" would reach 50 MeV. Unfortunately, bound tightly with the old theories, they explained this ray as a γ ray.

James Chadwick was assigned the task of tracking down evidence of Rutherford's tightly bound "proton-electron pair," or neutron. In 1932, he proposed that this particle was Rutherford's neutron. Using kinematics, $s = vt$, Chadwick was able to determine the velocity of the protons. Then, through conservation of momentum techniques, he was able to determine that the mass of the neutral radiation was almost exactly the same as that of a proton. This is Chadwick's equation:

$$^{4}_{2}\alpha + ^{9}_{4}\mathrm{Be} \rightarrow ^{12}_{6}\mathrm{C} + ^{1}_{0}n.$$

In 1935, he was awarded the Nobel Prize in Physics for his discovery. With Chadwick's announcement, Heisenberg then proposed the proton-neutron model for the nucleus.

3.1.3 *Neutron imaging [3.3]*

There are two types of neutron images. One is neutron radiography, and the other is neutron tomography.

(1) Neutron radiography

Neutron radiography is a non-destructive method based on the traditional x-ray radiography. It can determine the distribution of different elements or structures that have different neutron scattering cross sections.

The universal law of attenuation of radiation passing through matter is applied in neutron radiography.

$$I = I_0 \exp\left(-\sum d\right)$$

where I_0 and I are the neutron intensity before and after incident, d is the sample thickness, and $\sum$ is the macroscopic cross section of the sample. Different materials have different attenuation behavior, and the neutron beam passing through a sample can be interpreted as signals carrying information about the composition and structure of the sample.

Both thermal and fast neutrons can be used in neutron radiography, but thermal neutrons are more widely used. This is because thermal neutrons have larger cross sections and larger cross-section differences than fast neutrons. However, since fast neutrons can penetrate thick metallic or organic walls, they are used in neutron radiography with samples located in high-pressure cells, low-temperature dewars or other complicated containers.

Since neutrons carry no charge, they cannot be directly detected and the scattered neutrons usually have to go through a convector such as ^{6}LiFZnS(Ag) to produce high energy α particles and then visible light, which are finally detected with a film or a digitalized CCD (charge coupled device) camera [3.4].

(2) Neutron tomography

Neutron tomography is a method which provides cross-sectional images of an object from transmission data, measured by irradiating it from many different directions. Tomography imaging in a mathematical sense deals with reconstructing an image from its projections. Here, the projection at a given angle represents the integral of the image in the direction specified by that angle.

The intensity of the scattered neutron beam can be expressed as attenuation along the "line of response"

$$I(\rho,\theta) = I_0 e^{-\int_{-\infty}^{\infty} \sum(x,y)ds} . \tag{3.1}$$

The projection is the Radon transform of $\sum(x,y)$

$$\begin{aligned} P(\rho,\theta) = \ln\left(\frac{I_0}{I(\rho,\theta)}\right) &= \int_{-\infty}^{\infty} \sum(x,y)ds \\ &= \int_{-\infty}^{\infty} \sum(\rho\cos\theta - s\sin\theta, \rho\sin\theta + s\cos\theta)ds. \end{aligned} \tag{3.2}$$

$\sum(x,y)$ can then be obtained from the inverse Radon transform

$$\sum(x,y) = B(x,y) = \int_0^\infty P(x\cos\theta + y\sin\theta, \theta)d\theta\,. \tag{3.3}$$

The non-destructive analysis of an object by neutron tomography is mostly done by taking one or more two-dimensional parallel projections. In some cases, however, the transmission properties of the object seen from any angle are looked for. This can be achieved by rotating the object in small angular steps over 180° and calculating tomographic slices using the inverse Radon transform.

3.2 Neutron Diffraction

3.2.1 *Diffraction of radiation*

The scattering of radiation by condensed matter will be related to the distribution of atomic position if the wavelength is of the same order of magnitude as the interatomic spacing. If a beam of radiation falls on a target and wavelets scattered by different atoms have similar amplitudes and phases, then the scattered waves will interfere and the target will act as a diffraction grating. In this case, the distribution of scattered intensity contains information on the distribution of atoms.

For continuous radial distribution of scattering centers $\rho_{(2)}(r)$, bulk liquid density ρ_L and structure $g_{(2)}(r)$, scattering intensity is

$$I(\theta) = I_0\left\{1 + 4\pi\rho_\mathrm{L}\int_0^R g_{(2)}(r)\left(\frac{\sin sr}{sr}\right)r^2\mathrm{d}r\right\}. \tag{3.4}$$

The Fourier inversion of scattering data $I(\theta)$ gives distribution $g_{(2)}(r)$. For the pair theory of liquids, two-particle distribution $g_{(2)}(r)$ plays a central role.

The *correlation functions* are the central concepts in the study of scattering problems. A correlation function refers to the probability of finding object A at coordinates $\boldsymbol{x}$ at time t given object B is at $\boldsymbol{x}_0$ at time t_0. Suppose that A and B denote particular atoms or molecular species. If $A = B$, one is referring to autocorrelation (or self-correlation) functions. If $A \neq B$, one has to deal with *coherent scattering* phenomena, which is related to the momentum transfer. On the other hand, stochastic motion, diffusion, and migration phenomena, as well as collective motions in polycrystalline samples (seen as the density of phonon states) will all be viewed through *incoherent scattering.*

The interaction between the incident beam and the samples may involve the energy transfer between them. When this does happen, the scattering is *inelastic.* If the energy of the incident beam (namely the wavelength of the incident beam) is conserved, the scattering is *elastic.*

3.2.2 *Wave properties of neutrons*

A neutron has a mass of hydrogen atom, but it is electrically neutral. If a neutron passes an electron cloud and approaches the nucleus, then strong short-range force applies which can be approximated with a hard sphere potential. If a is the nuclear radius, it is also called scattering length, or (scattering) amplitude, or form factor on the order of a Fermi (10^{-15} m). One may use a^2 as the cross section of the nucleus, typically on the order of $10^{-28}\,\mathrm{m}^2$. The wavelength of neutrons is about 0.1 nm.

In cases where the energy is too small to excite the internal nuclear levels, elastic nuclear scattering is a major process.

The phase difference (see Fig. 3.1) between the incident and scattered beams is $(\boldsymbol{k}_0 - k) \cdot \boldsymbol{r}$. The wave scattered by the volume element dV at position $\boldsymbol{r}$, relative to the wave scattered by $\mathrm{d}V$ at the origin 0, has a phase factor $\exp[i(\boldsymbol{k}_0 - k) \cdot \boldsymbol{r}]$. For elastic scattering (a ball scattered by a wall), since energy conservation, $\hbar\omega = \hbar\omega_0$,

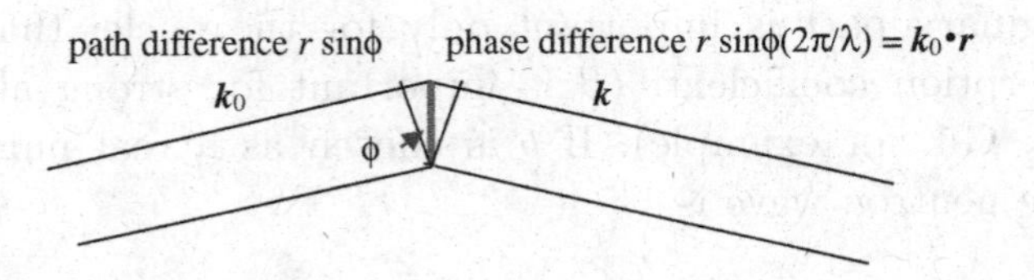

Fig. 3.1 Phase change in a scattering.

the outgoing frequency $\omega =$ incident frequency ω_0, or $ck = ck_0$, namely $k = k_0$. Laue's diffraction condition $\boldsymbol{k} - \boldsymbol{k}_0 = \boldsymbol{G}$, where $\boldsymbol{G}$ is a reciprocal lattice vector, leads to $(\boldsymbol{k}_0 + \boldsymbol{G})^2 = k^2$. One has then $2\boldsymbol{k} \cdot \boldsymbol{G} + G^2 = 0$, which is equivalent to the Bragg diffraction condition, $2dsin\theta = n\lambda$ [3.4].

3.2.3 *Neutron elastic scattering*

For simplicity, we will first consider the case of infinite nuclear mass to treat the coherent effect of three-dimensional nuclei set. This is equivalent to a classical elastic collision between a particle and a wall, in which case there is no energy transfer between neutrons and target nuclei.

A neutron wave satisfies the following equation

$$(\Delta + k^2)\psi(r) = (2m/h)V(r)\psi(r) \tag{3.5}$$

where $m = M/(A+1)$ is the reduced mass, M — mass of atom, $A = M/$neutron mass.

The incident neutron plane wave is

$$\Psi = e^{i\kappa x}, \tag{3.6}$$

where $\kappa = 2\pi/\lambda$ is a wave number. The scattering wave of this incident wave from a nucleus is of spherical form

$$\Psi = -(b/r)e^{i\kappa x}, \tag{3.7}$$

where r is the distance from the measuring point to the origin where the nucleus is located. Quantity b has the dimension of length, and is called "*scattering length*". It is a complex number $b = \alpha + i\beta$. The imaginary part is important only for the nuclei that have a high absorption coefficient. (β is important for strong absorption materials; Gd, for example). If b is chosen as a real number, the composite neutron wave is

$$\Psi = e^{i\kappa x} - (b/r)e^{i\kappa x}. \tag{3.8}$$

The scattering cross section is defined as

$$\sigma = \frac{\text{Outward scattered neutron flux}}{\text{Incident neutron flux}}$$
$$= 4\pi r^2 \frac{|(b/r)e^{i\kappa r}|^2}{v|e^{i\kappa x}|^2}$$
$$= 4\pi b^2 \quad (3.9)$$

where v is the neutron speed.

The composite wave function after scattering by an atom system set is

$$\Psi = e^{i\kappa x} - \sum_{\rho} (b_\rho/r)e^{i\kappa r} - e^{i\rho(\kappa-\kappa')} \quad (3.10)$$

where ρ is the vector pointing to the nucleus from the coordinate origin, and κ, κ' are the wave vectors before and after scattering, respectively.

Expression $\exp[i\rho(\kappa-\kappa')]$ is the *phase difference* after considering all nuclei, and it can be expressed as $\exp\{2\pi i(hx/a_0+ky/b_0+lz/c_0\}$, where x, y and z are the Cartesian of the nuclei, a_0, b_0, c_0, the unit lengths of a unit cell of the nuclei, and h, k, l, the Miller indices in the direction of $(\kappa - \kappa')$.

At unit distance from the nuclei, the amplitude of the scattering neutron wave is

$$A = -\sum_{\rho} b_\rho \exp\{2\pi i(hx/a_0 + ky/b_0 + lz/c_0)\}. \quad (3.11)$$

The scattering cross section of a nucleus at a unit solid angle of a certain direction, or differential cross section is

$$G_{hkl} = (1/N_0)|A|^2, \quad (3.12)$$

where N_0 is the total number of nuclei in the crystal of soft matter material under study, and it is supposed to be sufficiently small for absorption effects to be neglected.

In the case of elastic scattering, neutron energy does not change; κ_0 and κ have the same length, or their kinetic energy

$$E_k = \frac{1}{2}mv^2 = \frac{\hbar^2\kappa^2}{2m}. \tag{3.13}$$

However, the momentum has changed, and it is absorbed by the crystal at a quantity given by

$$mv_1 - mv_2 = \hbar(\kappa_0 - \kappa). \tag{3.14}$$

It is shown in the above figure that the momentum conservation equation can be expressed as

$$(\kappa_0 - \kappa) = 2\pi\tau. \tag{3.15}$$

3.2.4 *Neutron inelastic scattering*

In the last section of neutron scattering, we supposed an infinitely heavy scatter mass of nuclei and concentrated on the scattering effect of the three-dimensional periodic structure of crystals. However, if nuclei are combined into a crystal form, and if the nuclei are not infinitely heavy, nuclei will absorb energy from incident neutrons and pass it on to crystals for vibration, or vice versa; they absorb energy from crystal vibration and pass it on to neutrons to increase neutron speed. So for finite nuclei mass, neutron scattering may be non-elastic. In this case, the scattered neutron wavelength would be longer or shorter than the incident neutron wavelength, and the neutron wavelength change is measurable.

In inelastic neutron scattering, the quantities κ_0, κ are no longer equal to each other. The neutron energy changes, and a phonon of frequency ω and wave-vector $\boldsymbol{q}$ is produced. The corresponding energy and momentum equations are

$$\frac{\hbar^2\kappa_0^2}{2m} - \frac{\hbar^2\kappa^2}{2m} = \hbar\omega, \tag{3.16}$$

$$\boldsymbol{Q} = \kappa - \kappa_0 = 2\pi\tau + \boldsymbol{q}, \tag{3.17}$$

where $\boldsymbol{Q}$ is the momentum-transferring vector. $\hbar 2\pi\tau$ is the momentum given by neutrons to the scatter, but there still exist phonons with momentum $\hbar \boldsymbol{q}$. Phonons have to satisfy the dispersion law as well,

$$\omega = \omega_i(\boldsymbol{q}), \tag{3.18}$$

which gives a relation of ω and wave-vector $\boldsymbol{q}$ that is related to polarization i.

The dispersion relation $\omega(q)$ of the acoustical branch is correspondent to the case where the mass center of two atoms oscillates with atoms; when the mass center does not move with oscillating atoms, the dispersion relation $\omega(q)$ is called optical branch since such oscillations can be stimulated with the electric field of an optical wave.

The scattering amplitude U is determined by the integral

$$U = \int \mathrm{d}V n(\boldsymbol{r}) \exp(-i\boldsymbol{Q} \cdot r), \tag{3.19}$$

where $\kappa_0 + \boldsymbol{Q} = \kappa$, and $\boldsymbol{Q}$ measures the wave-vector change, and is called the scattering vector [3.4].

If the Fourier components of $n(\boldsymbol{r})$ are inserted into Eq.(3.19), one has

$$U = \sum \int dV n_G \exp(i(\boldsymbol{G} - \boldsymbol{Q}) \cdot \boldsymbol{r}). \tag{3.20}$$

When the scattering vector $\boldsymbol{Q}$ is equal to a vector of reciprocal lattice $\boldsymbol{G}$, the angle of the exponent becomes zero, and $U = V n_G$. On the other hand, it can be proven that when the difference of $\boldsymbol{Q}$ and $\boldsymbol{G}$ is sufficiently large, U is so small that it can be neglected.

Major applications of neutron scattering can be classified into five categories: (1) the study of the structure of condensed matter to determine the position of light atoms, especially hydrogen atoms; (2) the identification of atoms with atomic numbers that are very close, of which x-ray scattering amplitudes are very similar; (3) the study of magnetic materials, in which there is an additional scattering in the atoms that have magnetic moments; (4) the vibration study using non-elastic scattering, including magnetic vibrations; and (5) the study of non-ideal arrangements–liquid, gas and defects.

3.3 Structure Determination of Soft Matters

3.3.1 *Neutron scattering of light elements*

In x-ray measurement, scattering amplitude is proportional to atomic number, so it is not an effective method to detect materials consisting mainly of light elements. In neutron scattering, light elements are not a serious problem. For example, the neutron scattering amplitude of oxygen is 0.58×10^{-15} m, and it is quite easy to detect when it is with tungsten, gold or lead, which have b values of 0.48×10^{-15} m, 0.76×10^{-15} m and 0.94×10^{-15} m, respectively. The scattering amplitude of hydrogen, 0.37×10^{-15} m, is not too small in comparison to other elements.

The previous discussion of neutron scattering of three-dimensional nuclear lattices concentrated on how to determine $\bar{b}_r$ effective "coherent scattering amplitude". $\bar{b}_r$ is the mean value of scattering length b over various isotopes and their parallel and anti-parallel spin states. The "coherent cross section" of elements

$$\Omega = 4\pi\bar{b}_r^2, \tag{3.21}$$

and the disordered scattering cross section

$$\varsigma = 4\pi\{\bar{b}_r^2 - (\bar{b}_r)^2\}, \tag{3.22}$$

are defined. In many areas of soft matter study, the positioning of hydrogen atoms (or atoms of other light elements) is a common issue, and this is where neutron scattering can find its role. The positioning resolution of light atoms using neutron diffraction is complete with that of heavy atoms using either x-ray or neutron diffraction.

3.3.2 *The neutron scattering of liquid*

The atomic structure of liquid has a nature between the disordered gas and highly ordered solid structures. The immediate neighboring surroundings of an atom in a liquid is not so much different from that in a solid, and the liquid packing density is only a few percent lower than that of a solid. However, there is only very local order in liquid, as opposed to the long-range order in solid. This brings a difficulty in interpreting the liquid experimental data of neutron scattering, and it is not easy to determine liquid structure.

One of the important advantages of neutron scattering over liquid is its low absorption coefficient of most materials, which permits one to study samples using a transmitted method, while x-ray study is limited to a reflected method for its strong absorption coefficient. Reflected method only allows for the study of liquid surface, which is not representative of bulk properties, and, in many cases, is seriously contaminated with external atoms. Low neutron absorption also allows for the study of liquid in a heavy sample cell of high temperature and/or high pressure.

Liquid atoms are fixed in certain positions as in a solid, so elastic scattering becomes important.

At long wavelengths, the peak is due to the diffusion of quasi-elastic atoms; at short wavelengths, the curve is from non-elastic scattering due to molecular rotation and the vibration of atoms in the molecules. This figure gives a general description of dynamic characteristics of liquids.

3.3.3 *Radial distribution function g(r) of liquid*

In soft matter study, the radial (pair) distribution function $g(r)$ is often used to describe the local atomic or colloidal structure of amorphous materials. The radial distribution function is important, since once it is determined, it can be used to derive many thermodynamic quantities such as energy, pressure and compressibility.

The distribution function of molecules ${n_{\mathrm{N}}}^{(1)}(\boldsymbol{r}_1)\mathrm{d}\boldsymbol{r}_1$ is proportional to the probability of finding a molecule at $\boldsymbol{r}_1$ in a volume element $\mathrm{d}\mathbf{r}_1$; ${n_{\mathrm{N}}}^{(2)}(\boldsymbol{r}_1, \boldsymbol{r}_2)\mathrm{d}\boldsymbol{r}_1\mathrm{d}\boldsymbol{r}_2$ is proportional to the probability of finding a molecule at $\boldsymbol{r}_2$ in a volume element $\mathrm{d}\boldsymbol{r}_2$ if, at the same time, there is a molecule at $\mathbf{r}_1$ in volume element $\mathrm{d}\boldsymbol{r}_1$ (irrespective of where the remaining molecules are). The pair distribution function, $g(r)$, is the molecule distribution function divided by density $g_m(\boldsymbol{r}^m) = \rho^{-m}{n_{\mathrm{N}}}^{(m)}(\boldsymbol{r}^m)$, or

$$g(r) = \exp(-\phi(r)/k_B T), \tag{3.23}$$

where $\phi(r)$ is called the "potential of mean force".

The general form of the radial distribution function $g(r)$ of high-density liquid is as follows: $g(r)$ is almost zero at small r because the energy needed for atoms to be superposed is extremely large. When r is equal to the radius of an atom, there is a peak, which means the nearest neighboring atom. At distances greater than the atomic radius, oscillations appear, which represent further neighbors. When r further increases, the amplitude of oscillations gradually decreases, and $g(r)$ eventually reaches the mean density of the system.

Bernal pointed out in 1960 that the general form of $g(r)$ of simple liquids can be understood from the geometric packing of particles. Just like the hard ball model, atoms are packed together but cannot reach any distance smaller than r_1. Bernal calculated $g(r)$. The measurement of $g(r)$ for atomic fluids can be found in [3.5].

The radial distribution function can be calculated from experimental data of optical images, light scattering, x-ray diffraction or neutron scattering.

3.3.4 *Form factor and structure factor of neutron scattering spectrum*

In the practical term, the scattered neutron intensity $I_{\mathrm{N}}(q)$ is proportional to the differential scattering cross section $\mathrm{d}\sigma(q)/\mathrm{d}\Omega$ [3.1]. The proportional constant depends on the neutron wavelength, detector efficiency, sample size and its transition, and they are specific to a given experimental setup.

$$I_{\mathrm{N}}(q) \approx \frac{\mathrm{d}\sigma_{coh}}{\mathrm{d}\Omega}(q) \approx CP(q)S(q). \tag{3.24}$$

Cylindrical symmetry is assumed in the above equation for simplicity. $S(q)$ is the interparticle *structure factor*.

$P(q)$ is the *form factor*, which carries the information on the size and shape of the scattering object. The form factor $P(q)$ is a function that describes how a differential cross section is modulated by interference effects between radiation scattered by different parts of the same scattering center, and it is sensitive to the shape of the scattering center. With a static light scattering instrument, one can measure the form factor of highly diluted colloidal dispersions.

The form factor $P(q)$ for monodisperse homogeneous hard spheres is

$$P(q) = (n_p - n_s)^2 \frac{9}{(qR)^6} [\sin(qR) - qR\cos(qR)]^2, \quad (3.25)$$

where q is the wave-vector, R is the radius of the sphere and n_P and n_S are the refractive indices of the spheres and the solvent. Independent of the contrast $(n_P - n_S)$, the minimum of the form factor is at $qR = 4.493$. Based on the position of the minimum we determine the size of the colloids with this relation. In reality, the colloids are not monodisperse and the minimum of the measured form factor is broader. From the width of the minimum we can determine the polydispersity σ of the colloids.

The form factor for a Gaussian distribution of segment density about a center-of-mass characterized by a radius-of-gyration R_g is [3.7]

$$P(q) = \frac{2(\exp(-q^2R_g^2) + q^2R_g^2 - 1)}{(q^2R_g^2)^2}. \quad (3.26)$$

The general form of the form factor is

$$P(q) = \frac{1}{V_p^2} \left| \int_0^{V_p} \exp[if(q\alpha)]dV_p \right|^2, \quad (3.27)$$

where α is the shape parameter, and it can be, for example, the length or the radius of a gyration in a polymer solution. Form factors of some particular geometries, such as sphere, disc, rod or Gaussian random coil, can be found in books on neutron diffraction.

The interparticle structure factor $S(q)$ is a function that tells how a differential cross section is modulated by interference effects between the radiation scattered by different scattering centers. It is dependent on the degree of local order in the sample (e.g. dispersion of charged particles interaction via an electrostatic potential). Its general expression is

$$S(q) = 1 + \frac{4\pi N_p}{qV} \int_0^{\infty} [g(r) - 1] r \sin(qr) dr. \quad (3.28)$$

Once the density distribution $g(r)$ is obtained from the Fourier inversion of the above expression, the radial distribution function r.d.f. $= \frac{4\pi N_p r^2}{V} g(r)$, and one may gain information on the relative position of scattering centers. This is the key issue in this chapter.

It is interesting to study the relation between the distribution function and the neutron scattering spectrum. The differential cross section per unit solid angle in any direction for a set system of N_0 can be written as

$$\frac{\mathrm{d}\sigma}{\mathrm{d}\Omega} = \left| \sum_n b_n \exp(i\boldsymbol{Q} \cdot \boldsymbol{R}_n) \right|^2 , \tag{3.29}$$

where $b_n, \boldsymbol{R}_n$ represents the scattering length and position of the n-th nucleus, and $Q = k - k'$. It can be proven that $\boldsymbol{Q} \cdot \boldsymbol{R}_n = 2\pi(hx/a_0 + ky/b_0 + lz/c_0)$, which provides the phase difference of the scattering waves of atoms at points 0 and P.

The above equation can be expanded as

$$\frac{\mathrm{d}\sigma}{\mathrm{d}\Omega} = N_0\{\bar{b}_r^2 - (\bar{b}_r)^2\} + (\bar{b}_r)^2 \left| \sum_n \exp(i\boldsymbol{Q} \cdot \boldsymbol{R}_n) \right|^2 . \tag{3.30}$$

The first term represents the isotropic incoherent scattering, and the second, coherent scattering.

In the case of inelastic scattering, scattering probability depends on the amount of energy exchange as well, and the second order differential cross section, $\mathrm{d}^2\sigma/\mathrm{d}\Omega\mathrm{d}E$, which means the cross section per unit solid angle per energy exchange range, may need to be considered. It has been proven by Van Hove in 1954 that

$$\frac{d^2\sigma}{d\Omega dE} = N\frac{\kappa}{\kappa_0}\frac{1}{2\pi\hbar}\int_{-\infty}^{\infty} dte^{-i\omega t} \sum e^{iQ\cdot r}\left[(\bar{b})^2 G(r,t) + \left\{(\bar{b}^2) - (\bar{b})^2\right\} G_0(r,t)\right]. \tag{3.31}$$

In inelastic scattering measurements, dynamic structure factors $S(Q,\omega)$ can be obtained for different frequencies ω of incoming neutrons. The integration of $S(Q,\omega)$ over ω, namely the area under the curve of $S(Q,\omega)$, provides a point of a curve of $S(Q)$. To find the liquid structure factor $S(Q)$ is the primary purpose of liquid diffraction experiments. The curve $S(Q)$ is very similar to that of the radial distribution function $g(r)$. This is natural, as the two are closely related as in Eq. (3.28) and both can be used to determine the structures of soft matters.

3.3.5 *Small angle neutron scattering*

Small angle neutron scattering is an experimental configuration where the scattering angle is less than 10°. Why do we need to use small angle neutron scattering? First, there is a limit for large angle (10° – 180°) neutron scattering, and that is 180°, while small angle (< 10°) has no such limitation — you may go down to 0.001° or even smaller. Secondly, this is because the length scales are correct. If the scattering vector $q = (4\pi/\lambda)\sin\theta$, and neutron wavelength $\lambda = 0.5\,\mathrm{nm}$, from $0.001° < \theta < 10°$, one may have $0.0002\,\mathrm{nm}^{-1} < q < 2\,\mathrm{nm}^{-1}$, or $0.5\,\mathrm{nm} < 1/q < 5\mu\mathrm{m}$, where $1/q$ is a real size which is compatible with many requirements of soft matter study [3.6].

Similar circumstances also occur with small angle x-ray analysis. Colloidal particles usually have a size between a few to a few hundred nanometers, which are much larger than the wavelengths of either x-rays or neutrons (about 0.1 nm).

As Figs. 3.2 and 3.3 depict, no matter what the particle size is, the waves scattered from the two points to an angle 2θ have a path difference of 1λ, leading to no scattering in the direction of 2θ as a result of destructive interference. Also, no matter what the particle size is, the scattering maximum always occurs at the direction of zero scattering angle, where all waves are exactly in phase. When the particle size is small (Fig. 3.2 (a)) 2θ is pretty large, and the observable intensity is like curve 1 in Fig. 3.3. However, when the particle size is much larger than the wavelength of the probe wave, one sees a large change of intensity when 2θ is small (Fig. 3.2 (b) and Fig. 3.3, curve 2). This is the case of small angle x-ray scattering (SAXS), which is used in studying structural features of colloidal size. X-rays are primarily scattered by electrons, and the electrons resonate and emit coherent secondary waves which interfere with each

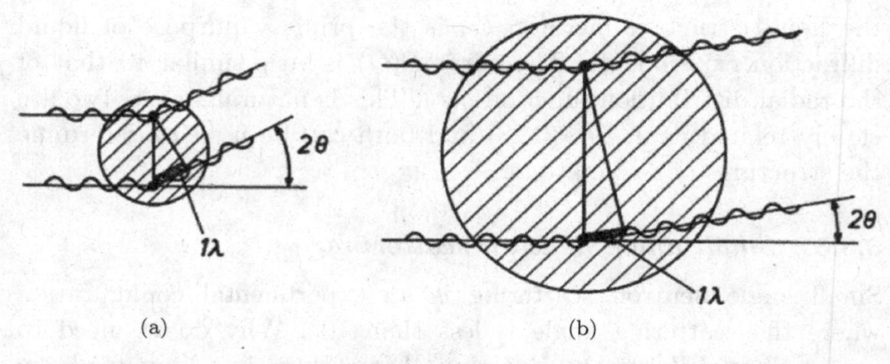

Fig. 3.2 Different particle sizes as compared to the wavelength of the probe wave lead to different 2θ, where the scattering wave is out of phase [3.13].

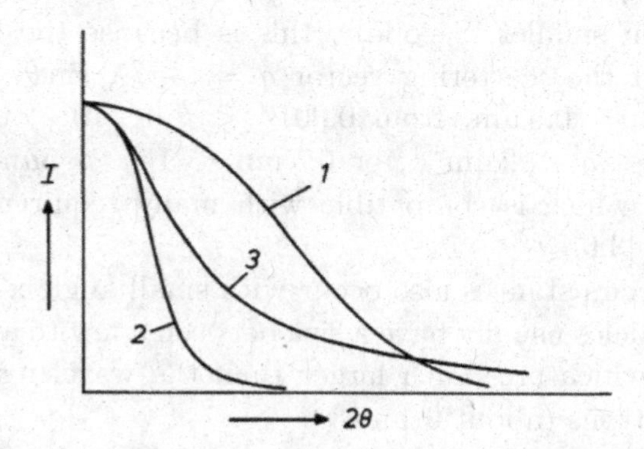

Fig. 3.3 Intensity of the scattering wave versus 2θ for small (curve 1) and large (curve 2) particle sizes as compared to use of the wavelength. Curve 3 is for the intermediate particle size [3.13].

other. It has been proven that SAXS can always be observed, and only be observed, when electron density inhomogeneities of colloidal size exist in the sample [3.13].

The flux of neutrons captured by the detector located L away from the sample at scattering angle θ is related to the number of transmitted, absorbed and scattered neutrons. If the detector size is $\mathrm{d}x\mathrm{d}y$, and the solid angle element is $\mathrm{d}\Omega = \mathrm{d}x\mathrm{d}y/L^2$, then the flux is

$$I(\lambda, \theta) = I_0(\lambda)\Delta\Omega\eta(\lambda T(\lambda)V d\sigma'(Q)/d\Omega, \tag{3.32}$$

where $I_0(\lambda)$ is an incident flux, $\eta(\lambda)$, the detector efficiency, $T(\lambda)$, the sample transmission, and $\mathrm{d}\sigma'(Q)/\mathrm{d}\Omega$, the differential scattering cross section.

The scattering vector is $Q = |\boldsymbol{Q}| = |\boldsymbol{k} - \boldsymbol{k}_0| = (4\pi n/\lambda)\sin\theta$, and since the refraction of the fluid n is slightly less than one, the scattering vector Q $\sim (4\pi/\lambda)\sin\theta$. Using Bragg's law of diffraction, $\lambda = 2d\sin\theta$, one has $d = 2\pi/Q$, which is important in small angle neutron scattering.

The differential scattering cross section is

$$d\sigma'(Q)/d\Omega = N_p V_p^2 (\Delta\rho)^2 P()S(Q) + B, \tag{3.33}$$

where N_p is the number concentration of scattering centers, V_p is the volume of one scattering center, $(\Delta\rho)^2$ is the contrast match factor, $P(Q)$ the form or shape factor, $S(Q)$ the interparticle structure factor and B, the background signal. The unit of the differential scattering cross section is 1/length [3.7].

The contrast factor is $(\Delta\rho)2 = (\rho_p - \rho_m)^2$, where ρ_p and ρ_m are the neutron scattering length densities of a sample and a medium, respectively. The neutron scattering length density is defined as

$$\rho = \sum_i b_i \frac{\delta N_A}{m},$$

where δ is the bulk density of the molecule, m is the relative molar mass, N is the number density of the scattering center, and b_i is the coherent neutron scattering length of neutron i. When the scattering length densities of the particles and media are exactly the same, $(\Delta\rho)^2 = 0$, the neutron beams are not able to distinguish them, so the differential cross section and, consequently, the flux is zero. This is called the scattering match, and no small angle neutron scattering can be detected. This is why the parameter $(\Delta\rho)^2$ is called the contrast factor.

Since large unit cells give small angle reflection at a given wavelength $\lambda = 2d\sin\theta$, if the neutron wavelength is 0.153 nm, the distance between planes is 0.15 nm, and a diffraction peak appears at 3°.

The following is an example of how neutron scattering can clarify the structure of a chemical [3.8]. Figure 3.4(a) is a graph of neutron

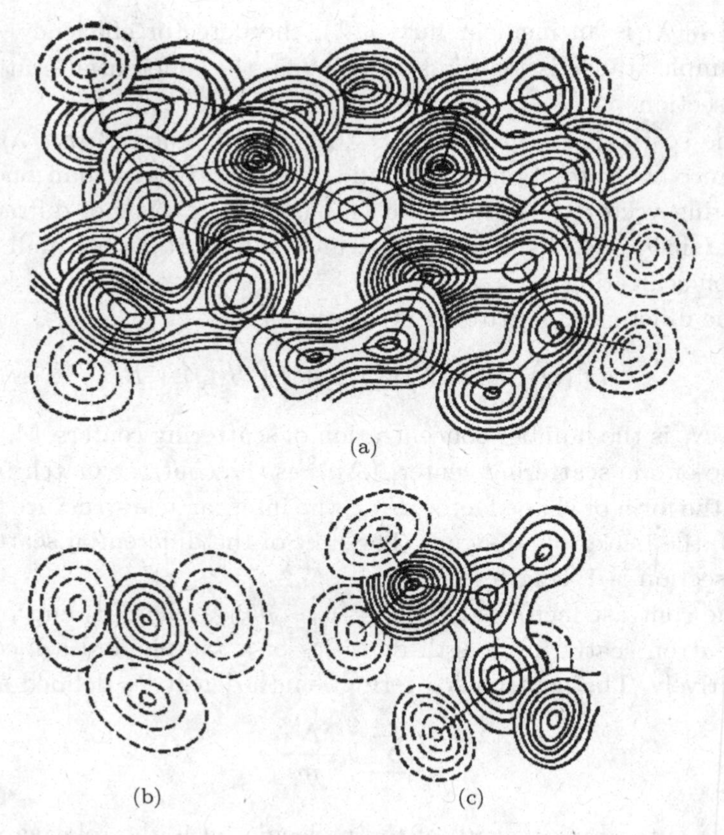

Fig. 3.4 Projections of neutron scattering density of a derivant of B_{12} [3.8].

scattering density. It is interesting to see that the centered atom of cobalt has only very few constant value lines, magnitude contours, meaning that it is a light atom, while it is shown to be a heavy atom in x-ray diffraction. In Fig. 3.4(b), since three hydrogen atoms can be clearly identified, it can then be definitely confirmed that methyl CH_3 does not rotate, while from previous studies conducted without neutron scattering it was not able to identify whether or not it rotates. Figure 3.4(c) shows an atom group for which, with only x-ray diffraction, it could not be determined whether it was $CO{\cdot}NH_2$ or COOH. However, in this neutron diffraction pattern, two divided hydrogen atoms can be clearly identified, and nitrogen atoms can also be identified since the nitrogen atom has larger scattering than that of oxygen. The conclusion is that this is $CO{\cdot}NH_2$.

3.4 Optical Microscopy and Light Scattering

3.4.1 *Structure determination with optical microscopy*

If particle size is quite large, say, above several hundred nanometers, optical microscopy can often be used to determine the pair correlation function or distribution function. The principle is as follows: since $4\pi r^2 \rho_0 g(r)$ is the number of particles whose centers are in a shell between r and $r + \mathrm{d}r$, where ρ_0 is the global particle density, the radial distribution function $g(r)$ is found to be the probability of finding particles in the shell.

Typical colloidal configuration is a hard-sphere fluid. The pair distribution function can be obtained by choosing an arbitrary particle as the origin and counting the atoms whose centers lie within a distance $\mathrm{d}r$ of a circle of radius r from the origin.

In the limit of infinite dilution, the *pair correlation function* $g(r)$ is related to the *pair interaction energy* $u(r)$ through the Boltzmann distribution [3.9]

$$\lim_{n \to 0} g(r) = \exp[-u(r)/k_B T],$$

where n is the areal density of spheres.

For finite concentration, the *radial distribution function* also reflects the influence of neighboring particles via the following formula:

$$w(r) = -k_B T \ln g(r),$$

where $w(r)$ is the *potential of mean force.*

Strictly speaking, the method of $U(r)$ converted from $g(r)$ is only valid in the dilute limit [3.10].

The pair correlation function $g(r)$ measures the mean number $ng(r)$ of particles per unit area separated from any given sphere by displacement r. This average is calculated in practice by counting the number of r pairs in a recorded image and normalizing by the number of particles actually tested for a partner at separation r.

Given $N(r)$ pairs at separation r in a snapshot, the pair correlation function is

$$g(r) = N(r)/(\pi n^2 r \delta A)$$

or

$$g(r) = \frac{2N(r)}{An^2 2\pi r dr - n\sum_i^{edge} \delta A_i(r)},$$

where $N(r)$ is the number of particle pairs at separation r in each image, $2\pi r dr$ is the bin area, n is the number density of the particles in the image, and A is the area of the image.

All these parameters can be gathered from optical microscopic pictures.

Once the curve of the pair correlation function is obtained, the pair interaction energy can be found through the Boltzmann distribution in the dilute limit. When particle concentration is finite, the Percus–Yevick correction must be used. This method does not work for the case of high concentration [3.10].

3.4.2 *Static and dynamic light scattering*

In traditional dynamic light scattering (DLS) experiments, non-interacting colloidal spheres in suspension systems experience Brownian diffusion. If the composition of the colloidal spheres is dilute enough, there can be only one scattering. Suppose that the incident wave-vector is $\boldsymbol{k}_0$, outgoing wave-vector is $\boldsymbol{k}$, and scattering angle is θ. If the light electric field intensity scattered from a single particle is E_0, the total light electric field intensity scattered from all N particles in the scattering region is then

$$E(t) = \sum_{i=1}^{N} E_0 \exp[i\boldsymbol{q} \cdot \boldsymbol{r_i}(t)], \tag{3.34}$$

where $\boldsymbol{q} = \boldsymbol{k} - \boldsymbol{k}_0$, is the scattering wave-vector, $\boldsymbol{r}_i(t)$, is the position of the i-th particle, and $\exp[i\boldsymbol{q} \cdot \boldsymbol{r}_i(t)]$ is the phase factor, determined by the phases of scattering wave-vectors and particles. If particles are moving, the phase of the electric field intensity of the scattering light is also changed, making the light intensity fluctuate. This light fluctuation can be described by the instantaneous autocorrelation function

$$g_{(2)}(t) = \frac{1}{\beta}\left(\frac{\langle I(t)I(0)\rangle}{\langle I\rangle^2} - 1\right) = \left(\frac{\langle E(0)E^*(t)\rangle}{\langle |E|^2\rangle}\right)^2 \equiv |g_{(1)}(t)|^2, \tag{3.35}$$

where β is a constant, determined by the optical condition of the experiment. $g_{(1)}(t)$ is the autocorrelation function of the light electric field intensity, and can usually be expressed as

$$g_{(1)}(t) = \frac{\sum_{i=1}^{N}\sum_{j=1}^{N}\langle \exp\{i\boldsymbol{q}\cdot[\boldsymbol{r}_i(0) - \boldsymbol{r}_j(t)]\}\rangle}{\sum_{i=1}^{N}\sum_{j=1}^{N}\langle \exp\{i\boldsymbol{q}\cdot[\boldsymbol{r}_i(0) - \boldsymbol{r}_j(0)]\}\rangle}. \tag{3.36}$$

Since there is no interaction between different particles, or the correlation is zero when $i \neq j$, $g_{(1)}(t)$ can be simplified as

$$g_{(1)}(t) = \langle \exp[-i\boldsymbol{q}\cdot\Delta\boldsymbol{r}(t)]\rangle \tag{3.37}$$

where $\Delta\boldsymbol{r}(t) \equiv \sum_{i=1}^{N}[\boldsymbol{r}_i(t) - \boldsymbol{r}_i(0)]$. If $\Delta\boldsymbol{r}(t)$ is a Gaussian random variable,

$$g_{(1)}(t) = \exp[-q^2\langle\Delta r^2(t)\rangle/6]. \tag{3.38}$$

For a diffusion model,

$$\langle\Delta r^2(t)\rangle = 6Dt. \tag{3.39}$$

D is a self-diffusion coefficient of the particles. One then has

$$g_{(1)}(t) = \exp(-q^2 Dt), \tag{3.40}$$

meaning that the autocorrelation function of the electric field intensity decays exponentially with the characteristic decay time.

For spherical particles, according to the Einstein equation,

$$D = \frac{k_B T}{6\pi\eta a} \tag{3.41}$$

where k_B is the Boltzmann constant, T is the absolute temperature, η is the viscosity of the dispersion system and a is the particle radius. This connects the characteristic decay time and dynamic properties of scattering particles.

3.4.3 *Diffusing-wave spectroscopy [3.11]*

In comparing diffusing-wave spectroscopy and dynamic light scattering, one finds the following similarities: first, both measure the scattering intensity with time; secondly, both calculate the instantaneous autocorrelation functions; thirdly, both want to find the dynamic information of the scattering centers. However, there are some major differences in these two methods: in DLS, the sample must be very dilute (usually much less than 1%), it strongly desires that a photon is scattered only one time; while in DWS measurement, a dense sample is expected, and multiple scattering is needed; a statistical approximation is widely used.

(1) Two major parameters in multi-scattering

In multi-scattering, there are two major parameters that characterize light scattering and transportation: scattering mean free path l and transport mean free path l^*. Scattering mean free path is the distance between two adjacent scatterings. In a dilute suspension system, $l = 1/\rho\sigma$, where ρ is the number density of the particles, and σ is the total cross section of a single particle. The transport mean free path l^* is the length that a photon must travel before its direction is completely randomized. The relation between l and l^* is

$$l^* = \frac{l}{\langle 1 - \cos\theta \rangle}, \tag{3.42}$$

where θ is the scattering angle and $\langle\ \rangle$ means the ensemble average; an average over many scattering events. l and l^* can be measured with experiments. When the particle size is equivalent to the wavelengths, one finds the least l and l^*, and thus the strongest scattering; in contrast when the particle size is much smaller than the wavelengths, one has very large l and l^*, and multiple scattering is unlikely to happen.

(2) Two approximations

Similar to the case of DLS, the change of the intensity of DWS scattering light in a multiple scattering system is due to phase change. To calculate the phase change, it is necessary to introduce two approximations. The first approximation is that the propagating process of the light can be considered as photon diffusion. Each incident photon experiences a large number of scattering events, each of which can be considered as one random step of the photon. This is very similar to the diffusion process from high to low density. The light diffusion coefficient D_l in a dispersion system can thus be defined:

$$D_l = vl^*, \tag{3.43}$$

where v is the velocity of the light propagation in such media, and l^* is the transport mean free path.

To identify the diffusions of photons and particles, the particle diffusion may be estimated at $D_0 \sim 10^{-8}\,\mathrm{cm^2s^{-1}}$. However, the photon diffusion coefficient is $D_l \sim 10^{-8}\,\mathrm{cm^2s^{-1}}$ ($l^* \sim 100\,\mu$m). Comparing D_0 and D_l, particle diffusion can be neglected under this approximation.

The second approximation is that a single scattering event can be replaced by an "averaged scattering event". The total scattering effect of each scattering path can be expressed by the effect of the "averaged scattering event" multiplied by scattering number N, which is determined by the length of scattering path. These two approximations are the basic presuppositions in deriving the autocorrelation function of light intensity in multiple scattering.

(3) Autocorrelation functions

Suppose that the sample thickness $L \gg l^*$. Each photon has p choices of path, there are N scatterings in each path, and N is determined by the path length s. In the following diagram, only one path out of multiple scattering is depicted in Fig. 3.5.

$$s = \sum_{i=0}^{N} |\boldsymbol{r}_{i+1} - \boldsymbol{r}_i| = \sum_{i-0}^{N} \left(\frac{\boldsymbol{k}_i}{|\boldsymbol{k}_i|}\right) \cdot (\boldsymbol{r}_{i+1} - \boldsymbol{r}_i), \tag{3.44}$$

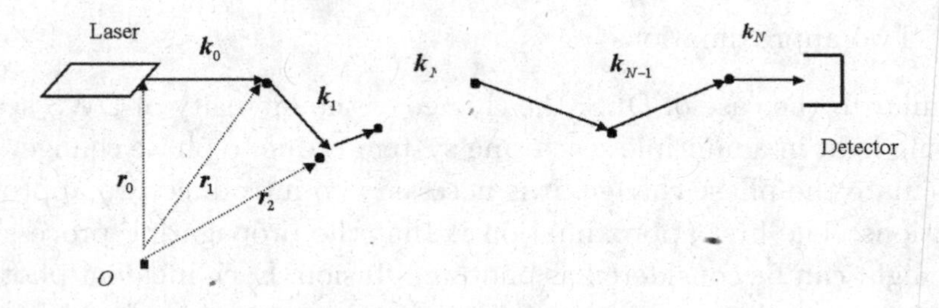

Fig. 3.5 N scatterings in a photon path [3.11].

where $\boldsymbol{k}_i$ is the wave-vector of i-th scattering, $\boldsymbol{r}_i$ is the position of the i-th particle, $1 \leq i \leq N$. $\boldsymbol{r}_0$ is the position of light source, and $\boldsymbol{r}_{\mathrm{N}+1}$ is the position of the detector. Since the scattering is quasi-elastic, $\boldsymbol{k}_i = \boldsymbol{k}_0$ (for any i).

Thus, the total phase change of photons in the process from incoming to leaving the sample is

$$\varphi(t) = k_0 s(t) = \sum_{i=0}^{N} \boldsymbol{k}_i(t) \cdot [\boldsymbol{r}_{i+1}(t) - \boldsymbol{r}_i(t)]. \tag{3.45}$$

The total electric field intensity includes all contributions from all paths

$$E(t) = \sum_p E_p e^{i\varphi_p(t)}, \tag{3.46}$$

where $\sum_p$ means the summation over all paths, and E_p is the electric field intensity of path p.

Substituting Eq. (3.45) into (3.46), noting the definition of Eq. (3.45) gives

$$\begin{aligned} g_{(1)}(t) &= \left(\frac{\langle E(0)E^*(t)\rangle}{\langle |E|^2 \rangle} \right) \\ &= \frac{1}{\langle I \rangle} \left\langle \left(\sum_p E_p e^{i\phi_p(0)} \right) \left(\sum_{p'} E^*_{p'} e^{-i\phi_{p'}(t)} \right) \right\rangle. \end{aligned} \tag{3.47}$$

$\langle I \rangle$ is the mean value of total light intensity in the detector. Since the scattering paths are independent, Eq. (3.47) can be simplified as

$$g_{(1)}(t) = \left\langle \sum_p \frac{|E_p|^2}{\langle I \rangle} e^{i[\phi_p(0)-\phi_p(t)]} \right\rangle$$
$$= \sum_p \frac{\langle I_p \rangle}{\langle I \rangle} \left\langle e^{i\Delta\phi_p} \right\rangle, \tag{3.48}$$

where$\langle I_p \rangle \equiv \left\langle |E_p|^2 \right\rangle$ is the mean light intensity of path p.

The phase change $\Delta\phi_p(t)$ is associated with particle motion

$$\Delta\phi_p(t) = \phi_p(t) - \phi_p(0)$$
$$= \sum_{i=0}^{N} \boldsymbol{k_i}(t) \cdot [\boldsymbol{r}_{i+1}(t) - \boldsymbol{r}_i(t)]$$
$$- \sum_{i=0}^{N} \boldsymbol{k_i}(0) \cdot [\boldsymbol{r}_{i+1}(0) - \boldsymbol{r}_i(0)]. \tag{3.49}$$

Defined as $\boldsymbol{q}_i \equiv \boldsymbol{k}_i(0) - \boldsymbol{k}_{i-1}(0)$, and $\Delta\boldsymbol{k_i}(t) \equiv \boldsymbol{k}_i(t) - \boldsymbol{k}_i(0)$, Eq. (3.49) is transformed as

$$\Delta\phi_p(t) = \sum_{i=0}^{N} \boldsymbol{q}_i \cdot \Delta\boldsymbol{r_i}(t) + \sum_{i=0}^{N} \Delta\boldsymbol{k}_{i(t)} \cdot [\boldsymbol{r}_{i+1}(t) - \boldsymbol{r}_i(t)] \tag{3.50}$$

where $\Delta\boldsymbol{r}_i = \boldsymbol{r}_i(t) - \boldsymbol{r}_i(0)$. In general, the second term in Eq. (3.50) is very small in comparison with the first term, so it can be neglected.

$$\Delta\phi_p(t) = \sum_{i=0}^{N} \boldsymbol{q}_i \cdot \Delta\boldsymbol{r_i}(t). \tag{3.51}$$

The quantity of $\boldsymbol{q}$ is related with scattering angle θ:

$$q = 2k_0 \sin\frac{\theta}{2}. \tag{3.52}$$

When N is very large, $\Delta\phi_p(t)$ is a Gaussian random variable

$$\left\langle e^{-i\Delta\phi_p(t)} \right\rangle = e^{-\langle \Delta\phi_p^2(t) \rangle/2}.$$

Using the result of Eq. (3.51),

$$\langle \Delta\phi_p^2(t)\rangle = \sum_{i=1}^{N} \left\langle [\boldsymbol{q}_i \cdot \Delta\boldsymbol{r}_i(t)]^2 \right\rangle = \frac{1}{3} N \langle q^2\rangle \langle \Delta r^2(t)\rangle. \tag{3.53}$$

q^2 can be expressed as mean free path

$$\langle q^2\rangle = \langle [2k_0 \sin(\theta/2)]^2\rangle = 2k_0^2 \langle 1 - \cos\theta\rangle = 2k_0^2 \frac{l}{l^*}. \tag{3.54}$$

When $N \gg 1$, $s = Nl$. Substituting (3.54) into (3.53), one has

$$\langle \Delta\phi_p^2(t)\rangle = \frac{1}{3}\frac{s}{l} 2k_0^2 \frac{l}{l^*} \left\langle \Delta r^2(t)\right\rangle = \frac{2}{3} k_0^2 \left\langle \Delta r^2(t)\right\rangle \frac{s}{l^*}. \tag{3.55}$$

l is deleted, while l^* is an important parameter related to $\langle \Delta\phi_p^2(t)\rangle$. For general diffusion systems, Eq. (3.39) exists, and then

$$\langle \Delta\phi_p^2(t)\rangle = 4k_0^2 D t \frac{s}{l^*}. \tag{3.56}$$

Equation (3.56) shows that $\langle \Delta\phi_p^2(t)\rangle$ depends only on the path length s. The summation over path p can be transformed into the summation over path length s. The scattering probability $P(s)$ of paths with path length s can be used to express the ratio of the mean light intensity of the path over the total mean light intensity $\frac{\langle I_p\rangle}{\langle I\rangle}$. The autocorrelation function can then be expressed as

$$g_{(1)}(t) = \sum_s P(s) \exp\left(-2k_0^2 D t \frac{s}{l^*}\right). \tag{3.57}$$

If s is continuous, the summation can be transformed as integral

$$g_{(1)}(t) = \int_0^{\infty} P(s) e^{-(2t/\tau)s/l^*} ds, \tag{3.58}$$

where $\tau = (k_0^2 D)^{-1}$.

Equation (3.58) is the basis of solving the light electric field autocorrelation function in multiple scattering. It is found that (1) for a path with length s, photons randomly walk (s/l^*) steps; on average, $g_{(1)}(t)$ decays $\exp(-2k_0^2 D t)$ at each step; the contribution of each path to $g_{(1)}(t)$ is expressed by the weight $P(s)$. (2) Paths of different lengths correspond to different characteristic decay times; the path of length s corresponds to the characteristic decay time of $(\tau l * /2s)$, during which the motion of the scattering particle changes the length by about a wavelength. Longer paths make faster decay, and shorter paths, slower.

3.4.4 *Applications of DWS*

(1) Particle sizing

The following equation

$$g_{(1)}(t) = \int_0^\infty P(s) e^{-(2t/\tau)s/l^*} ds$$

can be used for particle sizing. A characteristic decay time τ can be determined from the measured curve of the autocorrelation function. Since $\tau = (k_0^2 D)^{-1}$, and D is directly related to particle size, it can be considered a bridge between the microscopic values of measurements and the microscopic particle sizes. Backscattering geometry is easier for the measurements; this geometry has the advantage that only a single optical access is required. Furthermore, no additional knowledge of the scattering properties of the suspension is required, since the autocorrelation function does not depend on the value of l^*. Differentiation of the following equation $dg_{(1)}(t)/d\sqrt{t}$

$$g_{(1)}(t) = \frac{1}{1 - \gamma l^*/L} \frac{\sin h\left[(L/l^*)(6t/\tau)^{1/2}(1 - \gamma l^*/L)\right]}{\sin h\left[(L/l^*)(6t/\tau)^{1/2}\right]}, \quad (L \gg l^*)$$
$$\rightarrow \exp\left[-\gamma(6t/\tau)^{1/2}\right]$$

results in $\gamma/\sqrt{\tau}$, where γ is usually taken as 2.1, and then τ can be determined. Further, one may find particle radius R from the following Stokes–Einstein relation:

$$D_0 = \frac{1}{{k_0}^2 \tau_0} = \frac{kT}{6\pi\eta R}.$$

The results show that the particle sizes obtained using this method are statistical mean values which include information of whole particles since the outgoing light is scattered by many particles.

In comparison with DLS technique, the particle size measured with the DWS is more suitable for its time variation. DWS has been successfully applied by Horne to the study of aggregation processes in milk during the formation of cheese. [3.12]

(2) Particle motion on short length scales

This is one of the major applications of DWS that is different from DLS. $g_{(1)}(t)$ is associated with $\langle \Delta r^2(t) \rangle$ in the expression of the autocorrelation function. $\langle \Delta r^2(t) \rangle$ represents the mean square displacement of scattering particles at time t. In the characteristic decay time, the total length of a scattering path is changed by a wavelength, and this is due to the displacement of many particles in the scattering volume. In fact, the average displacement of each particle is very little, about a fraction of a nanometer. The particle displacement sensitivity of DWS is a major basis for observing the structure evolution in soft condensed matters.

References

[3.1] W. Zajac and B. J. Gabrys, *An introduction to neutron scattering, in Applications of neutron scattering of soft condensed matter*, ed. Gabrys, Barbara J. Amsterdam, (2000).

[3.2] Physics Laboratory Resource Lesson Discovery of the Neutron, http://online.cctt.org/physicslab/content/PhyAPB/lessonnotes/dualnature/Chadwick.asp.

[3.3] PSI Neutron Tomography, http://neutra.web.psi.ch/What/tomo.html

[3.4] Bin Tang, Xi-an Li, Song-bao Zang, Xia Ming, Xiang Dong, Wei Biao, *CT Theory Appl.*, **13** (2004), 20 (in Chinese).

[3.5] P. A. Egelstaff, An Introduction to the Liquid State. 2nd ed., Vol. 7 aus der Reihe: *Oxford Series on Neutron Scattering in Condensed Matter*, Oxford University Press, (1992).

[3.6] Mingyao Lin, NIST,*Lecture on Small angle Neutron Scattering* in Summer School on Neutron Scattering organized by Institute of Physics, CAS, Beijing, (2004).

[3.7] S. M. King, Chapter 7 in *Modern Techniques for Polymer Characterization* R A Pethrick & J V Dawkins (editors), John Wiley, (1999) (ISBN 0-471-96097-7).

[3.8] G. E. Bacon, Neutron diffraction. 3rd ed. Clarendon Press; Oxford (1975).

[3.9] Wei Chen, Susheng Tan, Tai-Kai Ng, Warren T. Ford, and Penger Tong, *Phys. Rev. Lett.* **95** (2005), 218301.

[3.10] Wei Chen, Private communication, October 10, (2007).
[3.11] D. A. Weitz and D. J. Pine, Diffusing Wave Spectroscopy, in *Dynamic Light Scattering*, ed. by W. Brown, Claredun Press, Oxford (1993).
[3.12] D. S. Horne, *J. Phys. D: Appl. Phys.*, **22** (1989), 1257.
[3.13] O. Kratky, in eds. O. Glatter and O. Kratky, *Small Angle X-ray Scattering*, Academic Press, London, (1982).

Chapter 4

Complexity of Soft Matters

Soft matters are usually called "complex fluids" because of their features of complexity. Self-organization is one of the most remarkable characteristics of soft matters, and it leads to self-similarity, the property of having a substructure analogous or identical to an overall structure. Soft matters manifest abundant behaviors of complexity such as fractal, chaos, percolation, self-organized criticality (SOC), complex network and so forth. Speaking of complexity of soft matters, P.G. de Gennes says, in a certain primitive sense, that modern biology has proceeded from studies on simple model systems (bacteria) to complex multicellular organisms (plants, invertebrates, vertebrates, ...). Similarly, from the explosion of atomic physics in the first half of this century, one of the outgrowths is soft matter, based on polymers, surfactants, liquid crystals, and also on colloidal grains [1.1].

4.1 Examples of Chaos in Soft Matters

As we have seen in Chapter 1, soft matters show abundant features of self-organization and self-similarity. In comparison with hard matters, soft matters have much more complicated characteristics, which come mainly from the interactions between the components of the soft matters. These interactions in soft matters are much stronger than those in air, but much weaker than those in solids as described in Chapter 2. In this chapter, we will examine the complexity of soft matters especially their properties of fractals and chaos on the basis of self-similarities and nonlinearity.

Outside of the scientific context, the term "chaos" is still used as the synonym of irregular, meaning great disorder, confusion, a disorderly mass or a jumble.

However, chaos is not simply a disordered and confused state, nor a complicated ordered state, but an apparently irregular, random state developed from a regular state and has a profoundly intrinsic regulation. Since chaos is an irregular movement out of deterministic equations, Haken calls it a "decisive chaos". Hao Bailin [4.1] points out that the physics the university students are studying is basically the physics of determinacy; at the other extreme is the physics of randomness such as Brownian motion; however the physics between the above two ends is the physics of decisive chaos.

We will start with three examples to get some feeling of what we mean by chaos.

4.1.1 *Rheochaos*

Rheochaos was first discovered in 2000 [4.2]. Rheochaos is an irregular time variation in the stress/shear rate at a constant shear rate/stress arising from nonlinearities in a system governed by the viscoelastic constitutive equation. In a shear-thinning wormlike micellar system, the small angle light scattering experiments and rheological measurements show the Rheochaos. In the experiments, [4.3, 4.4] the strength of flow-concentration coupling is tuned by the addition of NaCl. The time series of stress is plotted for different shear rates in Fig. 4.1. The power spectrum, a plot of the portion of a signal's power (energy per unit time) falling within given frequency bins, is presented in Fig.4.1(d). The broad and continuous band of the power spectrum shows that the time series of the signal is chaotic.

4.1.2 *Chaos in ECG*

The second example of the complexity of soft matter is a comparison of the electrocardiogram (ECG) of healthy older people and coronary artery disease (CAD) patients with sinus rhythms.

Is a "healthy" biological system of soft matters (a heart of a brain, for example) chaotic or periodic? This is still an open question. Some experimental studies claim that the healthy or normal

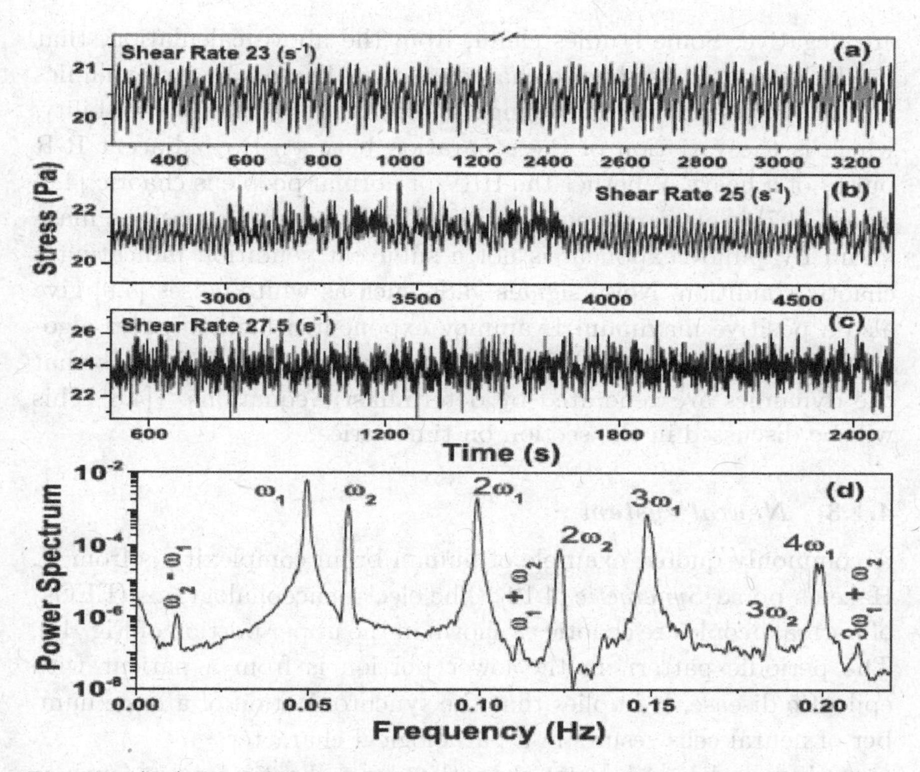

Fig. 4.1 (a-c) Stress relaxation dynamics for a shear-thinning wormlike micellar system with certain amount of NaCl for different shear rate; and (d) the broad band Fourier power spectrum of (a) shows the chaotic characteristics of the system [4.3].

hearts or brains are chaotic. One of the experiments [4.5] uses the data of synchronous 12-lead ECG signals using a multi-sensor (electrode) technique collected from 60 CAD patients with sinus rhythms and 60 healthy older people. Their Lyapunov exponents, parameters denoting the diverging or converging speed of two trajectories in a phase space, are compared. It is shown that the Lyapunov exponents computed from different locations on the body surface are not the same, but have a distribution characteristic for the ECG signals recorded from CAD patients with sinus rhythms and for signals from healthy older people. It is also shown that the maximum Lyapunov exponent λ_l of all signals is positive, while all the others are negative. Some studies claim, from the above calculation, that the ECG signal has chaotic characteristics. The nonlinear dynamics on ECG are often analyzed using the concept of heart rate variability, which is the variation of the separation between two adjacent R-R pulses of a heart. Whether the HRV of normal people is chaotic [4.6] or not [4.7] is also an open question. This is because a positive maximum Lyapunov exponent is not a sufficient condition indicating a chaotic condition. Noise signals [4.8], such as white noises [4.9] give also a positive maximum Lyapunov exponent using computer algorithms. As pointed by L. Glass, "the sine qua non of chaos is that the dynamics are generated by deterministic equations." [4.8] This will be discussed in the section on time series.

4.1.3 *Neural system*

A commonly quoted example of human brain complexity is from H. Haken's book, *Synergetic* [4.10]. The electroencephalograms (EEGs) of normal people are chaotic as shown in the upper portion of Fig. 4.2. The periodic pattern in the lower portion is from a patient with epileptic disease. It implies that the synchronization of a large number of neural cells results in a pathological character.

Other studies [4.11] also show that for a chaotic system (such as the brain), the maximum Lyapunov exponent L_1 implies the degree of chaos, or the degree of brain activity. They found that the maximum Lyapunov exponents in the brain center, where the portion governing memory and study are located, are larger than those in other regions.

It is noticed that the maximum Lyapunov exponents of human hearts are less than 0.1, while those of human brains range from

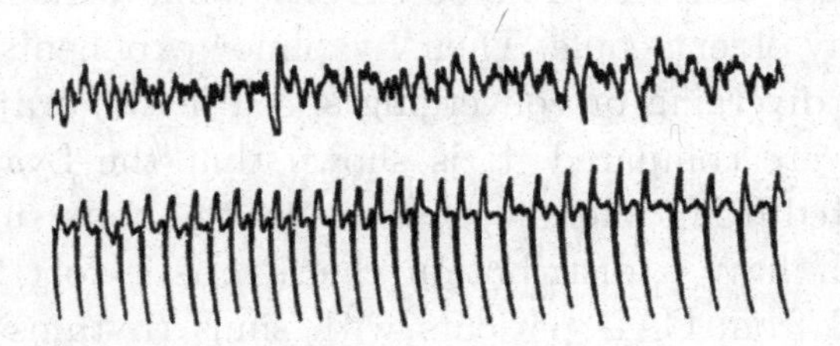

Fig. 4.2 The EEGs of normal people (upper) and a patient of epileptic disease (lower) [4.10].

about 20 to 30. This may imply that brains are much more complicated than hearts.

4.1.4 *Self-similarity*

In the soft matter study, quantitative analyses are necessary to understand many complex patterns. Figure 4.3 is taken in a process of a diffusion-limited monomer-aggregation starting with a seed near the center [4.12]. This is just an example of such complex patterns.

This pattern, and many others such as soft materials under a shear mentioned earlier, have one character trait in common — they are self-similar, and they can be analyzed with knowledge of fractals and fractal dimensions. Physicists are trying to find out the physics meanings behind the complex patterns.

As we mentioned in the first chapter, one of the major characteristics of soft matters is self-assembly. Many experiments have proven that during the process of self-assembly, soft matters would form certain geometric configurations that show self-similarity.

An example of trivial self-similarity is a straight line: any line segment looks the same as the whole line when magnified. Non-trivial examples include such things as the Sierpinski triangle and the Koch snowflake curve. It is interesting to form a Sierpinski triangle from "the chaos game" [4.13].

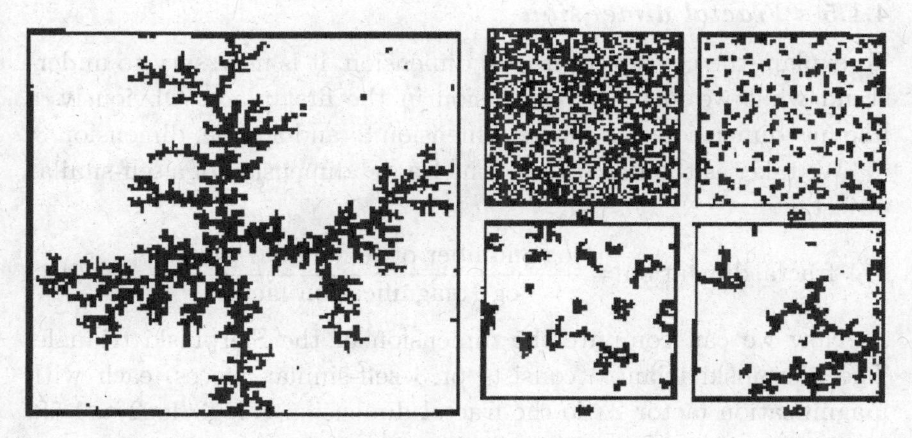

Fig. 4.3 A self-similar pattern grown from the branch center using a diffusion-limited monomer-aggregation [4.12].

Fig. 4.4 The Sierpinski triangle [4.13].

Any object that is self-similar in a non-trivial manner is a fractal. A fractal is a geometric figure or natural object that combines the following characteristics: **(a)** its parts have the same form or structure as the whole, except that they are at a different scale and may be slightly deformed; **(b)** its form is extremely irregular or fragmented, and remains so, whatever the scale of examination; **(c)** it contains "distinct elements" whose scales are very varied and cover a large range [4.14].

4.1.5 *Fractal dimension*

To explain the concept of fractal dimension, it is necessary to understand what we mean by dimension in the first place. Obviously, a line has dimension 1, a plane, dimension 2, and a cube, dimension 3.

We take, as the *definition* of the fractal dimension of a self-similar object,

$$\text{fractal dimention} = \frac{\log\,(\text{number of self-similar pieces})}{\log\,(\text{magnification factor})}. \tag{4.1}$$

Now we can compute the dimension of the Sierpinski triangle. The Sierpinski triangle consists of 3 self-similar pieces, each with magnification factor 2. So the fractal dimension is $\log 3/\log 2 \approx 1.58$.

So the dimension of Sierpinski triangle is somewhere between 1 and 2, just as our "eye" is telling us. A Sierpinski triangle breaks into $\mathbf{3}^N$ self-similar pieces with magnification factors $\mathbf{2}^N$, so we again have fractal dimention $= \log 3^{\mathrm{N}} / \log 2^{\mathrm{N}} \approx 1.58$.

Fractal dimension is a measure of how "complicated" a self-similar figure is. In a rough sense, it measures "how many points" lie in a given set. A plane is "larger" than a line, while a Sierpinski triangle sits somewhere in between these two sets.

For irregular fractals [4.14], the fractal dimension can be determined via a measurement of a fractal. A fractal is a point set inserted in Euclidian space, while a fractal dimension is a measurement of the size of a point set.

If a curve segment is covered using either linear, area or volume elements, there would be quite different results. A finite curve segment L can be completely covered with a linear element δ for $N(\delta)$ times of coverage, then its length $L = N(\delta)\delta \to L_0$ or $L_0\delta^0$ when $\delta \to 0$. When $\delta \to 0$, $L = L_0$. The curve segment can be covered with a surface element δ^2 for $N(\delta)$ times of covering. The area of the curve, $L_A = N(\delta)\delta^2 \to L_0\,\delta^1$ when $\delta \to 0$. If the curve segment is covered with a volume element δ^3, then the volume of the curve is $L_V = N(\delta)\delta^3 \to L_0\delta^2$ when $\delta \to 0$. Obviously, the area or volume of a curve segment is 0 when $\delta \to 0$.

Similarly, for a surface, its length A_L, area A and volume A_V can be expressed respectively as

$$A_L = N(\delta)\delta \xrightarrow{\delta\to 0} A_0\delta^{-1},$$

$$A = N(\delta)\delta^2 \xrightarrow{\delta\to 0} A_0\delta^{0},$$

$$A_v = N(\delta)\delta^3 \xrightarrow{\delta\to 0} A_0\delta^{1}.$$

When $\delta \to 0$, the corresponding length of a surface A_L is ∞, the area $A = A_0$, and the volume $\mathrm{A}_V = 0$ [4.16].

In general, a measurement of a point set S can be expressed as [4.15]

$$M_d = N(\delta)\delta^d. \tag{4.2}$$

$M_d \to \infty$ when $\delta \to 0$ and $d < d_f$; $M_d \to$ finite when $\delta \to 0$ and $d = d_f$; $M_d \to 0$ when $\delta \to 0$ and $d > d_f$. d_f is a critical dimension of measurement of M_d from 0 to ∞.

4.1.6 *Measurements of fractal dimension*

(1) Capacity dimension

The definition of fractals provides box counting as a method of measurement of fractal dimension. One may cover a fractal with boxes of side ε (see Fig. 4.6). Since there are holes and gaps in the fractal, some small boxes would be empty. Count the number of the boxes that are not empty, and denote the number as $N(\varepsilon)$. Reducing the box size ε, and repeating the above procedure, one may get series pairs of ε and $N(\varepsilon)$. The smaller the box size ε is, the larger the number of boxes $N(\varepsilon)$ needed to cover the fractal pattern. If $\ln N(\varepsilon)$ is plotted versus $\ln \varepsilon$, the negative value of the slope is the fractal dimension (Hausdorff dimension or capacity dimension), D_0,

$$D_0 = -\lim_{\varepsilon \to 0} \frac{\ln N(\varepsilon)}{\ln \varepsilon}. \tag{4.3}$$

From Eq. (4.2), one has, when $\delta \to 0$, $N(\delta) \propto 1/\delta^d = 1/\delta^{d_f}$ or

$$d_f = -\frac{\ln N(\delta)}{\ln \delta}. \tag{4.4}$$

Suppose a physical quantity (density, for example) linearly decreases with length on a log (density)-log (length) diagram, P is circumference, λ, length of a step and δ, fractal dimension 1.15 is determined as shown in Fig. 4.5.

However, the box-counting method is sometimes unrealistic, especially for high-dimensional cases where the $N(\varepsilon)$ counts increase rapidly. On the other hand, no matter how many points of the fractals are involved in a box, it is counted as a non-empty box, and has a similar contribution to $N(\varepsilon)$. So it cannot reflect the non-homogeneity inside a fractal.

(2) Scattering experiments

To understand the process of the measurement of fractal dimensions with light or neutron scattering, it is necessary to introduce a density-density correlation function (pair correlation function) [4.15]

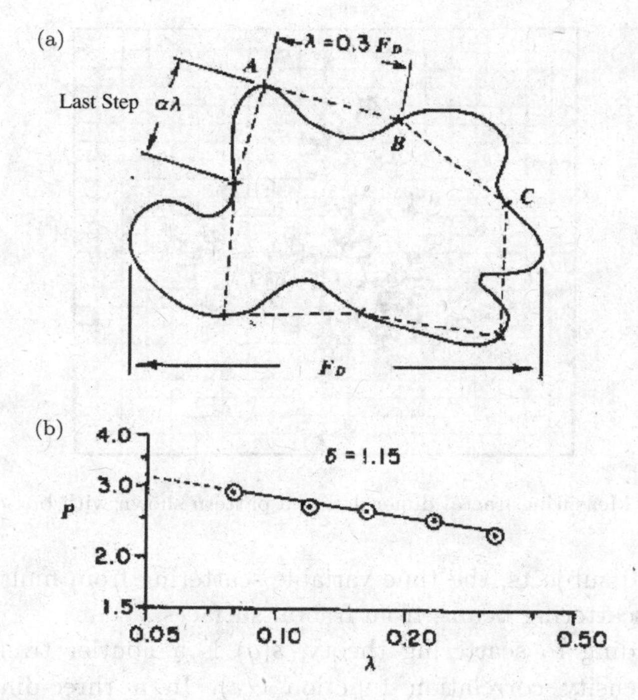

Fig. 4.5 An example of measuring fractal dimension of a pattern shown [4.15].

which is

$$c(\boldsymbol{r}) = \sum_{\boldsymbol{r}'} \rho(\boldsymbol{r}' + \boldsymbol{r})\rho(\boldsymbol{r}'). \tag{4.5}$$

The density correlation of two points (both within a cluster) that are $\boldsymbol{r}$ apart is the probability of finding a particle in $\boldsymbol{r} + \boldsymbol{r}'$, while there is a particle in $\boldsymbol{r}'$. $\rho(\boldsymbol{r}') = 1$ if $\boldsymbol{r}'$ belongs to the cluster, ($\rho(\boldsymbol{r}') = 0$, otherwise). If the fractal is isotropic, the density correlation function is independent of direction, $\mathrm{c}(\boldsymbol{r}) = \mathrm{c}(r)$. An irregular fractal should be scale invariant, namely one has

$$c(br) \propto b^{-a} c(r) \tag{4.6}$$

(where a is a non-integer, and $0 < a < d$). The above equation is satisfied only when $\mathrm{c}(r) \propto r^{-a}$.

Fractal dimension can be measured with light, x-ray or neutron scattering experiments. One should first find structure factors $S(q)$ of

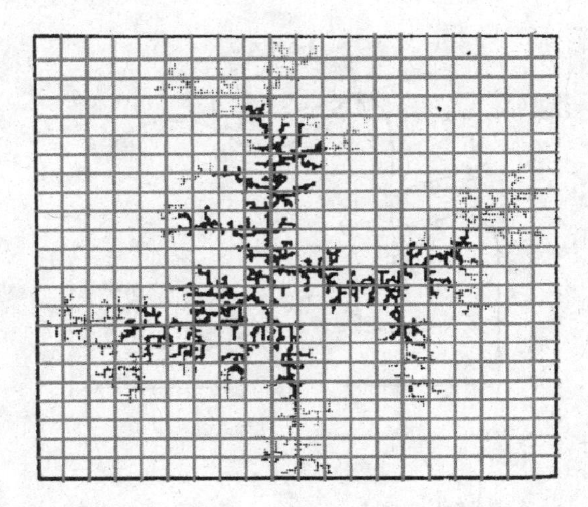

Fig. 4.6 Measuring fractal dimension of a pattern shown with box-counting.

the fractal subjects, the time variable scattering from multi-groups, and the scattering beams from fractal surfaces.

According to scattering theory, $s(q)$ is a Fourier transform of density-density correlation function $\mathrm{c}(r)$. In a three-dimensional isotropic system,

$$s(q) = 4\pi \int_0^\infty 0(r) r^2 \frac{\sin(qr)}{qr} dr, \tag{4.7}$$

$$c(r) \propto r^{dj-d}. \tag{4.8}$$

For a finite sample with radius R, $c(r) \to 0$ when $r > R$. For a three-dimensional sample,

$$\begin{aligned} &c(r) \propto r^{d_f-3} f(r/R), \\ &f(r/R) \approx \text{constant}, \qquad \text{if } r \ll R; \\ &f(r/R) \ll 1, \qquad\qquad \text{if } r \gg R. \end{aligned} \tag{4.9}$$

Substituting $c(r)$ into Eq. (4.7), and letting $r = z/q$, one has

$$s(q) \propto q^{-d_f} \int_0^\infty z^{d_f-2} f\left(\frac{z}{qR}\right) \sin\, zdz. \tag{4.10}$$

The scattering intensity $I(q)$ from experiments is given by

$$I(q) \approx s(q) \propto q^{-d_f}, \quad \text{for } \tfrac{1}{R} \ll q \ll \tfrac{1}{r_0}, \tag{4.11}$$

where r_0 is the particle radius.

Fractal dimension can be estimated from Eq. (4.11) with light and x-ray scattering for dilute aggregates [4.17]. For dense aggregates, small angle neutron scattering is used [4.18] with the intensity of the scattering neutrons $I(q)$ proportional to total cross section, $I(q) \sim d\Sigma/d\Omega$, measured as a function of the scattering wave-vector $q = (4\pi/\lambda)\sin(\theta/2)$, where λ is the incident wavelength and θ, the scattering angle. As mentioned in the last chapter, the scattering light intensity is proportional to the structure factor [4.19].

$$S(q,t) \sim q_m(t)^{-d_f}\tilde{S}(q/q_m(t)), \tag{4.12}$$

where a time-independent scaling function

$$\tilde{S}(X) = \frac{2(2 + X^{d_f})}{5 + X^{2d_f}} \tag{4.13}$$

was chosen in its curve fitting to determine the fractal dimension d_f.

4.2 Physical Mechanism of Fractals

It is said that the science of 20th century remembers only three concepts: relativity, quantum mechanics and chaos [4.20]. Relativity eliminates the Newtonian image of absolute space and time, quantum mechanics eliminates Newton's dream of the process of controllable measurement, while chaos eliminates the Laplacian fancy of decisive predictability. Not as the previous two, the third revolution can be directly used to study objects that we can touch, and that have sizes similar to human beings itself.

4.2.1 *Butterfly effect*

(1) Sensitivity to initial conditions

In 1979, E. Lorenz (1917–2008), a meteorologist of MIT, gave a talk at Washington DC, titled "Predictability: Does the flap of a butterfly's wings in Brazil set off a tornado in Texas?" The "flap of a

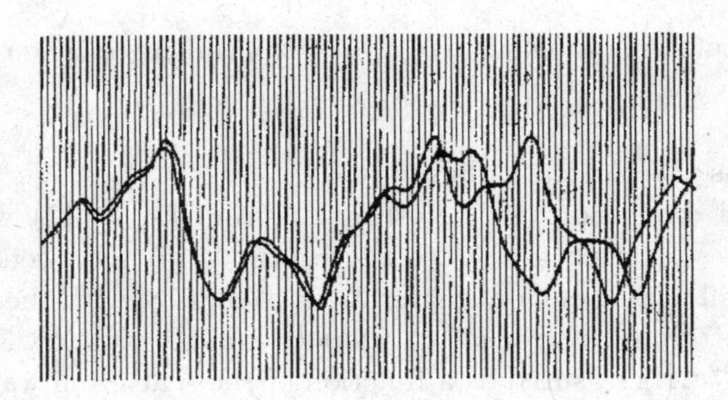

Fig. 4.7 E. Lorenz's curve of calculation of atmospheric convection: A little error would bring tremendous disaster [4.21].

butterfly's wings" comes from a neglect of 0.0001 — once he used a number 0.506 rather than 0.506127 as a middle input to ask his computer to repeat the computation of weather data according to his equation set. The second curve (see Fig. 4.7) started similar to the first one, but after a short while some difference becomes obvious, and at the time corresponding to the fourth peak on the second curve — the difference becomes very large — just like a sunny day with a gentle breeze on one curve, but a tornado on the other. A serious problem would be brought to a weatherman — "which curve should he use in his weather forecast?"

Poincare, a French mathematician, pointed out in his book *Science and Method* that a tiny difference in initial conditions leads to a resulting tremendous difference, a little difference of the former may bring large differences of the latter, and a prediction becomes impossible. Lorenz himself once said: "We have never made any long term forecast indeed, but now we found an excuse."

(2) Roles of nonlinearity

The Lorenz system is about an atmospheric convection, and it is governed by the following ternary nonlinear equation set

$$\begin{cases} \dot{X} = -10(X - Y) \\ \dot{Y} = -XY + YZ - Y \\ \dot{Z} = XY - \dfrac{8}{3}Z. \end{cases} \tag{4.14}$$

An attractor in a phase space is a final configuration of a set of orbits onto which all the orbits near it are converging. The attractor can be divided into two categories. A steady stationary solution is a 0-d attractor, a point in a phase space, which is called a fixed point; a stable periodic solution is a 1-d attractor, a closed curve in a phase space, which is called a limit loop. (Fig. 4.9- **v, x**)

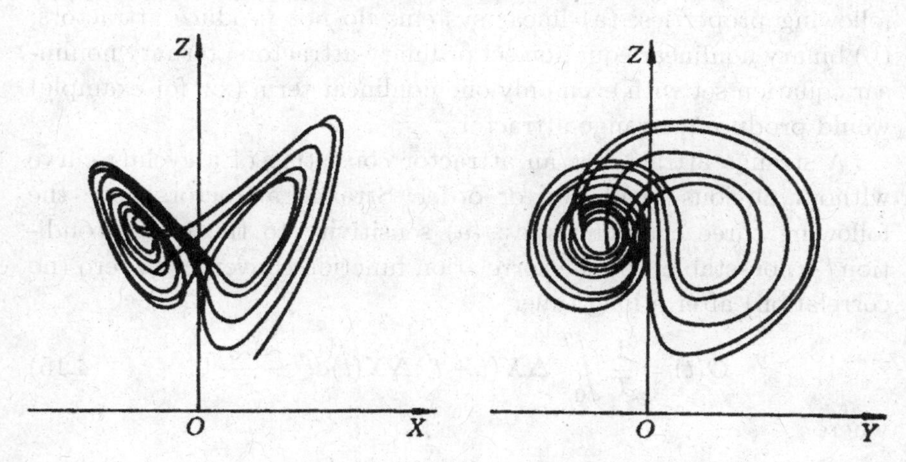

Fig. 4.8 Lorenz system solution projected onto X-Z and Y-Z planes [4.15].

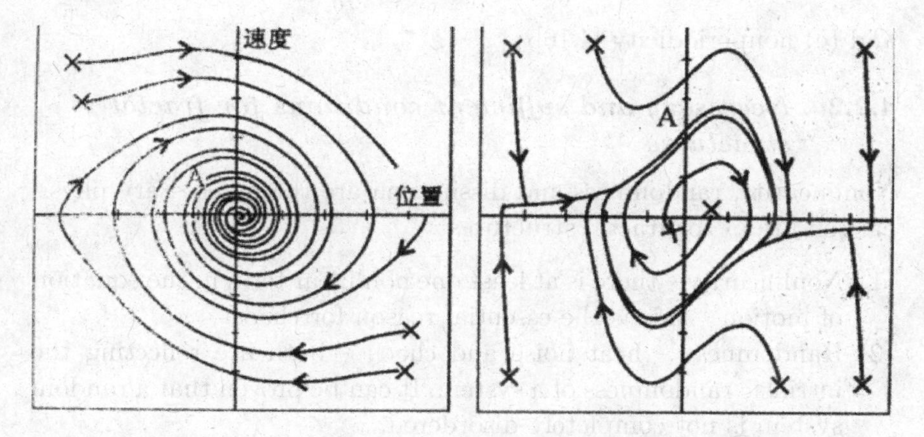

Fig. 4.9 The orbits and attractors of steady solutions. The left diagram shows the orbits of a damped pendulum. Four x's denote four different initial positions of the pendulum, and all the orbits come to an attractor of a steady point A (at the origin). The right diagram shows the orbits of a clock pendulum, and the attractor is a steady loop A [4.1].

Random movement solution is a strange attractor, and it is a fractal structure in phase space. A nonlinear and dissipative system may have a solution in a random motion in a certain limited area of phase space, and the random motion converges to an "attractor" — a fractal structure in phase space.

The solution of the Lorenz system shown in Fig. 4.6 has the following properties: (a) linear systems do not produce attractors; (b) binary nonlinear equation set ordinary attractors; ternary nonlinear equation set with even only one nonlinear term (xz, for example) would produce a strange attractor.

A strange attractor is an attractor consisting of a cycling curve without obvious regulation or order. Strange attractors have the following three characteristics: (a) sensitivity to the initial condition — non-stable; (b) the correlation function converges to zero (no correlation) after a long time;

$$C(t) = \frac{1}{T}\int_0^T \Delta X(t+t')\Delta X(t)dt' \xrightarrow{t\to\infty} 0, \tag{4.15}$$

where

$$\Delta X(t) = X(t) - \bar{X}, \bar{X} = \frac{1}{T}\int_0^T X(t)dt, \tag{4.16}$$

and (c) nonperiodicity [4.16].

4.2.2 *Necessary and sufficient conditions for fractal structures*

Nonlinearity, randomness and dissipation are three necessary physical conditions for fractal structures.

(1) Nonlinearity – there is at least one nonlinear term in the equation of motion – this is the essential reason for chaos.
(2) Randomness – heat noise and chaos – both are reflecting the intrinsic randomness of a system. It can be proven that a random system is not completely disordered.
(3) Dissipation — can be expressed as $\mathrm{d}E/\mathrm{d}t < 0$ (if $\mathrm{d}E/\mathrm{d}t > 0$, an energy-gained system), that leads to a disordered and entropy-increasing state. Dissipation of a system must be analyzed to study a fractal structure in both real and phase space.

The sufficient condition for fractal structure (presenting a strange attractor) is the appearance of instability in a dissipated system [4.15].

From the above solution categories and characteristics of strange attractor, one concludes that the precondition of fractal structure (or strange attractor) is the instability of a nonlinear, random and dissipative system.

4.3 Quantitative Analysis of Chaos

4.3.1 *The broad band power spectrum*

As we have seen in Section 4.1.1, the broad band power spectrum corresponds to chaos. A power spectrum is a plot of power distribution versus frequency. The power at a given frequency is the summation of the square of sinusoidal wave coefficients at the frequency. If $x_i(t)$ is a time series of a nonlinear process and there are N data $x_1, x_2, \ldots, x_N$, at a constant time interval τ, its Fourier coefficients [4.22] are

$$a_k = \left(\frac{1}{N}\right)\sum_{i=1}^{N} x_i \cos\left(\frac{\pi i k}{N}\right),$$

$$b_k = \left(\frac{1}{N}\right)\sum_{i=1}^{N} x_i \sin\left(\frac{\pi i k}{N}\right). \tag{4.17}$$

The power can be obtained from the following:

$$\bar{p}_k = a_k^2 + b_k^2. \tag{4.18}$$

The averaged value of several calculated values of p_k would approach the power spectrum.

The power spectrum of a period one orbit is dominated by one central peak, ω_1. It was pointed out [4.1] that the chaotic motion has a broad band power spectrum with a rich spectral structure. The broad-band nature of the chaotic power spectrum indicates the existence of a continuum of frequencies. A purely random or noisy process also has a broad band power spectrum, so it is necessary to develop methods to distinguish noise from chaos. In addition to its broad-band feature, a chaotic power spectrum has also many broad peaks at the nonlinear resonances of the system. These nonlinear resonances are directly related to the unstable periodic orbits embedded within the chaotic attractor. So the power spectrum of a chaotic attractor does provide some limited information concerning the dynamics of the system, namely, the existence of unstable periodic orbits (nonlinear resonances) that strongly influence the recurrence properties of the chaotic orbit.

Experimental power spectra can be obtained in at least three ways: (i) using a spectrum analyzer or signal analyzer, which displays real-time power spectra; (ii) using a signal analyzer card to collect signals; or (iii) digitizing data to take an FFT.

4.3.2 *The positive maximum Lyapunov exponents [4.20, 4.22]*

(1) Definition of the Lyapunov exponent

The basic characteristic property of chaos is the high sensitivity of an initial condition, and this can be quantitatively measured with the Lyapunov characteristic exponents. Simply speaking, the Lyapunov exponents express the averaged speed of the separation of the orbits in a phase space under certain initial conditions.

We start with a linear ordinary differential equation $\mathrm{d}x/\mathrm{d}t = ax$ with an initial condition of $x = x_0$ at $t = 0$. The solution is $x = x_0\mathrm{e}^{at}$. Two adjacent orbits at the initial time would separate from each other according to $x = x_0\mathrm{e}^{at}$ if $a > 0$; the separation of the two orbits would be decayed if $a < 0$; the two orbits would keep their separation only when $a = 0$. The exponent determines the separation of the two orbits — either stretched or compressed, or simply rotating, which can be identified only after a long time averaging process.

Consider two initial points: point x_0 and its neighboring point $x'_0 = x_0 + \Delta x$, namely the initial separation between the two points is $\Delta_0 = x_0 - x'_0 = \Delta x$. After the first and second iterations using a mapping or a function $* \ f(x)$, the separation of the two points becomes

$$\Delta_1 = x_1 - x'_1 = f(x_0 + \Delta x) - f(x_0) = f'(x_0)\Delta x$$

and

$$\begin{aligned}\Delta_2 = x_2 - x_2' &= f^{(2)}(x_0 + \Delta x) - f^{(2)}(x_0) \\ &\approx \left.\frac{\mathrm{d}f^{(2)}}{\mathrm{d}x}\right|_{x=x_0} \Delta x = \Delta x e^{2\lambda},\end{aligned} \tag{4.19}$$

respectively. After n iterations, the separation between the two point orbits will be

$$\Delta_n = x_n - x_n' = f^{(n)}(x_0 + \Delta x) - f^{(n)}(x_0) \approx \left.\frac{\mathrm{d}f^{(n)}}{\mathrm{d}x}\right|_{x=x_0} \Delta x = \Delta x e^{n\lambda}. \tag{4.20}$$

The last equation comes from the assumption of exponential growth or decay of the point orbit separation, and this leads to the definition of the Lyapunov exponent λ:

$$\lambda = \lim_{n\to\infty} \frac{1}{n} \ln \frac{\mathrm{d}f^{(n)}(x_0)}{\mathrm{d}x}. \tag{4.21}$$

The Lyapunov exponents describe the time-averaged characteristics of the derivation of a system's motion from the datum orbit.

The appearance of the maximum positive Lyapunov exponent signals the transition to chaos. The sign of the Lyapunov exponents $(\lambda_1, \lambda_2, \lambda_3)$ at different parameters provides a classification for attractors, e.g.,

(-, -, -) fixed point,
(0, -, -) periodic motion,
(0, 0, -) quasiperiodic motion on a two dimensional surface
(+, 0, -) chaos.

In comparison to steady motion, chaos is motion that is sensitive to even a little difference in initial conditions, and it corresponds to the positive Lyapunov exponents, a measure of the divergence of trajectories in phase space. If a fractal correlation dimension is larger than 2, it implies that the signal is chaotic rather than stochastic noise. The condition of chaos is the existence of both positive and negative Lyapunov exponents, which correspond to the sub-spaces of exponential growth and decay, respectively.

(2) Algorithm of the Lyapunov exponent from time series

(i) Reconstruction of phase space

In general, the description of a deterministic system is the description of the system in a phase space. A phase space is a space in which all possible states of a system are represented, with each possible state of the system corresponding to one unique point in the phase space. After a long time evolution, a deterministic chaotic system of dynamics is often attracted into a strange attractor with noninteger fractal dimension in a phase portrait.

Packard *et al.* propose a method to reconstruct a phase space from a single variable time series $X(t)$ [4.23]. Their basic idea is that an evolution of any component of a system is determined by other components that are interacting with it, and the information of the dynamic system can be explored by the evolution process of a component of the system.

A d-dimensional dynamic system can be expressed by a time series

$$X(t) = [x(t), x'(t), x^{2'}(t), \cdots, x^{(d-1)'}(t)].$$

It can be reconstructed in a d-dimensional phase space with the time delay method by representing $x(t)$, its derivatives, $x'(t)$ or $x^{(d-1)'}(t)$ with a time delay variable τ, and the time series becomes a vector of delayed coordinator,

$$X(t) = [x(t), x(t+\tau), \cdots, x(t+(d-1)\tau)],$$

where τ is a delay time (time lag). This vector forms a d-dimensional reconstructed phase space, and it represents also a point in the phase space, or a state of the system [4.5, 4.24, 4.25].

(ii) Lyapunov exponent

The general idea in studying the initial condition sensitivity of dynamic systems is to follow two nearby orbits and to calculate their average logarithmic rate of separation. First, find a point that is

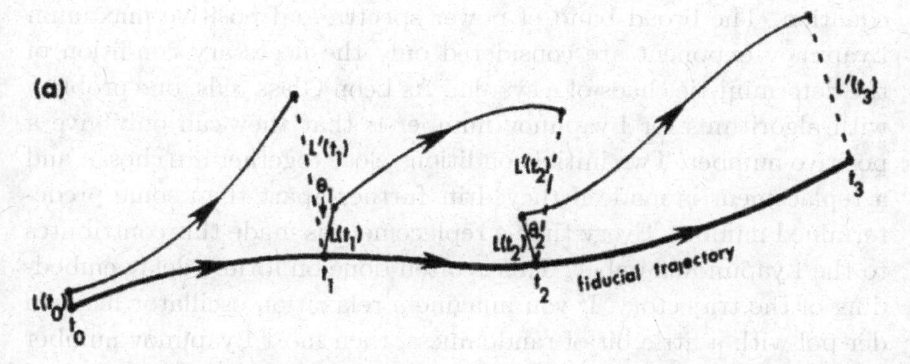

Fig. 4.10 A schematic representation of the evolution and replacement procedure used to estimate Lyapunov exponents from experimental data. The largest Lyapunov exponent is computed from the growth of length elements [4.24].

closest to the initial point $[X(t_0), X(t_0+\tau), \cdots, X(t_0+(d-1)\tau)]$, denoting the separation between the two points as $L(t_0)$ (Fig. 4.10); the separation $L(t_0)$ becomes $L'(t_1)$ at t_1 during the evolution. When the separation of the vector between two points becomes large, a new point is chosen near the fiducial trajectory (or reference orbit), minimizing both the replacement length L and the orientation change θ. This process is repeated till all data points are covered. Lyapunov exponent λ_1 is defined as

$$\lambda_1 = \frac{1}{t_N - t_0} \sum_{k=1}^{N} \log_2 \frac{L'(t_k)}{L(t_{k-1})} \tag{4.22}$$

where N is the total number of iterations of the length element. When λ_1 approaches a steady value, it is then the Lyapunov exponent of the time series [4.24,4.26]. By convention, the natural logarithm (base-e) is usually used ($\lambda \sim \ln(L'(t_1)/L(t_0))$), but for maps, the Lyapunov exponent is often quoted in bits per iteration, in which case you would need to use base-2. (Note that $log_2 x = 1.4427 log_e x$.) [4.27]

4.3.3 *Conditions for deterministic chaos of time series*

To determine whether a time series is deterministic chaotic or random, one needs to judge whether the time series is a result of dynamic equation. The broad band of power spectra and positive maximum Lyapunov exponent are considered only the necessary condition of the deterministic chaos of a system. As Leon Glass tells, one problem with algorithms for Lyapunov number is that they can only give a positive number. Two initial conditions close together are chosen and a replacement is made if they drift further apart than some predetermined number. Every time a replacement is made this contributes to the Lyapunov number. This is often done on a time delay embedding of the trajectory. If you imagine a relaxation oscillator like van der pol with a little bit of randomness, then most Lyapunov number algorithms would give a positive Lyapunov number [4.28]. Nicolis and Nicolis [4.29], and Kaplan and Glass [4.30] propose methods to characterize a time series that is directed towards the analysis of whether the time series is generated by a deterministic system.

(1) Decorrelation time criteria

C. Nicolis and G. Nicolis propose that the criterion of deterministically chaotic time series is that the decorrelation time, T_d, (see Fig. 4.9) the first zero of the autocorrelation function

$$AC(I) = AC(t' - t) = cov\frac{[X(t'), X(t)]}{E[x(t) - M]^2} \tag{4.23}$$

is on the order of the inverse of Lyapunov exponent, namely

$$T_d \sim \lambda^{-1}. \tag{4.24}$$

If this is fulfilled, the time series is deterministically chaotic [4.9, 4.29].

(2) Weighted average of trajectory vector summation

As pointed out by Leon Glass, finding a positive Lyapunov number in a cardiac system using a computer algorithm does not necessarily mean that the underlying system displays deterministic chaos, because random systems also give a positive Lyapunov number. Finally, the *sine qua non* of chaos is that the dynamics are generated by deterministic equations [4.8]. He added that in some cases it is clear that there is a deterministic rule. For example, it is sometimes

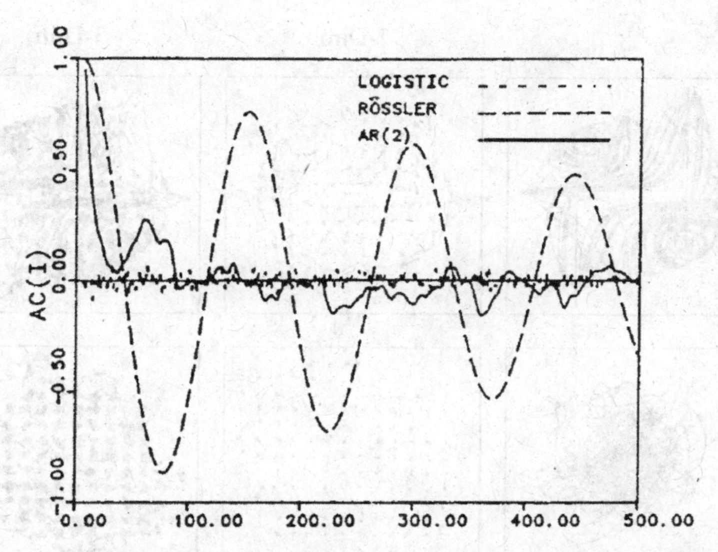

Fig. 4.11 Decorrelation time is the first zero of an autocorrelation function [4.9].

possible to construct a Poincare map and see that the points fall on a curve that is known to give chaos. An example is in Ref. [4.31].

A way to demonstrate determinism in continuous time series is given by Kaplan and Glass. It is based on the observation that the tangent to the trajectory (Fig. 4.12) generated by a deterministic system is a function of position in a $(x(t), x(t-\tau))$ phase space for a two-dimensional system, and in a $(x(t), x(t-\tau), x(t-2\tau))$ phase space for a three-dimensional system. Therefore all the tangents to the trajectory in a given region of the phase space will have similar orientations [4.30].

First, the embedding plots $x(t-\tau)$ vs $x(t)$ and coarse-grained flow averages for the x component of Lorenz equations and for a random signal with the identical power spectrum are shown in Fig. 4.12. The middle two figures of Fig. 4.12 are plotted in a 16×16 two-dimensional coarse-grained phase space, while the right two figures are plotted in a $16 \times 16 \times 16$ three-dimensional coarse-grained flow in the phase space $(x(t), (t-\tau), x(t-2\tau))$. The length of the arrows shows the alignment of the trajectory vectors $\boldsymbol{v}_{k,j}$ in the corresponding box, and the direction shows the direction of mean flow $\boldsymbol{V}_j$ in the boxes. Suppose that V_j is an arrow inputting to and exporting from

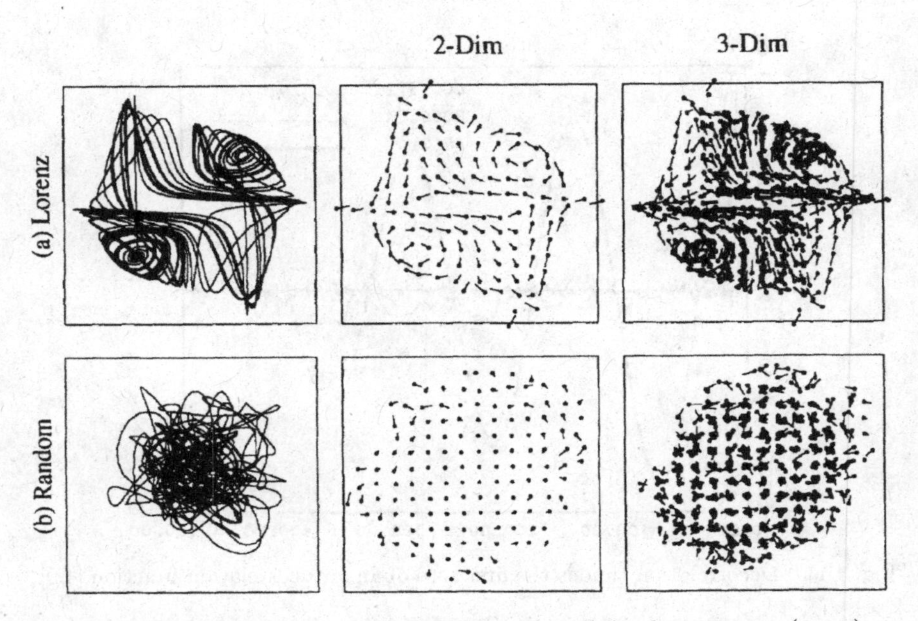

Fig. 4.12 The trajectories of the dynamics plotted in phase space $x(t-\tau)$ vs $x(\tau)$ for (a) Lorenz chaos and (b) randomness [4.30].

a box in the corresponding phase space (Fig. 4.12), and is also a trajectory vector summation over all n passes, k, in the box, normalized by passing number n_j. One has then [4.30]

$$V_j = \sum_k v_{k,j}/n_j, \tag{4.25}$$

where the mean flow vector V_j is a vector summation over all n passes (labeled by k) in a box, normalized by the total number of passes, n_j. This gives a coarse simulation of the dynamics of the system.

In a box, for each deterministic state dynamic flow has only one direction, while a random process may have more than one direction. Calculating averaged coarse-grained flow in each box and statistical quantities would be a good method to determine whether a short time series with some noises is deterministic chaos or random.

Secondly, the average displacement per step for large n is

$$\bar{R}_n^d = \left(\frac{2}{nd}\right)^{\frac{1}{2}} \frac{\Gamma((d+1)/2)}{\Gamma(d+2)}, \tag{4.26}$$

where Γ is the gamma function, $\Gamma(z) = \int_0^\infty t^{z-1}e^{-t}dt$. If z is a positive integer, $\Gamma(z) = z!$. One then has

$$\bar{R}_n^d = c_d n^{-\frac{1}{2}}, \tag{4.27}$$

where $c_d = \pi^{1/2}/2$ for $d = 2, c_d = 4/(6\pi)^{1/2}$ for $d = 3, c_d = 3\pi^{1/2}/32^{1/2}$ for $d = 4$, and $c_d = 1$ when d tends to infinity [4.30].

Finally, the weighted average of V_j over all the occupied boxes can be expressed as

$$\bar{\Lambda} = \left\langle \frac{(V_j)^2 - (R_{nj}^d)^2}{1 - (R_{nj}^d)^2} \right\rangle. \tag{4.28}$$

This is an indicator of a chaotic system since for a deterministic system, $\Lambda = 1$; for a random walk, $\Lambda = 0$ as depicted in Fig. 4.13.

When the first decorrelation time $T_d(D)$ is on the same order as the inverse maximum Lyopunov exponent, this is at least a sign that the system is of weak deterministic chaos [4.9]. The KG test [4.30] is a strict one and it gives strong evidence of chaos.

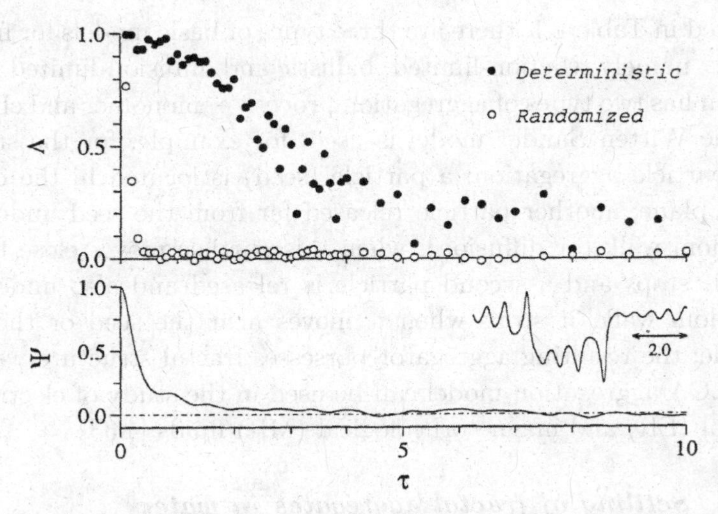

Fig. 4.13 Λ is the indicator to determine whether a time series is deterministically chaotic or randomized. A system is deterministic chaotic when the weighted average of V_j over all the occupied boxes $\Lambda = 1$, and random when $\Lambda = 0$ [4.30].

As pointed out by Ping Chen, this is because, in a box, for each deterministic state, dynamic flow has only one direction, while a random process may have more than one direction. Calculating averaged coarse-grained flow in each box and statistical quantities gives a good method for determining the complexity of short time series with some noises. It is also noted that there is no unique approach to identify deterministic chaos with certainty. Ping Chen pointed out that the word *chaos* is misleading. Chaotic motion has both regular and irregular characteristics. It is preferred to refer to continuous deterministic chaos as *imperfect periodic motion*, which has a stable fundamental period but an irregular wave shape and a changing amplitude. Although long term prediction of the chaotic orbit is impossible from the view point of nonlinear dynamics, a medium-term prediction of approximate period T can be made if we can identify strange attractors from the time series [4.9, 4.32].

4.4 Complexity Helps Better Understanding of Soft Matters

4.4.1 *Fractal growth in colloidal aggregation*

As listed in Table 4.1, there are three types of basic models for fractal growth, namely reaction-limited, ballistic and diffusion-limited. Each of them has two types of aggregation process — monomer and cluster.

The Witten–Sander model is used, for example, for the smoke-dust particle aggregation: a particle (seed) is located in the center of the plane; another particle released far from the seed undergoes a random walk (or diffusion); when this particle moves close to the seed, it stops and a second particle is released and also undergoes a random walk, it stops when it moves near the seed or the first particle; the resulting aggregator possesses fractal structure.

DLCA aggregation model can be used in the study of electrorheological (ER) and magnetorheological (MR) fluids [4.33].

4.4.2 *Settling of fractal aggregates in water*

The settling process of clay in water is an important issue in environmental science and technology [4.34, 4.35]. Quantitative descriptions

Table 4.1. Three basic models of fractal growth.

	Reaction-limited	Ballistic	Diffusion-limited
Monomer Aggregation	**Eden** Chemically-limited; Simulation of Tumor growth (Eden, 1961): a growth process in which the growth possibility is the same at all growing point.	**Vold** Ballistic happens when Mean Free Path of monomer > size of cluster in growth — monomers linearly move towards cluster.	**Witten-Sander** Monomer Brownian motion; Aggregation of smoke dust.
Cluster Aggregation	**RLCA** Random growth at stable interface	**Sutherland**	**DLCA** Clusters in Brownian motion; aggregation of particles onto water- air interface; The sedimentation of metal particles onto an electrode.

of such processes would benefit effective control of water cleaning process. Tambo and Watanabe measured settling velocity and diameter of a discrete floc in a quiescent water column. Floc density was then calculated by introducing the measured values and some constants such as water density and viscosity into a suitable settling velocity equation. It is found that the Stokes equation, which is basically used to describe the movement of fully solid particles of different densities in a flow, was not enough to understand the sedimentation process.

Since there is a permanent flow inside a floc, to analyze the free-settling test of activated sludge flocs, Brinkman equations were introduced [4.36]. In these works, the idea of fractal dimension of porous aggregates is used to estimate the floc permeability. Their theoretical analysis gives explicit calculations on the scaling behaviors of settling velocities, taking into account the fractality through a proper permeability k for fractal aggregates.

Under consideration of proper boundary condition and force analysis, u_a is given by the modified Stokes' law:

$$u_a = -\frac{2a^2 g(1-P)\Delta\rho}{9\mu} \times \frac{aa_0 \tanh(a/\sqrt{k}) + \sqrt{k}a}{\sqrt{k}(a_0 - a) + (k - aa_0)\tanh(a/\sqrt{k})}. \quad (4.29)$$

The aggregate velocity u_a is proportional to the difference of the aggregate and media densities, $\Delta\rho$. P is the aggregate porosity, μ is the kinetic viscosity of a fluid, and a is the radius of a spherical aggregate. The hydraulic permeability k, which reflects the fractality of aggregates, is a key parameter used to describe permanent flow in porous fractal-aggregates.

Using the theory of percolative metal/insulator composites and the scaling theory, k can be expressed as

$$k = d^2 \frac{(P - P_c)^t}{1 - P}, \quad P \geq P_c. \quad (4.30)$$

d is the characteristic length scale, $P_c = 1/3$, t is the conductivity exponent, $t = 1.264$ for $D_f = 1.79$ from the diffusion-limited cluster-cluster approximation, $t = 1.867$ for $D_f = 2.19$ and 2.25 from reaction-limited cluster-cluster approximation [4.37]. The calculated results for the scaling behaviors of settling velocities fit remarkably well with experimental data of the function of sedimentation speed u_a of aggregate size a.

This implies that knowledge of fractality gives a better understanding of sedimentation of fractal aggregates.

4.4.3 *Chaos helps mix microfluids*

Microfluidic mixing is important in the design and fabrication of lab-chips. However, since the Reynolds number R_{en} is much smaller than one, a viscous or lamellar flow is dominant, and the mixing of two different microfluids is difficult. An active mixer based on electrorheological (ER) valves is studied, where the chaotic state of the mixed microfluid is examined to ensure such microfluidic mixing [4.38, 4.39].

An ER flow is driven by a pump. The pressure of the main channel of the ER flow can be controlled by two ER valves when they are alternatively at the on or off state. At low (or high) pressure state, a diaphragm connected with the ER flow would pop up (or down) and open (or close) a microfluid side channel which is perpendicular to the main channel.

Having different side channel perturbation amplitude $A_p(= v_p/U)$ and frequency ω of the controlling signal in the main channel changes the mixing status of two input microfluids. The mixing intensity index is described by the mean Lyapunov exponent λ, which is defined as

$$\lambda = \lim_{t\to\infty} \lim_{d_0\to 0} \frac{1}{t} \ln \frac{d(t)}{d_0} = \frac{U}{L}\lambda^*, \tag{4.31}$$

where λ^* is the Lyapunov exponent of the normalized system, d_0 is the separation of the two fluid particles at $t = 0$, and $d(t)$ is the separation at a later time t. Shown in Fig. 4.14 is the mean

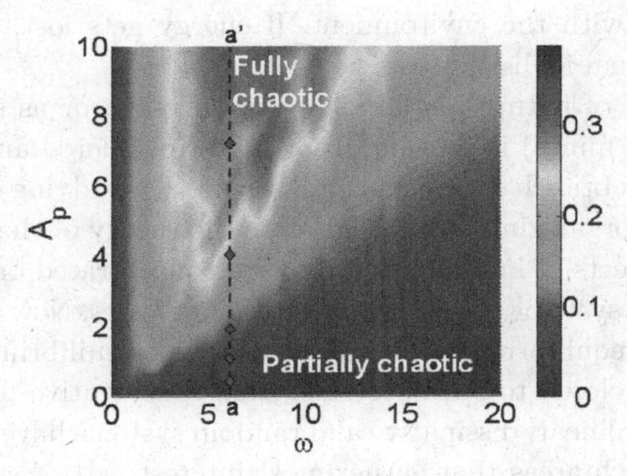

Fig. 4.14 Mean Lyapunov exponent distribution in the $A_p \sim \omega$(Hz) parameter space for the mixed microfluids [4.38].

Lyapunov exponent distribution in the $A_p \sim \omega$ parameter space for fluid particles fed into the main channel during one period of side channel pulsation. The phase diagram demonstrates that the perturbation of frequency of 2π (Hz) and side channel perturbation amplitude $A_p(= v_p/U)$ of 7 drives the two microfluids into a fully chaotic state, meaning a well-mixed state, while a signal of the same frequency but with $A_p = 1$ leads only to a partially chaotic state. This shows how knowledge of complexity benefits quantitative control of microfluid mixing.

4.4.4 *Life system is a dissipative structure*

Thermodynamics defines three different systems: isolated, closed and open systems. An isolated system refers to a system that has ideal isolation walls, and there is no exchange of matter and energy with the environment. In an isolated system, a natural process must be accompanied by an entropy increase, and it does not stop until the entropy reaches maximum and the system reaches an equilibrium state. This process is irreversible in an isolated system — namely, at a thermodynamic equilibrium state, the entropy does not reduce automatically, and it does not change from the most disordered state to an ordered state. A closed system exchanges energy, but not matter, with the environment. An open system exchanges both energy and matter with the environment. If energy gets lost, $dE/dt < 0$, and the system is dissipative.

A period of nature (as large as the universe, or as small as the formation of mines) is growing from chaos to ordered and organized states. Every branch of basic natural sciences is studying every aspect of this "self-organizing" process. Before the theory of chaos and fractal, the subjects of human research were concentrated on closed and conservative systems of equilibrium state processes. Now research has turned from equilibrium state processes to non-equilibrium state processes, from closed to open systems, from conservative to dissipative systems. Nonlinear, dissipative and random systems have become one of the research areas that garner most interest.

Prigogine developed the concept of dissipative structures to describe open systems, in which an exchange of matter and energy occurs between a system and its environment. Prigogine received the Nobel Prize in Chemistry in 1977 for his dissipative structure research and for his contributions to nonequilibrium thermodynamics.

One of the results of the study is Prigogine's "dissipative structure", which is an ordered and structured system created from non-equilibrium phase transitions. The characteristics of the dissipative structure are as follows:

(1) Dissipative structures are created in "open systems", where exists energy or mass transfer between the system and external world;
(2) Dissipative structures appear suddenly only when controlling parameters, such as temperature, flow speed, etc. are above certain threshold values;
(3) The symmetry of their spatiotemporal structures is lower than that before threshold values are achieved; and
(4) The dissipative structure may be a product when the former state is not stable, while it is quite stable once it is created, and it would not be destroyed by any small perturbation.

The first is the characterization that only dissipative states have, and other three are characteristics shared with equilibrium phase transitions.

Examples of dissipative structure are the Bénard cells, lasers, the Belousov–Zhabotinsky reaction, and even life itself [4.40, 4.41]. Lives possess all the above characteristics of dissipative structures. Far from equilibrium, nonlinearity is just a few of the necessary conditions for the appearance of dissipative structure. The appearance of life needs many other conditions to be satisfied simultaneously [4.42]. As an example, human consciousness was conceptualized as a dissipative structure reflecting relationships among time, space, and motion patterns and postulated to reflect a human field pattern with the potential for escape to a higher order and the expansion of consciousness [4.43]. This method helps researchers study consciousness quantitatively utilizing models of dissipative structure.

References

[4.1] Bailin Hao, *Chaos and Fractals: Collection of Essays of Popular Science*, Shanghai Science and Technology Publishing House, (2004), 46–68 (in Chinese).

[4.2] R. Bandyopadhyay *et al.*, *Phys. Rev. Lett.* **84** (2000), 2022.

[4.3] R. Ganapathy and A. K. Sood, *Phys. Rev. Lett.* **96** (2006), 108301.

[4.4] A.K. Sood, R. Ganapathy, *Pramana-J. Phys.* **67** (2006), 33–46.

[4.5] Zhenzhou Wang, Zhen Li, Xinbao Ning, Yuzheng Lin and Yixiang Wei, A comparison study of Lyapunov spectra of CAD patients' and healthy people's synchronous 12-lead ECG signals, *Chin. Sci. Bull.* **47** (2002), 1845 (in Chinese).

[4.6] Xiaolin Huang, Jianjun Zhuang, The Nonlinear dynamics study of ECG signals, www.nju.edu.cn.

[4.7] M. Costa, I. R. Pimentel, T. Santiago, P. Sarreira, J. Melo, and E. Ducla-Soares, *J. Cardiovasc. Electrophysio.* **10** (1999), 1350.

[4.8] L. Glass, *J. Cardiovasc. Electrophysio.* **10** (1999), 1358–1360.

[4.9] Ping Chen, Searching for Economic Chaos: A Challenge to Econometric Practice and Nonlinear Tests, in Richard Day and Ping Chen eds., *Nonlinear Dynamics and Evolutionary Economics*, Oxford University Press (1993), 217–253.

[4.10] H. Haken,*Erfolgsgeheimnisse der Natur — Synergetik: der Lehre vom Zusammenwirken*, Chinese edition translated by Lin Fuhua, Century Publishing Group, Shanghai Translation Publishing House, (1986), 148.

[4.11] Cong-bo Cai, Shen-chu Xu, Zhen-xiang Chen, Jin-shu Zhong, *J. Xiamen Univ.* (*Nat. Sci.*) **40** (2001), 1034.

[4.12] A. Aharony, Fractal Growth, in A. Bunde and S. Havlin, eds. *Fractals and Disordered Systems*, Springer-Verlag, Berlin Heidelberg, (1991), 151–173.

[4.13] M. Barnsley, *Fractals Everywhere*. Boston: Academic Press, (1989), http://math.bu.edu/DYSYS/chaos-game/node1.html.

[4.14] R. Devaney, *Chaos, Fractals, and Dynamics: Computer Experiments in Mathematics*. Menlo Park: Addison-Wesley, (1989).

[4.15] Zhanru Yang, *Fractal Physics*, Shanghia Publishing House of Science and Technology Education, Shanghai (1996) (in Chinese).

[4.16] Jizhong Zhang, *Fractals*, Tsinghua University Press, Beijing (in Chinese).

[4.17] D. W. Schaefer, J. E. Martin, P. Wiltzius and D. S. Cannel, Fractal geometry of colloidal aggregates, *Phys. Rev. Lett.* **52** (1984), 2371.

[4.18] B. D. Butler, C. D. Muzny and H. J. M. Hanley, Dynamic structure factor scaling in dense gelling silica suspensions,*J. Phys.: Condens. Matt.* **8** (1996), 9457.

[4.19] M. Carpineti and M. Giglio, Spinodal-Type Dynamics in Fractal Aggregation of Colloidal Clusters, *Phys. Rev. Lett.* **68** (1992), 3327.

[4.20] J. Gleick, *Chaos: Making a New Science*, Penguin Books, (1987).

[4.21] Internet.

[4.22] Zhong Chen, Yihua Sheng, *Modern System Scienology,* Shanghai Science and Technology Literature Publishing House, Shanghai, (2005), 262–266.

[4.23] N. H. Packard, J. P. Crutchfield, J. D. Fanner *et al.*, Geometry from a time series. *Phys. Rev. Lett.* **5** (1980), 712.

[4.24] A. Wolf, J. Swift, H. Swinney *et al.*, Determining Lyapunov exponents from a time series, *Physica D*, **16** (1985), 285–317.

[4.25] Nai Wang and Jiong Ruan, Principal component cluster analysis of ECG time series based on Lyapunov exponent spectrum, *Chin. Sci. Bull.* **49** (2004), 1980.

[4.26] J. S. Men *et al.*, Paper proposed for the application of the National Competition of Challenge Cup, Fudan University, (2007) (in Chinese).

[4.27] J. C. Sprott, *Chaos and Time-Series Analysis*, Oxford University Press, (2003). (The supplementary materials of the book can be found in http://sprott.physics.wisc.edu/chaostsa/)

[4.28] L. Glass, Private communication, October (2007).

[4.29] C. Nicolis and G. Nicolis, *Proc. Nat. Acad. Sci.* **83** (1986), 536.

[4.30] D. T. Kaplan and L. Glass, *Phys. Rev. Lett.* **68** (1992), 427.

[4.31] D. T. Kaplan, J. R. Clay, T. Manning, L. Glass, M. R. Guevara, *Phys. Rev. Lett.* **76** (1996), 4074–4077.

[4.32] Ping Chen, in *Civilization Bifurcation, Economic Chaos and Evolutionary Economic Dynamics*, Peking University Press, Beijing (2004), 285–319 (Chinese translation).

[4.33] M.-Carmen Miguel and R. Pastor-Satorras, *Phys. Rev. E* **59** (1999), 826.

[4.34] N. Tambo and Y. Watanabe, *Water Res.* **13** (1979), 409. The original versions in Japanese were published in *Suido Kyokai Zasshi*, **397** (1967) 2; **410** (1968) 14; **445** (1971) 2;

[4.35] W. J. Tian, T. Nakayama, J. P. Huang and K. W. Yu, *Europhys. Lett.* **78** (2007), 46001.

[4.36] D. J. Lee, G. W. Chen, Y. C. Liao and C. C. Hsieh, *Water Res.* **30** (1996), 541.

[4.37] T. Nakayama, K. Yakubo and R. Orbach, *Rev. Mod. Phys.* **66** (1994), 381.

[4.38] Xize Niu, Liyu Liu, Weijia Wen, and Ping Sheng, Hybrid Approach to High-Frequency Microfluidic Mixing, *Phys. Rew. Lett.* **97** (2006), 044501.

[4.39] Liyu Liu, Xiaoqing Chen, Xize Niu, Weijia Wen and Ping Sheng, Electrorheological fluid-actuated microfluidic pump, *Appl. Phys. Lett.* **89** (2006), 083505.

[4.40] Bai-lin Hao, *Elementary Symbolic Dynamics and Chaos in Dissipative Systems*, World Scientific, Singapore, (1989), 389–395.

[4.41] Shishi Liu, Shida Liu, Nonlinear Equations in Physics, Peking University Publishing House, Beijing, (2000) (in Chinese).

[4.42] I. Prigogine, Structure, Dissipation and Life. in *Theoretical Physics and Biology*, (Versailles, 1967) M. Marois, ed. 23–52, North-Holland Publ. Cie, Wiley Intersci. New York, (1969).
[4.43] J. A. Schorr and C. A. Schroeder, Consciousness as a dissipative structure: an extension of the Newman model, *Nurs. Sci. Q* **2** (1989), 183.

Chapter 5

Static Electrorheological Effect

5.1 Electrorheological Effect

5.1.1 *Basic phenomena*

Electrorheological (ER) fluids are a homogeneous mixture of miniature particles of high dielectric constant and oil with much lower dielectric constant. When the external electric field is lower than some critical value, the ER fluids are of liquid phase of suspensions; when the external electric field is higher than some critical value, the ER fluids are of a phase where liquid and solid phases are separated. The transition time between the two phases under high electric field is on the order of milliseconds, and the transition is reversible. The relative dielectric constants of various solid particles can vary from several to several thousand, and their size can vary from several nanometers to several microns. The critical values of the electric field are generally about several thousand volts per millimeter, and the working electric current density of good ER fluids are about 10^{-6}–10^{-5} A/cm^2. The shear stress of ER fluids when the electric field is above the critical value may exceed 20 kPa. The highest yield stress of the current best ER fluids may even exceed 100 kPa. The shear field of excellent ER fluids under high electric field may be several dozen, several hundred times higher or even higher than that under zero electric field. Near critical field, one may effectively control the shear stress of the suspension by changing the external electric field. It was predicted that "ER fluids may make revolutionary transformation in the field of industry and technology." [5.1]

In the industry of industrial automation equipment, ER fluids may be used in liquid valves, clutches, brakes, controlling equipment, servo machinery, milling machinery, computer-controlled torque measuring systems, damping systems and so forth. In the industry of general and special machinery, ER fluids can be used in machinery for construction, farming and oil fields. In the industry of oil pressure, they can be used in hydraulic oil cylinders, hydraulic pistons, hydraulic lifting, valves and closed-loop control systems. The most promising applications of ER fluids are in the transportation industry. This smart material can play an important role in various types of equipment, such as aerospace machines, torque transducers, liquid valves, clutches, dampers of motor bases, transmissions, stepless transmissions, hydraulic power steering systems, bump absorbers, electronic control systems for suspensions, coupling and braking systems, and so on. These wide applications are based on advantages that traditional machines do not have, such as computer control ability, high response speed, wide range of control speed, and accurate adjustment of damping systems. In some equipment, moving parts are reduced and even eliminated, so weight is reduced and energy efficiency is increased.

The development of ER fluids has a long history. At the end of the 19th century, it was noticed that the apparent viscosity of some suspensions would change under an electric field. The electrorheological phenomenon was first reported by Winslow [5.2] in 1949. He also studied suspensions consisting of silica gel, corn starch or gypsum particles distributed in dielectric oil, and found that the energy of an electric field can greatly increase the flowing resistance. He then thought that the resistance increment was due to fibers of suspended particles induced in the electric field. However, for a long period of time, the development of ER fluids was slow, mostly due to the fact that their application was not promising because the function of water was not recognized properly, and ER fluid with water content is not easily applied. In the 1980s, basic anhydrous synthesized materials with strong ER effects appeared, and progress was made in the area of basic research. Interest in ER fluids increased tremendously, and the area has gained new attraction since then.

In principle, the ER effect is the result of the interaction between particle chemical configuration, polarized structure and liquid environment. To make a breakthrough in the study of ER fluids, it is necessary to understand their mechanism.

It is found through computer simulation that when an electric field increases from zero, the state of ER fluids changes from liquid to liquid crystal, and then to a liquid-solid separating state. The mechanism of the ER effect has not been fully understood, even after the phenomena had been known for several decades, which caused the slow development of research on ER fluids. A good understanding of the mechanisms of solid particle interaction, ordered arrangement and field-induced liquid-solid phase transition of ER fluids would be beneficial to the design and synthesis of practical ER fluids with market value.

5.1.2 *Static particle structure of ER fluid*

Under zero electric field, an ER fluid is a Newtonian fluid: However, when the electric field is above a critical value, it becomes a Bingham fluid, and its shear stress τ, viscosity η_0 and shear rate $\dot{\gamma}$ satisfy the following equation

$$\tau = \tau(E) + \eta_0 \dot{\gamma}. \tag{5.1}$$

When the electric field is zero, dielectric particles are distributed randomly in base oil (a). When the electric field intensity increases gradually, the polarization intensity of solid particles in ER fluids also increases gradually; the particles in the ER fluids under either no shearing or under weak shearing form chains upon the effect of electric field (b,c). Further increasing the electric field enhances the interaction between the chains so that particle columns (d) are formed [5.3].

The thermodynamics of ERF has been studied, and it has been found that the column thickness is $r(d/r)^{2/3}$, where r is the particle radius, and d is the gap between electrodes [5.4]. After the careful study of laser diffraction experiment of glass beads and three-dimensional computer simulations, R. Tao *et al.* verified that the ER particles (glass beads) in columns under a certain electric field are in body-centered tetragonal (BCT) structure [5.3, 5.5–5.7].

In the experiment of Chen, Zitter and Tao (Fig. 5.1) [5.3], the ER fluid used consists of preprocessed glass beads and silicone oil. The distance of electrode L is 3 mm, the width of particle columns is about $R(L/R)^{2/3}$. The diameter of one type of glass bead is $2R = 20\,\mu\text{m} \pm 10\%$, and that of the other type of glass bead is $2R = 40.7\,\mu\text{m} \pm 10\%$. The electric field was kept on for several hours allowing for oil to be drained and the structure of glass beads to be stabilized. The laser beam was transmitted along the crystallographic axis. According to laser diffraction theory, the mode of nondivergent light may propagate along a periodic array of transparent spheres if a certain condition is satisfied.

Tao and Sun [5.6] propose that the ER-effected glass beads in columns are arranged in BCT structure from the diffraction patterns

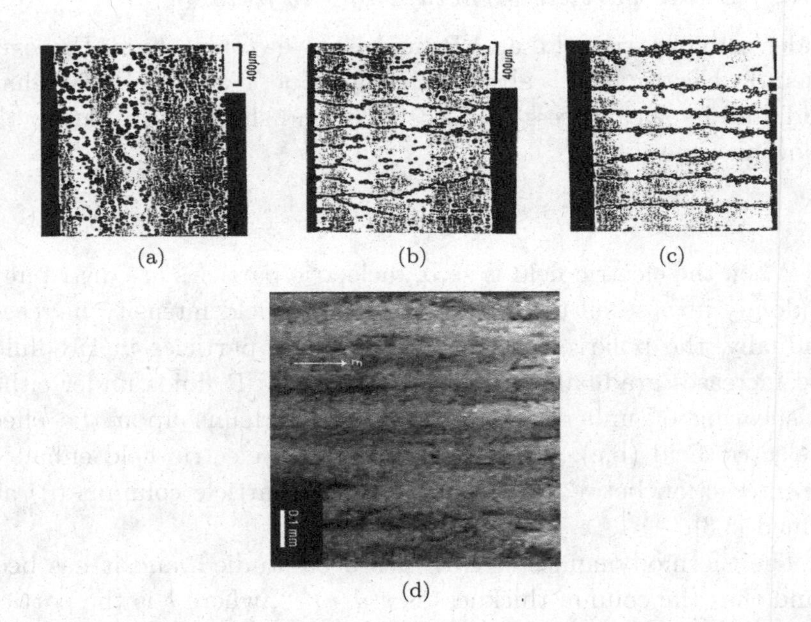

Fig. 5.1 Particles are distributed randomly under zero electric field (a), when the electric field increases, ER particle polarization increases, particles aggregate into "chains" (b, c) and "columns" (d). The black blocks at the right and left boundaries of each panel are electrodes [5.3].

of planes (110) and (100). On the (100) plane, there are rectangles with unit cell sides of $\sqrt{6}R \times 2R$; while on the (110) plane, there are rectangles with unit cell sides of $2\sqrt{3}R \times 2R$. Figure 5.2(b) is the diffraction pattern of 20-μm-diameter spheres, and this pattern is a reciprocal pattern of the (110) plane of BCT structure. Figure 5.2(c) is the diffraction pattern of 40.7-μm-diameter spheres, and this pattern is also a reciprocal pattern of (110) plane of BCT structure with a pattern density that is about double that for 20-μm-diameter spheres. This is because when the diameter of glass spheres is doubled, the distance of the pattern spots in the reciprocal space is reduced by half. Figure 5.2(d) is still the pattern for 40.7-μm-diameter spheres, but the pattern is more orthogonal than the previous one: this is just the pattern of the (100) plane. The authors have proven that the glass spheres range in the BCT structure (Fig. 5.3).

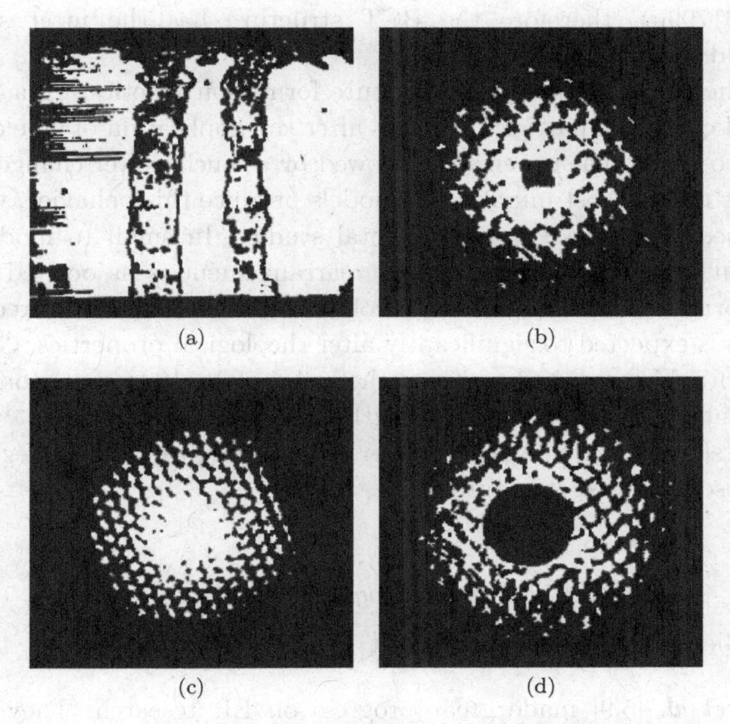

(a) (b) (c) (d)

Fig. 5.2 Laser diffraction pattern of glass spheres [5.6].

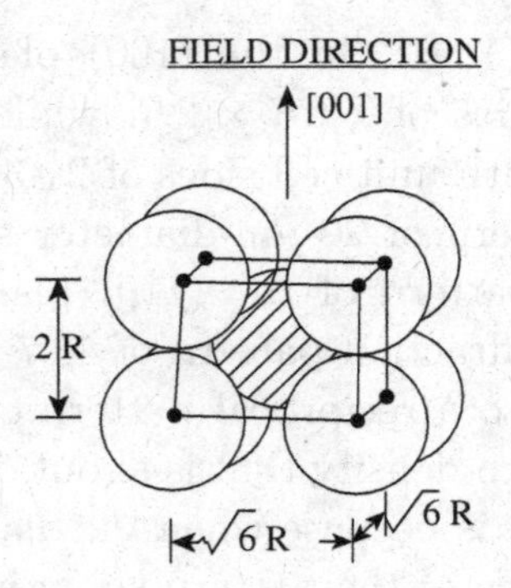

Fig. 5.3 The body-centered structure of glass spheres [5.8].

In different crystallographic arrangements, the dipole interaction of BCT structure has the least interaction energy (BCT = −0.381268u, FCC = −0.370224u, HCP = −0.3700284u, cubic lattice = −0.261799u), therefore the BCT structure has the most stable ground state.

The general features of structure formation kinetics are a rapid initial change in particle positions after the application of an electric field to a random suspension, followed by a much slower change over longer times. Most microscopic models produce this behavior, which has been described in experimental studies. In small to moderate concentrations, this initial rapid rearrangement is associated with the formation of chains that percolate in the electric field direction, which is expected to significantly alter rheological properties. Chaining is rapid, predicted to be on the order of 10–100 ms in moderate concentrations and for field strengths on the order of $1\,\text{kVmm}^{-1}$. The much slower coarsening over longer times arises from the aggregation of these extended structures.

5.1.3 *Colloidal electrorheological effect*

(1) Giant electrorheological effect

Wen *et al.* [5.9] made great progress on ER research. They have fabricated electrorheological suspensions of coated nanometer-sized particles that show giant, electrically controllable liquid-solid transitions. The solid state can reach a yield strength of 130 kPa, breaking the theoretical upper bound on conventional ER static yield stress that is derived from the general assumption that the dielectric and conductive responses of the component materials are linear. The giant electrorheological (GER) fluid is set apart from conventional ER fluids because it exceeds the theoretical upper limit, 1.38 $\sqrt{(R/\delta)}(\varepsilon_1 E^2/8\pi)$ in units of kPa. R in the above formula is the particle radius, δ, the particle gap, ε_1, the dielectric constant of the liquid, and E, the applied field in kV/mm. This limitation is derived from the general assumption of linear dielectric and electrical conductive responses of the component materials [5.10–5.12].

(2) Polar molecule electrorheological effect

Lu *et al.* [5.13, 5.14] developed a polar molecule electrorheological (PM-ER) fluid, which has a yield stress of 220 kPa at an external electric field of 4.5 kV/mm. The yield stress of the PM-ER fluids with proper adsorption of polar molecules increases linearly with the external electric field, and is proportional to a^{-1} (a = particle size). They found that the polar molecules adsorbed on the particles play a decisive role in PM-ER fluids. The leaking current density of the PM-ER fluids is on the order of μm/cm^2 or less, and it varies with E^2. The model proposed to explain the PM-ER effect will be introduced in the next section.

In colloidal ER fluids, wettability plays an important role [5.15–5.20], and it will be discussed in detail in Section 5.3.2.

5.1.4 *Polarization types and electric double layer*

(1) Polarization types

Before introducing any actual models, we should give a basic classification of five types of polarization. The polarization of dielectric particles can be classified into five categories [5.21–5.23].

a) Electronic polarization due to electron displacement. Under an electric field, the distribution of negative and positive of atoms is slightly deformed, and particle dielectric constant is directly affected, although the contribution to the whole polarization is small.

b) Polarization due to ion displacement. Under an electric field, charged atoms in a solid particle move slightly. This affects the dielectric constant of the material, especially those of inorganic materials.
c) Dipolar polarization due to dipole re-orientation under an electric field. This is often seen in ER fluids.
d) Nomadic polarization. This is due to the heat-induced charges that migrate among different lattices under an electric field. This polarization is stronger than both electron and atom polarizations.
e) Interfacial (or Maxwell–Wagner) polarization. Under an electric field, charges aggregated on the particle surface move freely from one side of a particle to another, forming a dipole moment. It is hard to distinguish interface polarization from electron and ion polarizations.

(2) Polarization of electric double layers

When a solid particle contacts a liquid, a thin layer of charges (potential ions) appears on the surface of the particle due to the strong surface energy of the particle. The charges with different signs in the liquid are adsorbed tightly on the particle surface, while the charges with the same sign are repulsed from the surface. The temperature effect tries to mix the charges with different signs. There are then two layers near the surface of the solid particle: one is called the Stern layer, the other, the pervasion layer. The Stern layer is very thin and closely contacts the particle surface. This layer does not slide even when there is a strong shear. The pervasion layer, on the other hand, is much thicker and can be shifted under shear.

Electric double layers on the surface of particles are deformed under an electric field; this is the polarization of the electric double layers [5.24]. This causes an unbalanced or asymmetric distribution of

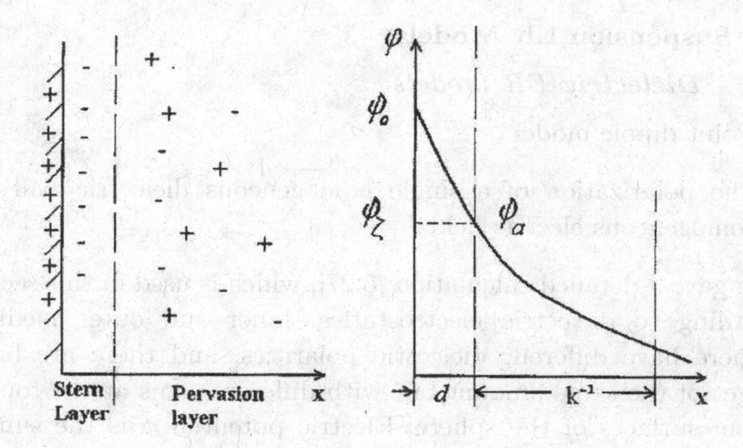

Fig. 5.4 Diagram of electric double layer.

charges, and causes attraction and repulsion between charges [5.25]. The interactions between the electric double layers of two spherical particles are derived [5.26] and can be expressed as

$$F_c = \frac{\pi n_0 R z^2 e^2}{12 k_B T}\left[\frac{48V_0}{k}(V_0 e^{-kL_0} + E_0 e^{-kL_0/2}(kL_0^2 - 8L_0)) - E_0^2 L_0^3\right], \tag{5.2}$$

where R is the particle radius, L_0 is the shortest distance between the two particles, 1/k is the effective thickness of the pervasion layer and $k^2 = 2n_0 z^2 e^2/\varepsilon_0 \varepsilon_f k_B T$, z is the absolute value of positive and negative ionic valence, and n_0 is the average ionic concentration. The first term is the static repulsion term, and the second and third terms are attractive terms. It is demonstrated that F_c is proportional to E_0^2.

However, there are also negatives for the model of electric double layer, for example, the effective thickness of the electric double layer is too large, and it is often larger than the distance between two electrodes.

5.2 Suspension ER Models

5.2.1 *Dielectric ER models*

(1) Point dipole model

(i) The polarization of a single homogeneous dielectric ball in a homogeneous electric field

Weng gave a detailed calculation [5.27], which is used in this section. According to dielectric electrostatics, inner and outer media of a sphere have different dielectric polarities, and there are bound charges of the same amount but with different signs at the top and bottom surfaces of the sphere. Electric potential ϕ is the summation of the potential due to the external field and surface polarized charges. In the case of no free charge, and only monodisperse and hard dielectric spheres with Newtonian continuous phase of zero conductivity, electrostatic potential ϕ should be the solution of Laplace's equation

$$\nabla^2\varphi = 0, \tag{5.3}$$

with certain boundary conditions (idealized electrostatic polarization model). Boundary conditions at each sphere/continuous phase interface:

a. electric potential φ is finite at $r = 0$;
b. the tangential components of electric field intensity vector and the normal components of electric displacement vector of the inner and outer sphere at the boundary of the sphere-oil interface (at $r = a$) are continuous, respectively, or

$$-\frac{1}{a}\frac{\partial\varphi_i}{\partial\theta}\bigg|_{r=a} = -\frac{1}{a}\frac{\partial\varphi_o}{\partial\theta}\bigg|_{r=a}, \tag{5.4}$$

$$\varepsilon_p\frac{\partial\varphi_i}{\partial r}\bigg|_{r=a} = \varepsilon_c\frac{\partial\varphi_o}{\partial r}\bigg|_{r=a}; \tag{5.5}$$

c. while when $r \to \infty$,

$$\varphi_o|_{r\to\infty} = -E_O z = -E_O r\cos\theta. \tag{5.6}$$

According to the axial symmetry of geometric shape, the Laplace equation has the following general solution:

$$\varphi(r,\theta) = \sum_{n=o}^{\infty}\left(A_n r^n + \frac{B_n}{r^{n+1}}\right)P_n(\cos\theta), \tag{5.7}$$

where $P_n(\cos\theta)$ is the Legendre polynomial, and

$$\begin{aligned} P_0(x) &= 1, \\ P_1(x) &= x, \\ P_2(x) &= \frac{1}{2}(3x^2 - 1). \end{aligned} \tag{5.8}$$

Considering condition (a), φ is finite at $r = 0$. Electric potential φ inside ($r < a$) and outside of a sphere ($r > a$) can be expressed as:

$$\varphi_i = \sum_{n=0}^{\infty} A_n r^n P_n(cos\theta), \quad r < a, \tag{5.9}$$

$$\varphi_o = \sum_{n=0}^{\infty}(B_n r^n + C_n/r^{n+1})P_n(\cos\theta), \quad r > a. \tag{5.10}$$

Using boundary conditions, Eqs. (5.4)–(5.6), and the orthogonality of the Legendre polynomial, one may find that

$$A_n = B_n = C_n = 0 \quad (n \neq 1), \tag{5.11}$$

$$A_1 = -\frac{3\varepsilon_f}{2\varepsilon_f + \varepsilon_p}\cdot E_0, \tag{5.12}$$

$$B_1 = -E_0, \tag{5.13}$$

$$C_1 = \frac{\varepsilon_p - \varepsilon_f}{2\varepsilon_f + \varepsilon_p}E_0 a^3. \tag{5.14}$$

Substituting Eqs. (5.11)–(5.14) in Eqs. (5.9) and (5.10), we have that

$$\varphi_i = -\frac{3\varepsilon_f}{2\varepsilon_f + \varepsilon_p}E_0 r\cos\theta, \tag{5.15}$$

$$\varphi_o = -E_0 r\cos\theta + \frac{\varepsilon_p - \varepsilon_f}{2\varepsilon_f + \varepsilon_p}\cdot(E_0 a^3/r^2)\cdot\cos\theta. \tag{5.16}$$

From Eqs. (5.15) and (5.16), it can be seen that the electric potential inside a dielectric sphere corresponds to the potential of a homogeneous electric field parallel to external field E_0 that is given by

$$E_i = -\frac{\partial \varphi_i}{\partial z} = \frac{3\varepsilon_f}{2\varepsilon_f + \varepsilon_p} \cdot E_0, \tag{5.17}$$

while the electric potential outside of a dielectric sphere, φ_o, corresponds to a summation of two potentials: one a dipole potential p produced in the outside of the sphere, positioned at the sphere's center, which is given by

$$P = 4\pi\varepsilon_f \frac{\varepsilon_p - \varepsilon_f}{2\varepsilon_f + \varepsilon_p} a^3 \cdot E_0, \tag{5.18}$$

and the other, the potential due to the homogeneous external field. The direction of the dipole is parallel to the external electric field.

Equations (5.15) and (5.16) imply that when a dielectric sphere, which has different dielectric constant from that outside of the sphere, is exposed under an electric field, there are bound charges on both hemispheres of the sphere surface with the same amount but different signs, and the charge density is

$$\begin{aligned}\sigma_{\text{pol}} &= -\frac{1}{4\pi}\left(\left.\frac{\partial \varphi_o}{\partial r}\right|_{r=a} - \left.\frac{\partial \varphi_i}{\partial r}\right|_{r=a}\right) \\ &= \frac{3(\varepsilon_p - \varepsilon_f)}{4\pi(2\varepsilon_f + \varepsilon_p)} E_0 \cos\theta. \end{aligned} \tag{5.19}$$

Obviously, on both poles of the sphere, or when $\theta = 0$ and $\theta = \pi$, charge density approaches its maximum.

On the other hand, when dielectric constants are almost the same, $\varepsilon_p \simeq \varepsilon_f$, that is, the electric field inside the sphere is the same as the external field; while when $\varepsilon_p \to \infty$, the electric field inside the sphere is zero, and the dipole moment P of the point dipole reaches its maximum value of $4\pi\varepsilon_f a^3 E_0$.

(ii) Interaction between particles under point-dipole approximation

ER fluids consist of solid many-body particles suspended in an insulating medium, and the study of dielectric polarization under an electric field and the induced interaction is very important.

From the discussion in the previous section, since ER fluid consists of multi-particles, each solid particle is polarized under a homogeneous electric field and produces a non-homogeneous field surrounding the particle. This non-homogeneous field then makes the particles near that particle polarized. This makes the electric field polarization problem of multi-particles due to mutual polarization quite complicated.

However, when the particle distance is sufficiently large for particle interaction to become negligible in comparison with particle polarization due to external field, one can then use an ideal case discussed in the previous section, namely, replacing each actual particle with a point-dipole located in the particle center. Let us examine particle interaction under this hypothesis.

As shown in Fig. 5.5, two spheres with the same size and same properties under electric field $\boldsymbol{E}_0$ and in homogeneous media of dielectric constant ε_f will be polarized. The two spheres can be approximately replaced with two point dipoles of the same amount

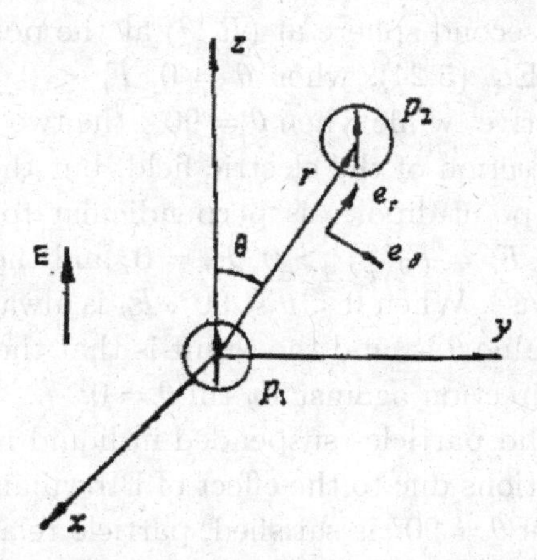

Fig. 5.5 Interaction between two spherical particles due to electric field polarization [5.27].

and same direction, and they can be expressed as

$$\boldsymbol{P}_1 = \boldsymbol{P}_2 = 4\pi\varepsilon_f \frac{\varepsilon_p - \varepsilon_f}{2\varepsilon_f + \varepsilon_p} a^3 \boldsymbol{E}_0. \tag{5.20}$$

If the coordinate system as shown in Fig. 5.5 is used, the interaction potential can be expressed as

$$U = \frac{\boldsymbol{P}_1 \cdot \boldsymbol{P}_2 - 3\,(\boldsymbol{n} \cdot \boldsymbol{P}_1)\,(\boldsymbol{n} \cdot \boldsymbol{P}_2)}{4\pi\varepsilon_f r^3} = \frac{(1 - 3\cos^2\theta)}{4\pi\varepsilon_f r^3} P^2, \tag{5.21}$$

where $P = 4\pi\varepsilon_f \frac{\varepsilon_p - \varepsilon_f}{2\varepsilon_f + \varepsilon_p} a^3 E_0$, with r being the distance between two spheres and $\boldsymbol{n}$, the unit vector of $\boldsymbol{r}$. The interaction components between the two spheres are

$$F_r = -\frac{\partial U}{\partial r} = 12\pi\varepsilon_f a^2 \beta^2 E_0^2 \left(\frac{a}{r}\right)^4 (-3\cos^2\theta + 1), \tag{5.22}$$

$$F_\theta = -\frac{1}{r}\frac{\partial U}{\partial \theta} = -12\pi\varepsilon_f a^2 \beta^2 E_0^2 \left(\frac{a}{r}\right)^4 \sin 2\theta, \tag{5.23}$$

where $\beta = \frac{\varepsilon_p - \varepsilon_f}{2\varepsilon_f + \varepsilon_p}$. If it is written in vector form, and making $F_0 = 12\pi\varepsilon_f a^2 \beta^2 E_0^2$, one has

$$\boldsymbol{F} = F_0 \left(\frac{a}{r}\right)^4 \left\{(1 - 3\cos^2\theta)\, \boldsymbol{e}_r - \sin 2\theta\, \boldsymbol{e}_\theta\right\}. \tag{5.24}$$

We can now discuss the time averaged force on a sphere at the origin due to the second sphere at (R, θ) at the point-dipole limits.

According to Eq. (5.24), when $\theta = 0$, $F_r < 0$, $F_\theta = 0$, the two spheres are attractive, while when $\theta = 90°$, the two point-dipoles are parallel to the direction of the electric field, but the connection line between the two point-dipoles is perpendicular to the direction of the external field, $F_r = F_0(\frac{a}{r})^4 > 0$, $F_\theta = 0$, and the two spheres are repulsive in this case. When $0 < \theta < 90°$, F_θ is always less than zero, no matter what value θ is, and the result is that the two spheres will move along the direction against $\boldsymbol{e}_\theta$ till $\theta = 0$.

In fact, the solid particles suspended in liquid media will change their relative positions due to the effect of Brownian motion. As long as the condition of $\theta \neq 90°$ is satisfied, particle relative rotation and attraction occurs, eventually leading to a particle chaining arrangement along the external electric field. Both ends of these particle chains attract other polarized particles. If the intensity of the external electric field is strong enough and particle concentration is high enough, particle chains form a chain structure across two electrodes. This is what has been found in numerous experimental studies.

Since it is the interaction between point dipoles that causes the formation of chaining structure and the ER effect, the strength of point-dipole interaction should be proportional to the intensity of ER effect. From Eq. (5.24), one may find that (i) the intensity of ER effect should be proportional to E^2; (ii) the intensity of ER effect should be proportional to β^2; the conclusion (i) is consistent with the relation that yield stress is proportional to the square of electric field intensity obtained from the experimental study mentioned previously. For (ii), if $K_{pf} = \varepsilon_p/\varepsilon_f$, then

$$\beta = \frac{\varepsilon_p - \varepsilon_f}{2\varepsilon_f + \varepsilon_p} = \frac{K_{pf} - 1}{K_{pf} + 2}. \tag{5.25}$$

It is obvious that β^2 increases with increasing K_{pf}, so increasing the ratio of dielectric constant of particles to that of base liquid increases the ER effect. This is consistent with the experimental fact that increasing particle dielectric polarizability increases the ER effect.

From the above analysis, (1) a dielectric sphere in a homogeneous electric field produces an inhomogeneous electric field, and its effect is equivalent to an electric field outside of a sphere produced by a dipole located at the center of the sphere. (2) When the distance of two particles is quite large, one may estimate the electric field polarization force of a spherical particle by using a point dipole to replace the spherical particle. In the case that there is no shear flow, particles form chains along the external electric field due to Browning motion and the polarizing force of electric field.

(2) The multipole-moment, multiple-image and many-body effect [5.28–5.31]

(i) The multipole-moment

Equation (5.24) is not accurate when highly polarizable spheres are close to each other; the disturbance fields created by the spheres further polarize each other, resulting in electrostatic moments beyond the dipole. These moments can have an enormous effect on the magnitude of the electrostatic force.

Similar to the last section about the dipole moment model, in the case that the two dielectric spheres are close to each other, the scalar electrostatic potentials governed by Laplace's equation Eq. (5.3) are quite different from those shown in Eqs. (5.9) and (5.10) [5.31]. The two spheres in vacuum have the same dielectric constant ε_p, radius a, and a center distance d. The two spheres A and B are in two spherical polar coordinate systems (r, θ) and (R, Θ), respectively (Fig. 5.6).

The Laplace equation (Eq. (5.3)) is solved under two conditions: $\phi = -Ez$ when $r \to \infty$, and the perturbation approaches 0 when $r \to \infty$. The terms of r^n in the external potential (Eq. (5.10)) are eliminated. The external potential becomes [5.31, 5.32]

$$\phi_o = \sum_{n=0}^{\infty} \left[a_n \left(\frac{a}{r}\right)^{n+1} \mathbf{P}_n(\lambda) + A_n \left(\frac{a}{R}\right)^{n+1} \mathbf{P}_n(\Lambda) \right] - Ez \qquad (5.26)$$

where $\lambda = \cos\theta$, $\Lambda = \cos\Theta$, and $\mathbf{P}_n$ is the Legendre polynomial of order n. The potentials inside spheres A and B are, respectively,

$$\phi_i = \sum_{n=0}^{\infty} c_n \left(\frac{r}{a}\right)^n \mathbf{P}_n(\lambda) \qquad (5.27)$$

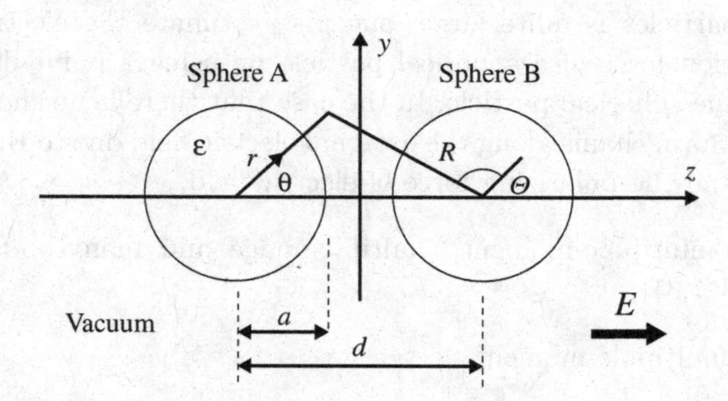

Fig. 5.6 Two closed dielectric spheres A and B in spherical coordinate systems (r, θ) and (R, Θ), respectively [5.31].

and

$$\Phi_i = \sum_{n=0}^{\infty} C_n \left(\frac{R}{a}\right)^n \mathbf{P}_n(\Lambda). \tag{5.28}$$

The terms of the Legendre expansion of an axially symmetric potential in one spherical coordinate system (Eq. (5.28), for example) can be re-expanded at another point on the axis of symmetry [5.31].

Consider two series of the spherical harmonics with the singularity at $r = 0$ and $r = \infty$ [5.32].

$$\phi_n = r^n \mathbf{P}_n(\cos\theta), \tag{5.29}$$

$$\Phi_n = R^{-n-1} \mathbf{P}_n(\cos\Theta), \tag{5.30}$$

$$\frac{\partial \phi_n}{\partial x} = n\phi_{n-1}, \tag{5.31}$$

$$\frac{\partial \Phi_n}{\partial x} = -(n+1)\Phi_{n+1}. \tag{5.32}$$

The zero order potential, $\Phi_0 = (1/R)P_0(\cos\Theta)$, around sphere B can be re-expanded with the Legendre terms around sphere A. According to the cosine law from Fig. 5.6,

$$R^2 = r^2 + d^2 - 2rd\cos\theta.$$

The relation among the parameters in the coordinates is then

$$\frac{d}{R} = \frac{1}{\sqrt{1 - 2(r/d)\cos\theta + (r/d)^2}} \tag{5.33}$$

which is the generating function of P_n. The re-expansion of Φ_0 is given then as

$$\begin{aligned}\Phi_0 &= \frac{1}{R} = \frac{1}{d}\sum_{k=0}^{\infty} \left(\frac{r}{d}\right)^k P_k(\cos\theta) \\ &= \sum_{k=0}^{\infty} d^{-k-1} r^k P_k(\cos\theta) = \sum_{k=0}^{\infty} d^{-k-1}\phi_k.\end{aligned} \tag{5.34}$$

To apply this re-expansion into the higher-order potentials, nth derivative of Φ_0 with respect to x is taken by using Eqs. (5.31) and

(5.32) recurrently,

$$\Phi_n = \frac{(-1)^n}{n!}\frac{\partial^n}{\partial x^n}\Phi_0 = \frac{(-1)^n}{n!}\frac{\partial^n}{\partial x^n}\sum_{k=0}^{\infty} d^{-k-1}\phi_k$$

$$= \frac{(-1)^n}{n!}\sum_{k=n}^{\infty} d^{-k-1}k(k-1)\cdots(k-n+1)\phi_{k-n}$$

$$= \sum_{k=0}^{\infty}(-1)^n d^{-k-n-1}\frac{(k+n)!}{k!n!}\phi_k. \tag{5.35}$$

One has then a re-expansion formula of [5.31, 5.32]

$$\left(\frac{a}{R}\right)^{n+1}\mathbf{P}_n(\Lambda) = (-1)^n\left(\frac{a}{d}\right)^{n+1} = \sum_{m=0}^{\infty}\frac{(m+n)!}{m!n!}\left(\frac{r}{d}\right)^m\mathbf{P}_m(\lambda). \tag{5.36}$$

Equation (5.36) only converges in the domain $r < d$.

The re-expansion of the external potential Eq. (5.26) is then

$$\phi_o = \sum_{n=0}^{\infty}\left[a_n\left(\frac{a}{r}\right)^{n+1} + b_n\left(\frac{r}{a}\right)^n\right]\mathbf{P}_n(\lambda) + Er\mathbf{P}_1(\lambda) + \frac{Ed}{2}\mathbf{P}_0(\lambda) \tag{5.37}$$

with

$$b_n = -\left(\frac{a}{d}\right)^n\sum_{m=0}^{\infty}\left(\frac{a}{d}\right)^{m+1}\frac{(m+n)!}{m!n!}a_m, \tag{5.38}$$

noting that $\mathbf{P}_0(\lambda) = 1$ and $\mathbf{P}_1(\lambda) = \lambda$, and in particular when $n = 0$, $b_0 = -\sum_{m=0}^{\infty}(a/d)^{m+1}a_m$.

The continuity boundary conditions of the electrostatic potential at the surface of the spheres are $\phi_i\phi_o$ and $k\partial\phi_i/\partial r = \partial\phi_o/\partial r$, at $r = a$. These boundary conditions lead to $a_1 = (Ea-b_1)(\varepsilon-1)/(\varepsilon+2)$, and for $n > 1$

$$a_n = \frac{n(k-1)}{n(k+1)+1}\left(\frac{a}{d}\right)^n\sum_{m=0}^{\infty}\left(\frac{a}{d}\right)^{m+1}\frac{(m+n)!}{m!n!}a_m, \tag{5.39}$$

and

$$c_n = -a_n(2n+1)/[n(k-1)] \quad \text{for } n > 0. \tag{5.40}$$

The electrostatic potential outside both spheres A and B can be expressed as

$$\phi_o = \sum_{n=1}^{\infty} a_n \left[\left(\frac{a}{r}\right)^{n+1} \mathbf{P}_n(\lambda) + \left(-\frac{a}{R}\right)^{n+1} \mathbf{P}_n(\Lambda) \right] - Ez \quad (\text{for } r, R \geq a) \tag{5.41}$$

$$\phi_o = \frac{Ed}{2} + \sum_{n=1}^{\infty} a_n \left\{ \left[\left(\frac{a}{r}\right)^{n+1} - \frac{n(k+1)+1}{n(k-1)} \left(\frac{r}{a}\right)^n \right] \times \mathbf{P}_n(\lambda) - \left(\frac{a}{d}\right)^{n+1} \right\} \quad (\text{for } a \leq r < d). \tag{5.42}$$

The potential inside spheres A and B are respectively

$$\phi_i = \frac{Ed}{2} - \sum_{n=1}^{\infty} a_n \left[\left(\frac{a}{d}\right)^{n+1} + \frac{2n+1}{n(k-1)} \left(\frac{r}{a}\right)^n \mathbf{P}_n(\lambda) \right] \tag{5.43}$$

and

$$\Phi_i = -\frac{Ed}{2} + \sum_{n=1}^{\infty} a_n \left[\left(\frac{a}{d}\right)^{n+1} + \frac{2n+1}{n(k-1)} \left(-\frac{R}{a}\right)^n \mathbf{P}_n(\Lambda) \right]. \tag{5.44}$$

Applying Gauss' flux theorem in a vacuum to a very thin volume enclosing a small portion of the surface of sphere A gives the surface charge density

$$\rho = \epsilon_0 (\partial\phi_i/\partial r - \partial\phi_o/\partial r)_{r=a}. \tag{5.45}$$

Applying Eqs. (5.42) and (5.43) yields

$$\rho = (\epsilon_0/a) \sum_{n=1}^{\infty} (2n+1) a_n \mathbf{P}_n(\lambda). \tag{5.46}$$

The electrostatic force is the interaction of this surface charge with the electrostatic field outside of the sphere which is *not* a result of this surface charge. That is, the potential to be used is that from

Eq. (5.42) but excluding the terms resulting from the polarization of sphere A. This potential is

$$\phi_o^* = \frac{Ed}{2} - \sum_{n=1}^{\infty} a_n \left[\frac{n(k+1)+1}{n(k-1)} \left(\frac{r}{a}\right)^n \mathbf{P}_n(\lambda) + \left(\frac{a}{d}\right)^{n+1} \right]. \quad (5.47)$$

Since the solution is symmetric around the z axis, only the z components of the field are required for the calculation.

$$E_z = -\partial \phi_o^* / \partial z, \quad \text{at } r = a. \quad (5.48)$$

Using $\partial/\partial z = \cos\theta \partial/\partial r - (\sin\theta/r)\partial/\partial\theta$ and the identity

$$(1 - x^2)\mathbf{P}'_n(x) + nx\mathbf{P}_n(x) = n\mathbf{P}_{n-1}(x) \quad (5.49)$$

in Eqs. (5.47) and (5.48), one has

$$E_z = \frac{1}{a(k-1)} \sum_{n=0}^{\infty} [(n+1)(k+1)+1] a_{n+1} \mathbf{P}_n(\lambda). \quad (5.50)$$

Substituting Eqs. (5.46) and (5.50) into the following integral for the force

$$F_e = \int_0^{\pi} E_z \rho 2\pi a^2 \sin\theta d\theta, \quad (5.51)$$

and applying the orthogonality condition for Legendre polynomials yield

$$F_e = \frac{4\pi\epsilon_0}{k-1} \sum_{n=1}^{\infty} [(n+1)(k+1)+1] a_n a_{n+1} \quad (5.52)$$

where

$$a_n = \frac{n(k-1)}{n(k+1)+1} \left(\frac{a}{d}\right)^n \sum_{m=0}^{\infty} \left(\frac{a}{d}\right)^{m+1} \frac{(m+n)!}{m!n!} a_m. \quad (5.53)$$

This force is plotted in Fig. 5.7 together with other theoretical and experimental [5.33] results for a comparison.

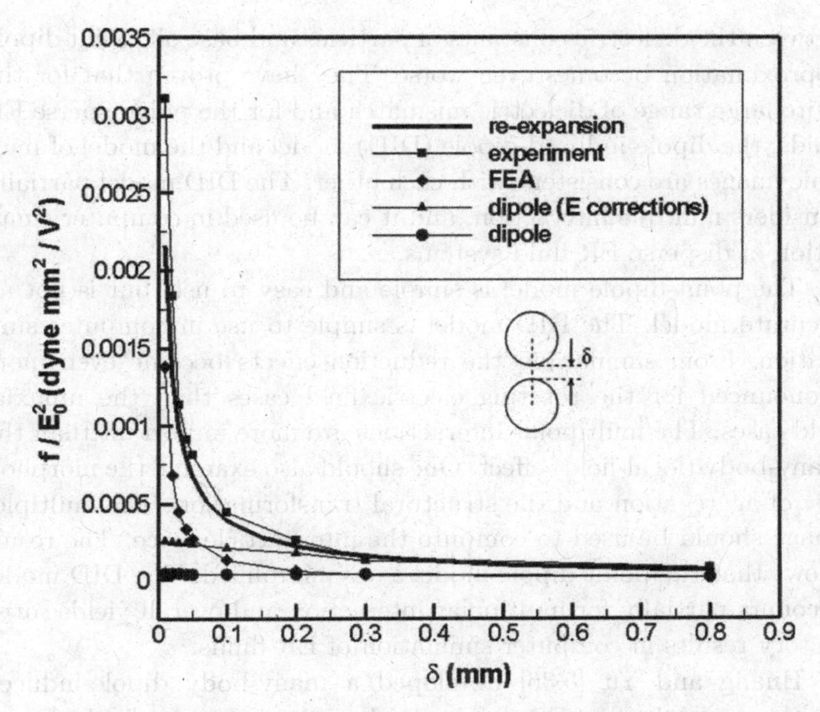

Fig. 5.7 Comparison between the experimental results and theoretical calculations based on the different modeling [5.33].

Wang *et al.* [5.33] measured the interaction between two millimeter-sized spheres and found that when the distance between the two spheres is very small, the interaction deviates severely from the point dipole model.

(ii) Dipole-induced-dipole model

Siu, Wan and Yu [5.35] developed a method of multiple-image to calculate the interaction between polydisperse ER particles. They developed a set of formulae and used them in dielectric spheres with different dielectric constants. They calculated the relation between particle interaction and particle distance. This proves that point dipole approximation would have quite a large error if multiple and many-body effects are neglected. In the case of a large difference between the dielectric constants of particles and base oil, point dipole approximation becomes even worse. They have proven that for the quite large range of dielectric mismatch and for the polydisperse ER fluids, the dipole-induced-dipole (DID) model and the model of multiple images are consistent with each other. The DID model partially considers multiple interaction, and it can be used in computer simulation of disperse ER fluid systems.

The point-dipole model is simple and easy to use, but is not an accurate model. The DID model is simple to use in computer simulation. From simulation, the reduction effects become even more pronounced for the rotating electric field cases than the uniaxial field cases. The multipolar interactions are more important than the many-body (local-field) effect. One should also examine the morphology of aggregation and the structural transformation. The multiple-image should be used to compute the interparticle force. The result shows that the point dipole model is oversimplified. The DID model accounts partially for multipolar interaction and overall yields satisfactory results in computer simulation of ER fluids.

Huang and Yu [5.36] developed a many-body dipole-induced dipole model for an ER solid, the lattice structure of which can be changed due to the application of external fields. They take into account both local-field effects and multipolar interactions. The results show that the multipolar interaction can be dominant over the dipolar interaction, while the local-field effect may yield an important correction.

(iii) Many-body effect

Brunner *et al.* studied three-body interactions amongst three charged colloidal particles in a deionized aqueous solution [5.37]. Two of the particles are confined to an optical line trap while the third one is approached by means of a focused laser beam. From the observed particle configurations we extract the three-body potential which is found to be attractive and roughly of the same magnitude and range as the pair interactions. In addition, numerical calculations are performed which show qualitative agreement with the experimental results.

(3) High field ground state of dielectric ER fluids

The theory of dielectric electrorheological (DER) fluids is well established [5.10–5.12]. The DER fluid system is in the long wavelength, or electrostatic limit. The response of the system to an external electric field which defines the z-axis can be captured by a complex number $\bar{\varepsilon}_{zz}$, the zz component of dielectric constant matrix, with the real part denoting dielectric constant and the imaginary part denoting the conductivity or relaxation effect. When spheres in DER fluids are larger than a few microns and/or the electric field is large, the entropy contribution TS can be omitted, and the *Gibbs free energy density* f of the system is

$$f = -(1/2)\mathrm{Re}(\varepsilon_{zz})E^2. \tag{5.54}$$

The stability of a physical system is governed by the principle of minimum free energy, which requires the maximization of the real part of $\bar{\varepsilon}_{zz}$ with respect to the positions of the solid spheres under certain electric field intensity E. In calculating $\bar{\varepsilon}_{zz}$, the body-centered tetragonal (BCT) structure is considered as a high field ground state which is experimentally confirmed as described in Section 5.1.

(i) *Determination of electric potential*

The *static electric potential* is the solution of the following equation:

$$\nabla \cdot (\varepsilon(\boldsymbol{r})\nabla\phi(\boldsymbol{r})) = 0. \tag{5.55}$$

The boundary condition of the problem is given by (with unit electric field)

$$\begin{aligned} \phi(x, y, z = 0) &= 0, \\ \phi(x, y, z = l) &= l. \end{aligned} \tag{5.56}$$

In calculating Eq. (5.55), it is crucial to find an effective dielectric constant which is closely related to the detailed microstructure of ER particles. The exact theory of the Berman–Milton representation [5.11] can be used to calculate $\bar{\varepsilon}_{zz}$. When a DER fluid consists of only one type of particle with dielectric constant ε_p suspended in a liquid with dielectric constant ε_l, and both ε_p and ε_l are real numbers, the spatially varying dielectric constant of the system can be written as

$$\varepsilon(\boldsymbol{r}) = \varepsilon_l\left(1 - \frac{1}{s}\eta(\boldsymbol{r})\right), \tag{5.57}$$

where the material characteristics of the problem are contained in the complex parameter

$$s = \frac{\varepsilon_l}{\varepsilon_l - \varepsilon_p}, \tag{5.58}$$

$s < 0$ or > 1, whereas the microstructure is implied by the indicator function $\eta(\boldsymbol{r})$.

$$\begin{aligned} &\eta(\boldsymbol{r}) = 1 \text{ in the region of } \varepsilon_p, \\ &\eta(\boldsymbol{r}) = 0 \text{ otherwise.} \end{aligned} \tag{5.59}$$

Substituting Eq. (5.57) into Eq. (5.55) yields

$$\nabla^2\phi(\boldsymbol{r}) = \frac{1}{s}\nabla\cdot(\eta(\boldsymbol{r})\nabla\phi(\boldsymbol{r})); \tag{5.60}$$

the right hand side of which is a source term.

Technically, a *Green's function*, $G(\boldsymbol{r}, \boldsymbol{r}')$, of a linear operator L acting on distributions over a manifold M, at a point $\boldsymbol{r}'$, is any solution of

$$LG(\boldsymbol{r}, \boldsymbol{r}') = \delta(\boldsymbol{r} - \boldsymbol{r}'), \tag{5.61}$$

where δ is the Dirac delta function. This technique can be used to solve differential equations of the form

$$L\phi(\boldsymbol{r}) = f(\boldsymbol{r}). \tag{5.62}$$

In practice, some combination of symmetry, boundary conditions and/or other externally imposed criteria will give a unique Green's function. If such a function G can be found for the operator L, then if we multiply Eq. (5.61) for Green's function by $f(\boldsymbol{r}')$, and then perform an integration in the $\boldsymbol{r}'$ variable, we obtain;

$$\int LG(\boldsymbol{r},\boldsymbol{r}')f(\boldsymbol{r}')d\boldsymbol{r}' = \int \delta(\boldsymbol{r}-\boldsymbol{r}')f(\boldsymbol{r}')d\boldsymbol{r}'' = f(\boldsymbol{r}).$$

The right hand side is now given by Eq. (5.62) to be equal to $L\phi(\boldsymbol{r})$, thus:

$$L\phi(\boldsymbol{r}) = \int LG(\boldsymbol{r},\boldsymbol{r}')f(\boldsymbol{r}')d\boldsymbol{r}'.$$

Because the operator L is linear and acts on the variable x alone (not on the variable of integration s), we can take the operator L outside of the integration on the right hand side, obtaining

$$L\phi(\boldsymbol{r}) = L\left(\int G(\boldsymbol{r},\boldsymbol{r}')f(\boldsymbol{r}')d\boldsymbol{r}'\right).$$

And this implies:

$$\phi(\boldsymbol{r}) = \int G(\boldsymbol{r},\boldsymbol{r}')f(\boldsymbol{r}')d\boldsymbol{r}'. \tag{5.63}$$

To solve Eq. (5.60), one can use Green's function [5.11, 5.12, 5.38]

$$G(\boldsymbol{r},\boldsymbol{r}')\frac{1}{4\pi|\boldsymbol{r}-\boldsymbol{r}'|}. \tag{5.64}$$

The solution can be written as a sum of two parts: the solution of the homogeneous Laplace equation, which satisfies the boundary conditions ($\phi_0 = E_0 z = z$ as $E_0 = 1$ by definition), plus the integral of Green's function of the Laplacian (Eq. (5.63)). The solution of Eq. (5.60) can be written as

$$\phi(\boldsymbol{r}) = z - \frac{1}{s}\int \mathrm{d}V' G(\boldsymbol{r},\boldsymbol{r}')\nabla' \cdot \underline{(\eta(\boldsymbol{r}')\nabla'\phi(\boldsymbol{r}'))}. \tag{5.65}$$

According to the integration by parts, one has

$$\phi = z + \frac{1}{s}\int \mathrm{d}V'\eta(\boldsymbol{r}')\nabla' G(\boldsymbol{r},\boldsymbol{r}') \cdot \nabla'\phi(\boldsymbol{r}'). \tag{5.66}$$

By introducing a linear integral operator $\hat{\Gamma}$ defined as

$$\hat{\Gamma}\phi(\boldsymbol{r}) \equiv \int \mathrm{d}V'\eta(\boldsymbol{r}')\nabla' G(\boldsymbol{r},\boldsymbol{r}') \cdot \nabla'\phi(\boldsymbol{r}'), \tag{5.67}$$

Eq. (5.66) can be written in a more compact form as

$$\phi = z + \frac{1}{s}\hat{\Gamma}\phi. \tag{5.68}$$

The eigenvalues of the operator $\hat{\Gamma}$ are real and bounded in the interval [0,1], while the material parameter s cannot be in the interval [0,1]. The potential thus has a formal solution of

$$\phi = \left(1 - \frac{1}{s}\hat{\Gamma}\right)^{-1} z. \tag{5.69}$$

(ii) Determination of effective dielectric constant of BCT-structured and dielectrically mixed material

With this formal solution, it is possible to write an expression for the *effective dielectric constant* which, by definition, is obtained as the ratio of the spatial average of the displacement vector $\boldsymbol{D} = \varepsilon\boldsymbol{E}$ to the applied electric field ($E_0 = 1$). Since the applied electric field is in the z direction, the zz component of the effective dielectric matrix can be written as [5.12]

$$\begin{aligned}\bar{\varepsilon}_{zz} &= \frac{1}{V}\int \mathrm{d}V[\varepsilon_l(1-\eta(\boldsymbol{r})) + \varepsilon_p\eta(\boldsymbol{r})]\frac{\partial\varphi(\boldsymbol{r})}{\partial z} \\ &= \varepsilon_l\left(1 - \frac{1}{V}\int \mathrm{d}V\frac{1}{s}n(\vec{r})\nabla\varphi\cdot\nabla z\right) \\ &= \varepsilon_l\left(1 - \frac{s^{-1}}{V}\langle z|\varphi\rangle\right).\end{aligned} \tag{5.70}$$

Inserting Eq. (5.69) into Eq. (5.70) yields

$$\begin{aligned}\bar{\varepsilon}_{zz} &= \varepsilon_l\left(1 - \frac{1}{V}\langle z|(s-\hat{\Gamma})^{-1}|z\rangle\right) \\ &= \varepsilon_l\left(1 - \frac{1}{V}\sum_u \langle z|(s-\hat{\Gamma})^{-1}|\varphi_u\rangle\langle\varphi_u|z\rangle\right)\end{aligned} \tag{5.71}$$

where ϕ_u is the eigenfunction of $\hat{\Gamma}$ (with associated eigenvalue s_u), namely $\hat{\Gamma}\phi_u = s_u\phi_u$, with $\sum_u |\varphi_u\rangle\langle\varphi_u| = 1$. The spectral representation of $\bar{\varepsilon}_{zz}$ is then

$$\bar{\varepsilon}_{zz} = \varepsilon_l \left(1 - \frac{1}{V}\sum_u \frac{|\langle z|\varphi_u\rangle|^2}{s - s_u}\right). \tag{5.72}$$

Here, V is the sample volume, a normalization factor. From Eq. (5.72) we have $\bar{\varepsilon}_{zz}$

$$\begin{aligned}
\bar{\varepsilon}^{zz}_{\text{col}} &= \varepsilon_l \left(1 - \frac{1}{V}\sum_u \frac{|\langle z|\varphi_u\rangle|^2}{s - s_u}\right) \\
&= \varepsilon_l \left(1 - \frac{1}{V}\sum_{\vec{R},l,m}\sum_u \frac{|\langle z|\chi_{lm}\rangle\langle\chi_{lm}|\varphi_u^{(0)}\rangle|^2}{s - s_u}\right) \\
&= \varepsilon_l \left(1 - \frac{N}{V}\sum_u\sum_{lm} \frac{|z_{lm}\varphi_{lm}^{(u)}|^2}{s - s_u}\right) \\
&= \varepsilon_l \left(1 - p_{\text{col}}\sum_u \frac{|\varphi_{10}^{(u)}|^2}{s - s_u}\right)
\end{aligned} \tag{5.73}$$

where p_{col} is the volume fraction of column, and the summation over $\boldsymbol{R}$ is noted to give a factor of N since both z_{lm} and $\phi_{lm}^{(u)}$ are independent of $\boldsymbol{R}$. The resulting V/N yields the unit cell volume. By utilizing the relation $z_{lm} = (4\pi a^3/3)^{1/2} z_{lm}\delta_{l1}\delta_{m0}$, the final answer can be obtained.

The material parameter s is a complex number, while the free energy is real and is directly proportional to the real part of $\bar{\varepsilon}_{zz}$ (Eq. (5.54)). Hence the structure, stress and other physical effects (derivable from the free energy) are only obtained through the real part of the effective $\bar{\varepsilon}_{zz}$. The imaginary part of $\bar{\varepsilon}_{zz}$ gives the overall dissipation of the system. It is clear from Eq. (5.70) that the imaginary part of the component dielectric constants (or s) has a non-trivial contribution to the real part of (the effective) $\bar{\varepsilon}_{zz}$, and hence the ER effect.

The structure of the ground state of DER fluids has been determined by both experiments [5.6] and theory [5.11] to be body-centered tetragonal (BCT) structure.

(iii) Determination of yield stress

The axis of the column structure is originally along the external electric field when there is no shear. To find the yield stress of the DER system [5.11], the system has to be sheared. Shear force that is perpendicular to the applied field tilts the column structure, and the gap between the adjacent particles in the same chain increases, while that in the two neighboring chains is shortened. Such a shear distortion is characterized by the tilt angle θ, and involves a distortion in the lattice constants c and a given by $c/R = 2/\cos\theta$, $a/R = [8 - (c^2/2R^2)]^{1/2}$ and a change in unit cell volume given by

$$\frac{v}{v_0} = \frac{4\cos^2\theta - 1}{3\cos^3\theta} \tag{5.74}$$

where v_0 is the cell volume at $\theta = 0$ (Fig. 5.8).

Consequently, under shear the volume fraction of solid spheres in the BCT structure is also θ-dependent [5.10]. When particles form a structure in a liquid under an applied field, the effective dielectric

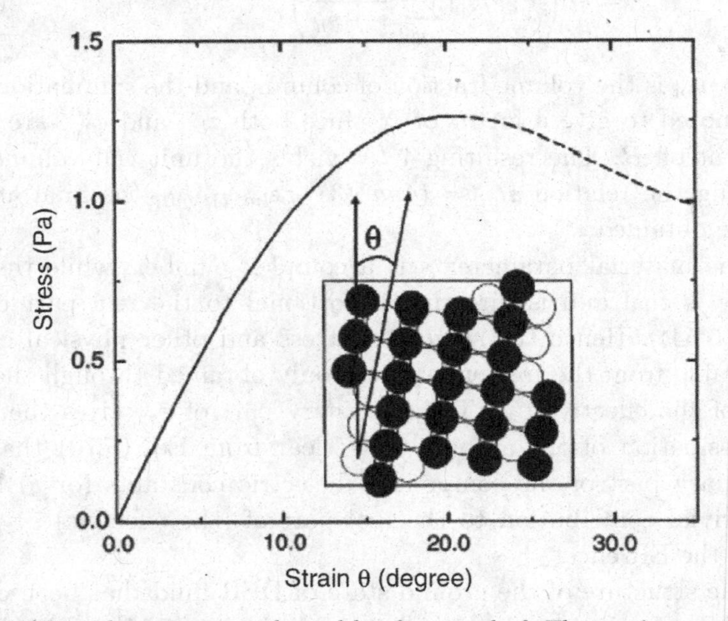

Fig. 5.8 Particle structure is sheared by shear angle θ. The maximum stress at a certain strain θ is defined as yield stress [5.10].

constant can be expressed as

$$\bar{\varepsilon}_{zz} = \frac{p}{p_{\text{str}}}\bar{\varepsilon}_{zz}^{\text{inf}} + \left(1 - \frac{p}{p_{\text{str}}}\right)\varepsilon_l, \tag{5.75}$$

where $\varepsilon_{zz}^{\text{inf}}$ is the volume average of the column dielectric constant, p is the global volume fraction of the dispersed component with particle dielectric constant ε_p, p_{str} is the solid particle volume fraction inside the columns, which depends on the given periodic structure, and ε_l is the liquid dielectric constant. Equation (5.75) shows that $\bar{\varepsilon}_{zz}$ is also a function of strain θ.

For finite distortions, the effective dielectric constant can again be calculated using the spectral representation, leading to energy as a function of θ, $E(\theta)$. By definition, the stress is given by $\mathrm{d}E(\theta)/\mathrm{d}\theta$. To calculate yield stress, the electrostatic field is expanded in vector spherical harmonic functions [5.11], and an exact representation of the electrostatic energy as a function of structure and component dielectric constants is obtained.

In this case, the direction of the sheared column is z', while x' is perpendicular to z' and on the $z'-E_0$ plane. When the shear angle is small, and the separation between the columns tends to infinity, the component $\varepsilon_{x'x'}$ can be evaluated by first calculating the infinite column dielectric constant in the $x'x'$ direction, $\varepsilon_{x'x'}^{\text{inf}}$, and then obtaining $\varepsilon_{x'x'}$ by the Maxwell–Garnett formula [5.38]:

$$\frac{\bar{\varepsilon}_{x'x'} - 1}{\bar{\varepsilon}_{x'x'} + 1} = \frac{p\,\bar{\varepsilon}_{x'x'}^{\text{inf}} - 1}{p_{\text{str}}\,\bar{\varepsilon}_{x'x'}^{\text{inf}} + 1}, \tag{5.76}$$

where $\bar{\varepsilon}_{zz}$ is then obtained from $\varepsilon_{z'z'}$ and $\varepsilon_{x'x'}$ by the following formula:

$$\bar{\varepsilon}_{zz} = \bar{\varepsilon}_{z'z'}\cos^2\theta + \bar{\varepsilon}_{x'x'}\sin^2\theta \tag{5.77}$$

through the rotational transformation of a second rank tensor [5.11].

By varying θ, Ma *et al.* [5.10] have numerically calculated the free energy density as a function of strain; the stress is obtained simply by differentiation. A typical stress–strain relation, shown in Figure 5.8, is noted to start at zero and is linear at small strain. It reaches a maximum and decreases with strain when sheared further.

The initial slope gives the shear modulus, and the maximum of the stress–strain relation is defined as the static yield stress. The part after the maximum is plotted as a dashed line, since decreasing stress with increasing strain indicates instability.

A numerically evaluated stress $\mathrm{Re}\{|\varepsilon_l|^{-1}\partial\bar{\varepsilon}_{zz}(\theta)/\partial\theta\}$ versus strain θ curve (in the unit of $|\varepsilon_l|E^2/8\pi$) is shown in Fig. 5.8. It is seen that at a small strain, the stress varies linearly with strain. The slope of the linear variation is precisely the shear modulus. The maximum stress at certain strain θ is defined as the yield stress, beyond which the system becomes unstable.

The upper bound of the yield stress is estimated assuming 3 points: that (1) $\varepsilon_p/\varepsilon_l \to \infty$; (2) the surfaces of the spheres cannot approach each other closer than a small distance 2δ, e.g. the lower limit on atomic separation (so the lowest s_n is slightly greater than 0); and (3) $p \le p_{\mathrm{BCT}} = 2\pi/9$. $\varepsilon_{zz}/\varepsilon_l = -3.95\ln(\delta/R) - 5.35$. In units of $|\varepsilon_l|E^2/8\pi$, the fitted upper limit of yield stress is $1.38\sqrt{(R/\delta)}$ for the static yield stress. Taking $R = 20\,\mu\mathrm{m}$, $\delta = 1\,\text{Å}$, $E = 1\,\mathrm{kV/mm}$ and $\varepsilon_l = 2.95$, the upper limit of yield stress is 8 kPa.

It has been pointed out that cases of coated microspheres can be treated with only slightly modified computational formalisms [5.12].

5.2.2 *Conduction ER models*

(1) Interfacial polarization (Maxwell–Wagner model) [5.39]

(i) Maxwell–Wagner model

All ER suspensions possess some level of conductivity. Anderson [5.40] and Davis [5.41] pointed out that for DC and low-frequency AC electric fields, particle polarization and particle interactions are controlled not by the particle and fluid permittivities as described above, but rather by the particle and fluid conductivities. According to Anderson and Davis, the conductivity in the bulk of both phases results in free charge accumulation at the particle/fluid interface. The migration of free charges to the interface prompts the alternative names "migration" and "interfacial" polarization. In a DC field, mobile charges accumulating at the interface screen the field within a particle, and particle polarization is completely determined by conductivities. In a high-frequency AC field, the mobile charges have insufficient time to respond, leading to polarization that is dominated solely by permittivities, unaffected by conductivities. At intermediate frequencies, both permittivity and conductivity play a role.

The Maxwell–Wagner model [5.42] is the simplest description of particle polarization accounting for both particle and fluid bulk conductivities, as well as their permittivities. In this theory, the permittivities and conductivities of the individual phases are assumed to be constants, independent of frequency. The complex dielectric constants of the disperse and continuous phases are written as

$$\varepsilon_p^*(\omega_e) = \varepsilon_p - j(\sigma_p/\varepsilon_0\omega_f), \tag{5.78}$$

where p = particle, f = fluid, $j = \sqrt{-1}$ and the asterisk represents complex quantities.

Consider again an isolated sphere in a uniform AC electric field:

$$E_0 = \mathrm{Re}\{E_o e^{j\omega_e t}\boldsymbol{e}_2\}. \tag{5.79}$$

The complex potential will still satisfy Laplace's equation $\nabla^2\varphi^* = 0$ in the bulk phases [5.42], subject to the boundary conditions at the interface,

$$\varphi^{*i} = \varphi^{*o}, \tag{5.80}$$

$$\varepsilon_p^*\nabla\varphi^{*i}\cdot\boldsymbol{n} = \varepsilon_f^*\nabla\varphi^{*o}\cdot\boldsymbol{n}. \tag{5.81}$$

Similar to an ideal case, the solution of complex potential is

$$\varphi^{*i} = -E_0 r\frac{3\varepsilon_f^*}{\varepsilon_p^* + 2\varepsilon_f^*}\cos\theta e^{j\omega_e t}, \tag{5.82}$$

$$\varphi^{*o} = -E_0 r\left[1 - \beta^*\left(\frac{a}{r}\right)^3\right]\cos\theta e^{j\omega_e t}, \tag{5.83}$$

where

$$\beta^* = (\varepsilon_p^* - \varepsilon_f^*)/(\varepsilon_p^* + 2\varepsilon_f^*). \tag{5.84}$$

The field external to the sphere is again equivalent to that of a dipole with moment

$$\boldsymbol{p}^{\text{eff}} = \frac{\pi}{2}\varepsilon_0\varepsilon_f\sigma^3\text{Re}\{\beta^* e^{j\omega_e t}\}E_0\boldsymbol{e}_r. \tag{5.85}$$

The time-averaged force on a sphere at the origin due to a second sphere at (R, θ) may be determined easily in the point-dipole limit as before.

$$\boldsymbol{F}_{ij}^{el\cdot\text{PD}}(R_{ij}, \theta_{ij}) = \frac{3}{16}\pi\varepsilon_0\varepsilon_f\sigma^2\beta_{\text{eff}}^2(\omega_e)E_{\text{rms}}^2\left(\frac{\sigma}{R_{ij}}\right)^2 \times \{[3\cos^2\theta_{ij} - 1]\boldsymbol{e}_r + [\sin 2\theta_{ij}]\boldsymbol{e}_\theta\} \tag{5.86}$$

where

$$E_{\text{rms}} = E_o/\sqrt{2} \tag{5.87}$$

and "effective relative polarization" is

$$\beta_{\text{eff}}^2(\omega_e) = \beta_d^2\frac{\left[(\omega_e t_{\text{MW}})^2 + \frac{\beta_f}{\beta_d}\right]^2 + (\omega_e t_{\text{MW}})^2\left[1 - \frac{\beta_f}{\beta_d}\right]^2}{[1 + (\omega_e t_{\text{MW}})^2]^2}, \tag{5.88}$$

where

$$\beta_d = \frac{\varepsilon_p - \varepsilon_f}{\varepsilon_p + 2\varepsilon_f}; \quad \beta_f = \frac{\sigma_p - \sigma_f}{\sigma_p + 2\sigma_f}, \tag{5.89}$$

$$t_{\text{MW}} = \varepsilon_0\frac{\varepsilon_p + 2\varepsilon_f}{\sigma_p + 2\sigma_f}. \tag{5.90}$$

Except for the fact that the effective polarizability is a function of field frequency, particle and liquid permittivities, this force is equivalent to the ideal case.

(ii) Interaction between particles

a. Low particle concentration ($R/a \gg 1$)

Based on Maxwell–Wagner–Sillars interface, the point dipole polarization model gives the particle interacting force:

$$f_{p\text{-}d} = 24\pi a^2\varepsilon_0^2 K_f'(\beta E_0)^2/(R/a)^4 \tag{5.91}$$

where a is the particle radius, ε_0 the vacuum dielectric constant, $\beta = (K'_p - K'_f)/(K'_p + 2K'_f)$ the dielectric mismatch, $K'_f = \varepsilon'_f/\varepsilon_0$ the real part of the relative dielectric constant of base liquid, $K'_p = \varepsilon'_p/\varepsilon_0$ the real part of the relative dielectric constant of particles, R the center distance of particles, and E_0 — the intensity of external electric field [5.28, 5.34].

b. High particle concentration

The above equation is modified when the effect of particle conductivity, base liquid conductivity, and electric field frequency are considered. The complex dielectric constant,

$$K^* = K' - \varepsilon_0\omega, \tag{5.92}$$

can be used to calculate dielectric mismatch [5.39, 5.43–5.46].

The stress can be calculated from $\tau = f_{p\text{-}d}/R^2$, then

$$\tau = f_{p\text{-}d}/R^2 = 24\pi\varepsilon_0^2 K'_f(\beta E_0)^2(a/R)^6. \tag{5.93}$$

The value of β_{eff}, and thus the pair force, depends on the frequency relative to the polarization time constant t_{mw}. In the limit of large frequencies, permittivities dominate the response.

$$\lim_{\omega_e t_{\text{MW}}\to\infty} \beta^2_{\text{eff}}(\omega_e) = \beta^2_d, \tag{5.94}$$

and at the DC limit

$$\lim_{\omega_e t_{\text{MW}}\to 0} \beta^2_{\text{eff}}(\omega_e) = \beta^2_f, \tag{5.95}$$

and thus conductivities control particle polarization forces, *regardless of the permittivities.* The square of the effective relative polarizability, β^2_{eff}, is plotted in Fig. 5.9 as a function of dimensionless frequency $\omega_e t_{\text{MW}}$ in the Maxwell–Wagner formalism for different ratios of β_d/β_f. ω_e is the frequency of the applied electric field and t_{MW} is the polarization time constant. As we will see below, this likely explains, qualitatively, why high dielectric constant materials such as barium titanate ($\varepsilon_p \approx \theta(10^3)$) do not show a very large ER effect in DC fields, and why many systems exhibit a decrease in apparent viscosity with increasing electric field frequency [5.42, 5.47–5.49].

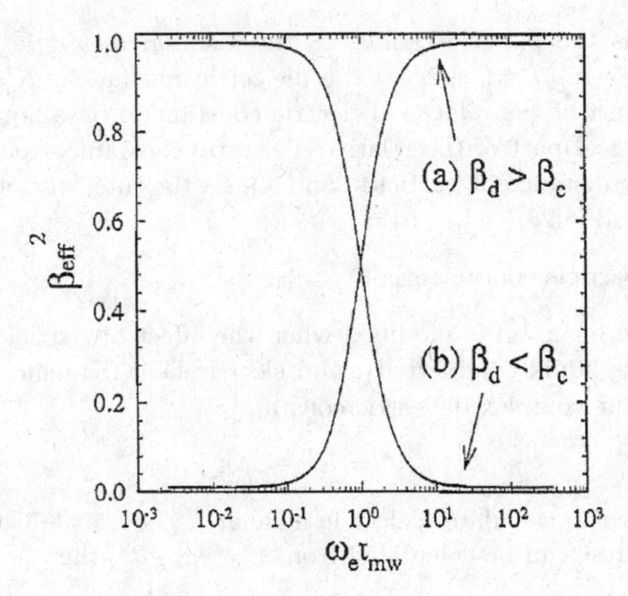

Fig. 5.9 β_{eff}^2 as a function of $\omega_e t_{\text{MW}}$ for (a) $\beta_d = 10\beta_f$ and (b) $\beta_d = 0.1\beta_f$ [5.42].

For particles with a surface conductivity, λ_s, polarization and interactions in the point-dipole limit are described by the Maxwell–Wagner model as described above, provided the particle conductivity is replaced by the apparent conductivity $\sigma_p + 2\lambda_s/a$ [5.42, 5.50].

(2) Equipotential model

When the particle surface is conductive, one may use a method of equipotential spheres to calculate polarization of ER fluids [5.25]. We will first consider the particle interaction of equipotential surface.

Wu and Conrad [5.51] and Felici [5.52] (also [5.53]) pointed out that at low external field and when particle fraction number is also low, solid particles can be treated as an equipotential sphere.

In Fig. 5.10, particle gap function $S = \delta/2a$, δ is the minimum distance between particles. From the hypothesis of equipotential spheres,

$$V(x) = V_0, \tag{5.96}$$

$$E(x)h(x) = E_0(2a + \delta). \tag{5.97}$$

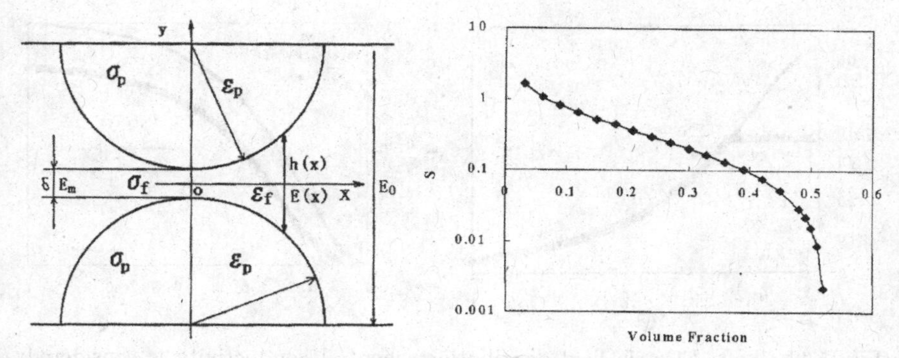

Fig. 5.10 Left: A particle configuration in calculation of electric potential between particles; Right: The electric potential between particles when the particles are treated as equipotential spheres [5.51].

At $x = 0$

$$E_{\max} = (1 + 1/S)E_0, \tag{5.98}$$

or, in general,

$$h(x) = \delta + 2a\{1 - [1 - (x/a)^2]^{1/2}\}, \tag{5.99}$$

$$E(x) = (1 + S)/[S + 1 - (1 - (x/a)^2)^{1/2}]E_0. \tag{5.100}$$

Thus, the reduced electric field E/E_0 at the center (point O in Fig. 5.10) of the joint line of the two hemispheres would be around 21 when $S = 0.1$, and about 160 when $S = 0.025$.

(3) Oil-conduction model

The conductive model deals with the interaction of particles in a conductive base liquid that is dielectric or obeys a nonlinear law of conductivity [5.42].

(i) An electric field between particles in cases of nonlinear conductivity of base liquid is considered.

Under a strong electric field, the conductance of base-liquid is nonlinear; namely, the relation between electric conductivity and electric field intensity does not obey Ohm's law, but, rather, the

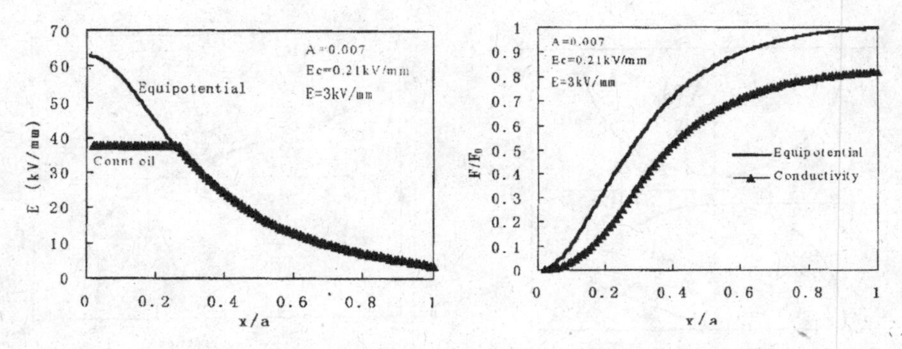

Fig. 5.11 Left: Electric field distribution when oil conductivity is considered; Right: Forces at different particle distances counted for different models [5.25].

following relation [5.50, 5.51]

$$\sigma_f(E) = \sigma_f(0)\{1 - A + A\exp[(E/E_c)^{1/2}]\}. \tag{5.101}$$

Thus, the electric field between particles changes when the nonlinear conductivity of base-liquid is counted.

The left hand side of Fig. 5.11 shows the function of electric field distributution between particles and particle distance when base-liquid conductivity σ_{oil} is counted. Different models of particle interaction give obviously different particle interactions at the same position. The right hand side of Fig. 5.11 shows the forces at different particle distances counted for different models [5.25].

H. Jie *et al.* [5.54] uses the technology of the optical Kerr effect to measure the local electric field between two stationary spheres. They proved that the interparticle local field is several times larger than the external field.

(ii) Electric potential and electric current near contact points

Particle interaction can be calculated using different models [5.52]. To calculate interaction, the local field in the liquid between the particles must be known; to calculate local field, the electric potential distribution V between the particles must be known, and V can be found by solving the Laplace equation under certain boundary conditions. However, this is difficult, and for simplicity, it is supposed that the electric field between particles is along particle connection

(y-direction in Fig. 5.10). Of course, this is precise only at the central cross section between the particles, or at the contact point. Under such hypothesis, ΔV is the electric potential difference between the upper and lower sphere at point x, Δy is the particle distance at those points, and the electric field in liquid can be obtained from $E(x) = \Delta V/\Delta y$. The problem now becomes how to solve the electric voltage $V(x)$ on the particle surface.

If the external electric field is E_0, V_0 is the voltage between the two hemispheres of the radius, while the lower hemisphere is grounded. For simplicity, two solid particles can be replaced with two solid plates (thickness t_s and area S), with liquid (thickness t_{L}) between the two plates, as shown in Fig. 5.11.

From the current continuity,

$$\sigma_s E_s = \sigma_{\mathrm{L}} E_{\mathrm{L}}, \tag{5.102}$$

where σ_s, σ_{L} are the conductivity of solid particles and base-liquid, respectively, and E_s and E_{L} the electric field of solid particles and base-liquid, respectively. The relation between electric field and the applied voltage V_0 is

$$V_0 = 2t_s E_s + t_{\mathrm{L}} E_{\mathrm{L}}. \tag{5.103}$$

t_s and t_{L} are the thicknesses of the solid plates and liquid region, respectively, as shown in Fig. 5.12.

$$E_{\mathrm{L}} = V_0/(t_{\mathrm{L}} + 2t_s\sigma_{\mathrm{L}}/\sigma_s). \tag{5.104}$$

When $t_{\mathrm{L}} \to 0$, the upper limit of E_{L} is

$$E_m = V_0(\sigma_s/2t_s\sigma_{\mathrm{L}}), \quad E_m \propto V_0, \tag{5.105}$$

and

$$E_m = (V_0/t_{\mathrm{L}})(R_{\mathrm{L}}/2R_{\mathrm{S}}) = E_0(R_{\mathrm{L}}/2R_{\mathrm{S}}), \tag{5.106}$$

where the resistance of base-liquid is $R_{\mathrm{L}} = t_{\mathrm{L}}/\sigma_{\mathrm{L}} S$, the resistance of solid particles, $R_{\mathrm{S}} = t_{\mathrm{S}}/\sigma_{\mathrm{S}} S$, and E_0, an average field.

As long as the particle distance is small enough, there is a limitation of the electric field. The limitation of electric field between the

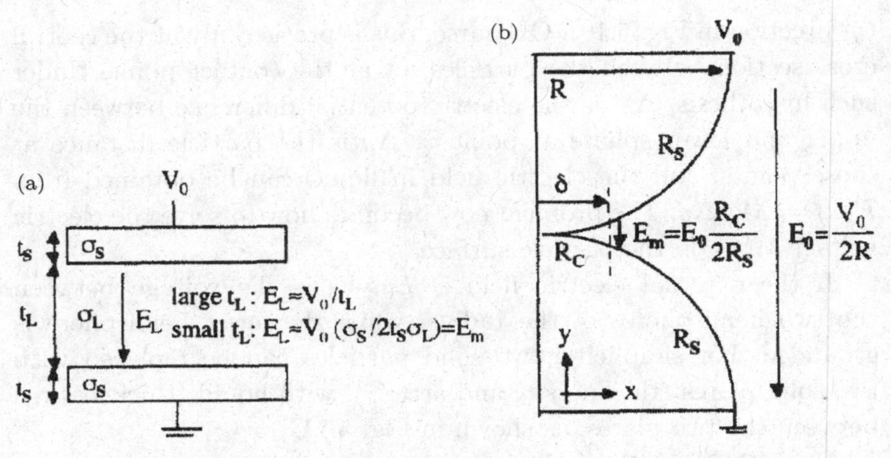

Fig. 5.12 (a) The two parallel plates are used to explain that when the distance between the two electrodes t_L is very small, there is a limitation in the field increment between particles; (b) spherical particle arrangement, where t_L is the particle distance, and t_L changes with position x. When $x < \delta$, the maximum value of electric field is E_m. If E_0 is the average external field ($E_0 = V_0/2R$), then the field enhancement factor E_m/E_0 is given by the ratio of the contact resistance $R_c(x < \delta)$ and surface resistance $R_s(x > \delta)$ [5.55].

two hemispheres near the contact region is

$$E_m = E_0(R_C/2R_S), \tag{5.107}$$

where R_C is the contact resistance, while the electric potential is

$$V = V_0/2 = (RE_m/2)(x/R)^2 \cdot (x < \delta). \tag{5.108}$$

The current density of solid particles and base-liquid, j_s and j_L are respectively

$$j_S = I_S/(2\pi x t_S), \quad I_S = (-dV/dx)(1/Z_s), \quad j_S = (E_m/r_S)(x/R). \tag{5.109}$$

Z_S is the distribution impedance of the solid plates;

$$j_L = I_L/(2\pi x t_L), \quad I_L = dI_S = (V_0 - 2V)(Y_L dx), \quad j_L = \sigma_L E_m. \tag{5.110}$$

Y_L is the distribution admittance of base-liquid. Note that the above equation is Ohm's law.

It can be obtained [5.51] that the electric field enhancement factor between the solid plates is

$$\frac{E_m}{E_0} \approx \frac{1}{\alpha \ln \frac{2R}{\delta}}. \tag{5.111}$$

When the approximate condition $\ln(R/\delta) \gg 1$ is valid, one has

$$\frac{\delta}{R} = \sqrt{\alpha}, \tag{5.112}$$

where

$$\alpha = \sqrt{2R\sigma_L \rho_s}. \tag{5.113}$$

For an insulating film, $\alpha = 7 \times 10^{-2}$, $\delta/R = 0.2$, and for a conducting film $\alpha = 3 \times 10^{-5}$, $\delta/R = 0.005$.

The electric field enhancement factor between the solid plates is

$$\frac{E_m}{E_0} \approx \frac{2}{\alpha \ln \frac{4}{\alpha}}. \tag{5.114}$$

For an insulating film, $E_m/E_0 \sim 7$; for a conducting film, $E_m/E_0 = 5.5 \times 10^3$.

The interaction force between solid particles is

$$F = \int_0^R \frac{1}{2}\varepsilon_L E^2 (2\pi x) dx \tag{5.115}$$

$$= \pi \varepsilon_{\mathrm{L}} \delta^2 E_m^2, \tag{5.116}$$

and particle interaction can be written as

$$F = \pi \varepsilon_L \frac{1}{\alpha \left(\ln \frac{4}{\alpha}\right)^2} V_0^2. \tag{5.117}$$

5.3 Colloidal ER Models

5.3.1 *Giant ER effect*

The GER [5.9] particles have an average size of 50–70 nm, each with a surface coating of urea ($< 3 \sim 10$ nm thick). The dielectric constant of GER suspension is around 50–60 (at 10 Hz) over the temperature

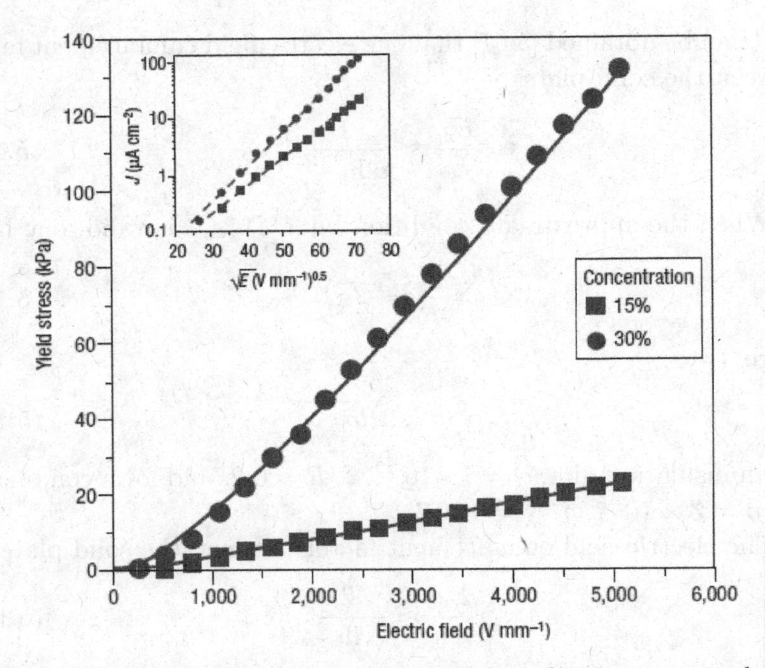

Fig. 5.13 The static yield stress curves for two volume fractions, measured under DC electric fields. The corresponding current densities are shown in the inset. The current density J is below $4\,\mu\,\mathrm{A\,cm}^{-2}$ at $E < 2\,\mathrm{kV\,mm}^{-1}$ for the 30% sample [5.9].

range of 10–120°C at 30% volume fraction of particles. Its effective dielectric constant is dominated by its urea coating. The molecular dipole moment of urea is $\mu = 4.6$ debye (and molecular number density of $1.3 \times 10^{22}\,\mathrm{cm}^{-3}$). An estimate of the polarizability $\alpha = \mu^2/kT$ based on the free-dipole model yields a dielectric susceptibility $\chi \approx 5$ (with a molecular volume of $\sim 100\,\text{Å}^3$). The dielectric constant $\varepsilon = 1 + 4\pi\chi$ of urea is about 60.

The static yield stress curves for two volume fractions, measured under DC electric fields, are shown in Fig. 5.12. In this GER effect, the static yield stress displays near-linear dependence on the electric field, in contrast to the quadratic variation usually observed. In general, the static yield stress is proportional to the energy density, $\mathbf{P}\cdot\mathbf{E}$, where $\mathbf{P}$ is the polarization density. A linear dependence of $\mathbf{P}$ on $\mathbf{E}$, that is, $\mathbf{P} = \chi\mathbf{E}$, implies a quadratic field dependence of the yield stress. The observed near-linear field dependence suggests that the

GER mechanism involves a constant $\mathbf{P}_0$, that is, a saturation polarization. The saturation surface polarization, in the contact region of the neighboring spheres, is responsible for the GER effect, while the saturation surface polarization is proposed to come from narrow gap (0.2 nm) due to the urea coating. The authors of Ref. [5.9] argue that the aligned configuration is possible because of (1) the high dielectric constant of the coating, which reduces the repulsive interaction between the aligned dipoles, (2) the favorable attractive interaction across the gap which considerably lowers the overall energy, and (3) the magnitude of the favorable interaction energy, which is sufficient to overcome the entropy effect.

Wen *et al.* [5.15] found a particle size effect of GER fluids that is different from the traditional one. In traditional ER fluids, the interaction force of ER particles is inversely proportional to a^4 (a being particle size). As indicated in Fig. 5.14, the GER fluid of 20 nm particles shows a yield stress above 250 kPa ($E = 5\,\text{kV/mm}$), which is much higher than that of the GER fluid of 50 nm particles (about 130 kPa at the same external electric field).

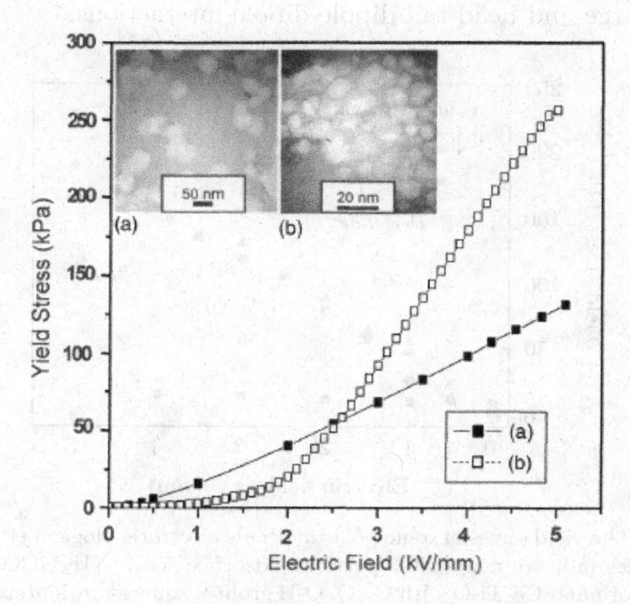

Fig. 5.14 Size effect of GER fluids [5.15].

5.3.2 *Polar molecule ER effect*

(1) Effect of a few polar molecules in the gap

As we have seen in Section 5.1.3, polar molecule electrorheological (PM-ER) fluids show giant ER effects (also see Fig. 5.14). Here we will introduce the polarization mechanism [5.13, 5.14].

The strong interaction between the polar molecules and charges under the enhanced electric field in the interparticle gap plays a key role in PM-ER fluids.

The polar molecules adsorbed on particle surfaces or floating around in the base liquid are randomly oriented when $E = 0$. When an electric field is applied, the polar molecules are attracted by the charges that appear on the particle surface near the interparticle gap and on the modified electrode (see the following for detail). The polar molecules in the interparticle gap are aligned along the field when E is high enough (see Fig. 5.16). In this case, a polar molecule becomes a bridge, while two "bridgeheads" are the charges either on two nearby particles or on an electrode and the nearby particle. In the enlarged image at the right, two kinds of interactions are shown: dipole-charge and head-tail dipole-dipole interactions.

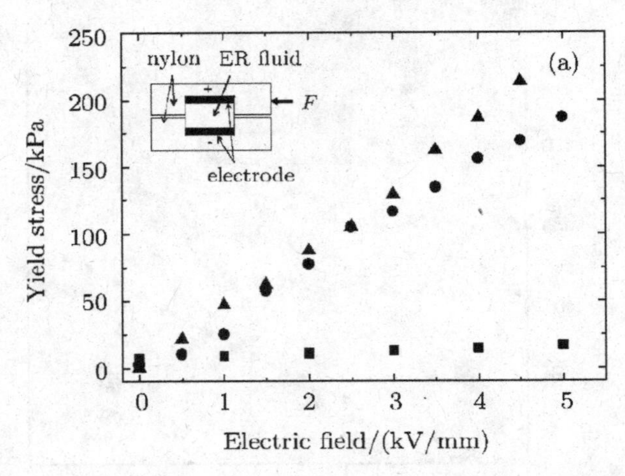

Fig. 5.15 The yield stress of some polar molecule electrorheological (PM-ER) fluids. Triangles indicate nano-TiO2 particles adsorbed with $(NH_2)_2CO$ molecules; circles are for nano-Ca-Ti-O with C=O, O-H groups; squares are sintered Ca-Ti-O particles [5.13].

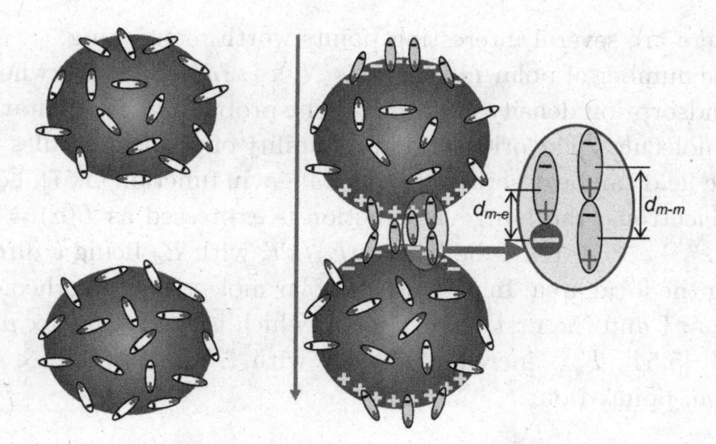

Fig. 5.16 The polar molecules aligned in the gap of the particles under zero field (left) and under an electric field (right) [5.13].

The attraction between a polar molecule and a charge on two nearby particles can be expressed as [5.13, 5.14, 5.56]

$$f_{m\text{-}e} = \frac{e\mu}{2\pi\varepsilon_0\varepsilon_f d_{m\text{-}e}^3}. \tag{5.118}$$

Here μ is the dipole moment of the polar molecule, and $d_{m\text{-}e}$ is the distance between charge and the molecule center. On the other hand, the attraction between head-tail aligned molecules on two nearby particles is [5.13]

$$f_{m\text{-}m} = \frac{3\mu^2}{2\pi\varepsilon_0\varepsilon_f d_{m\text{-}m}^4}, \tag{5.119}$$

where $d_{m\text{-}m}$ is the central distance between two head-tail aligned molecules. Since $d_{m\text{-}e} = d_{m\text{-}m}/2$, one has

$$\frac{f_{m\text{-}e}}{f_{m\text{-}m}} \approx \frac{16ed_{m\text{-}e}}{3\mu}. \tag{5.120}$$

For comparison, $f_{m\text{-}e}$ is about 25 times larger than $f_{m\text{-}m}$. This is the major source of the yield and shear stresses of such ER fluids.

There are several interesting points worth mentioning.

The number of polar molecules is $N \propto (\pi rd/2)\rho_m n(E)$, where ρ_m is the adsorption density and $n(E)$ is the probability of orientation of polar molecules. The orientation probability of polar molecules under electric field can be described by the Langevin function [5.57]. For free polar electrons, the Langevin function is expressed as $L(a) = (e^a + e^{-a})/(e^a - e^{-a}) - 1/a$, where $a = \mu E_d/kT$, with E_d being a directing field in the local area. In the case of polar molecule electrorheological fluid $a \approx 1$ and the first term of $L(a)$, which is $L(a) \approx a/3 \propto \mu E_d$, is enough [5.54]. $F_{m\text{-}e}$ increases linearly with E due to $n(E) \propto \mu E$ as Lu *et al.* pointed out:

$$\tau_y = \gamma F_{m\text{-}e} = \gamma \frac{3\varphi}{2\pi r^2} N f_{m\text{-}e} = A\gamma \frac{3\phi\rho_m e\mu^2 E}{\pi r \varepsilon_0 \varepsilon_f d^2}, \tag{5.121}$$

where $\gamma \approx \tan\theta \approx 1/3$ when $\theta \approx 20°$, which comes from the Langevin function. The maximum yield stress of the PM-ER fluids can also be estimated via the above equation. Taking $\mu = 2$ debye, $\phi = 35\%$, $d_{m\text{-}e} = 0.15\,\text{nm}$ ($d_{m\text{-}m} = 0.3\,\text{nm}$), $N = 3$, one has $\tau_y \approx 1700\,\text{kPa}$.

The polar molecules in a PM-ER fluid are easily reoriented under a rather low field since the polarization energy of a polar molecule μE is larger than $(3/2)\ k_B T = 6.2 \times 10^{-21}\,\text{J}$, which is less than adsorption energy U_{ad}.

The relation of current density J and E can be obtained from the Pool–Frenkel effect [5.58]. Since the external electric field E enhances the thermal excitation of charge carriers bounded by Coulomb potential, the conductivity can be expressed as

$$\sigma = \sigma_0 \exp[(e^3 E/\varepsilon)^{1/2}/k_B T]. \tag{5.122}$$

Electrodes become important in the giant ER effect. A smooth metallic electrode can hold only small polar molecules, and these molecules are difficult to be reoriented, so there is only a little molecule-charge interaction available, which leads to low shear stress. Electrodes modified with diamond sands exhibit the best results in holding enough polar molecules which can attract charges both on electrodes and the particles near the electrodes.

(2) Multiple scattering method

Gang Sun *et al.* [5.32] applied the multiple scattering method [5.59] to numerically calculate the electric field for a system composed of spheres. The key point of the method is to include the maximum order of polarization, $L_{\max}$, when it is needed in the calculation. The multiple scattering method is able to determine local electric and magnetic fields in a system composed of multi-scale spheres with arbitrary distribution, whereas the re-expansion method is restricted to the case of two spheres aligned along the field.

In the multiple scattering method, the field of a circularly polarized plane wave of wave-vector $\boldsymbol{k}$ is

$$\boldsymbol{E}_{\eta}^{(i)} = \sum_{\mathrm{LM}} W_{\eta\mathrm{LM}}(\hat{\boldsymbol{k}}) \left[j_{\mathrm{L}}(kr)\boldsymbol{X}_{\mathrm{LM}}(\hat{\boldsymbol{r}}) + \eta \frac{1}{k} \nabla \times j_{\mathrm{L}}(kr)\boldsymbol{X}_{\mathrm{LM}}(\hat{\boldsymbol{r}}) \right], \tag{5.123}$$

where

$$W_{\eta\mathrm{LM}}(\hat{\boldsymbol{k}}) = 4\pi i^{\mathrm{L}}(\boldsymbol{e}_1 + i\eta \boldsymbol{e}_2) \cdot \boldsymbol{X}_{\mathrm{LM}}^{*}(\hat{\boldsymbol{k}}), \tag{5.124}$$

and $\boldsymbol{e}_1$ and $\boldsymbol{e}_2$ are unit vectors orthogonal to $\boldsymbol{k}$ and to each other. $\boldsymbol{X}_{\mathrm{LM}}(\boldsymbol{r})$ is a vector spherical harmonic.

The field scattered by a cluster composed of N nonmagnetic spheres whose centers lie at R_α is expanded in a multicentered series of multipoles that includes only outgoing spherical waves at infinity:

$$E_{\eta}^{(s)} = \sum_{\alpha} \sum_{\mathrm{LM}} \Big[A_{\eta\mathrm{LM}}^{\alpha} h_{\mathrm{L}}(kr_\alpha)\boldsymbol{X}_{\mathrm{LM}}(\hat{\boldsymbol{r}}_\alpha) + B_{\eta\mathrm{LM}}^{\alpha} \frac{1}{k} \nabla \times h_{\mathrm{L}}(kr_\alpha)\boldsymbol{X}_{\mathrm{LM}}(\hat{\boldsymbol{r}}_\alpha) \Big], \tag{5.125}$$

where $\boldsymbol{r}_\alpha = \boldsymbol{r} - \boldsymbol{R}_\alpha$. The expansion coefficients A and B are uniquely determined by the boundary conditions for E at the surface of each of the spheres [5.59].

The local field of the numerical calculation greatly depends on $L_{\max}$ at first, and then tends to a stable value as $L_{\max}$ increases. For a smaller gap, larger $L_{\max}$ is needed to obtain stable results. When the gap is about one thousandth of the diameter of the spheres, which is the only limitation of the method, $L_{\max} = 30$ is used. It is found that the local electric field is a few hundred times larger than the external applied field if the gap is small, and it increases quickly as the gap decreases [5.60, 5.61].

(3) Superpolarization

It has been found that when the gap between a sphere and a plate is below a certain critical value at zero bias voltage, or when bias voltage is above a certain critical value at any value of the gap, water in the gap forms a bridge.

The energy barrier of forming a water bridge (see Fig. 5.17) consists of those due to surface, condensation and electric static energies, ΔU_s, ΔU_c and ΔU_e, respectively, where [5.62, 5.63]

$$\Delta U_s = (2\pi W_{\text{neck}} D)\gamma_{\text{LV}} - (2\pi W_{\text{neck}}^2)(\gamma_{\text{SV}} - \gamma_{\text{LS}}) \tag{5.126}$$

with $\gamma = \gamma_{\text{LV}} \approx (\gamma_{\text{SV}} - \gamma_{\text{LS}})$. The shape of a central cross section of the pillar is approximately circular, so it is replaced with the area of a circle in the calculation of ΔU.

$$\Delta U_c \approx (RT/\vartheta_m)\ln(1/H)\vartheta_o, \tag{5.127}$$

where W_{neck} is the center radius of the water bridge, $\vartheta_o = \pi W_{\text{neck}}^2 D$ is the bridge volume, ϑ_m is the molar volume, H is the relative humidity, $R = 8.31\,\text{J mol}^{-1}\,\text{K}^{-1}$, and T is the absolute temperature. The electrostatic energy associated with the polarization of the condensed liquid is $\Delta U_e = -(\varepsilon_0/2)(\varepsilon - 1))(V/D)^2\vartheta_0$. The total energy barrier

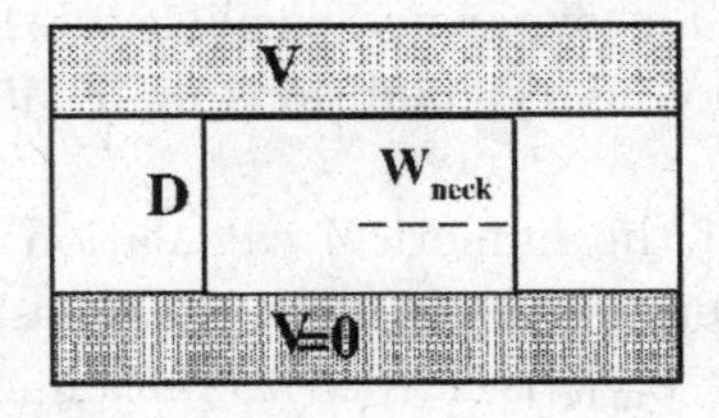

Fig. 5.17 A simplification of a water bridge between two plates. Gap thickness is D and the radius of the water bridge is W_{neck} [5.62].

can then be expressed as

$$\Delta U \approx \pi W_{\text{neck}}^2 D \left[\frac{RT}{\vartheta_m} \ln(1/H) - \frac{2\gamma}{D} - \frac{1}{2D^2}(\epsilon - 1)\epsilon_0 V^2 \right] + 2\pi\gamma W_{\text{neck}} D. \quad (5.128)$$

At a zero-bias ($V = 0$), only when the gap thickness is below a certain critical value D_c will a water bridge form:

$$D_C(V = 0) = 2\gamma/[(RT/\vartheta_m)\ln(1/H)]. \quad (5.129)$$

For any gap distance D there is a threshold bias V_{th} above which the most stable situation corresponds to the liquid filling the gap between the plates:

$$V_{\text{th}}^2 = \frac{1}{(\epsilon - 1)\epsilon_0} \left\{ 2D^2 \frac{RT}{\vartheta_m} \ln(1/H) - 4D_\gamma \right\}. \quad (5.130)$$

Experimentally, V_{th} can be determined by measuring the attractive force exerted by the water bridge on the tip using an amplitude modulation AFM, operated in noncontact mode at resonance (330 kHz). The above model is in quite good agreement with the experiment when sphere radius is 30 nm.

As the applied field reaches a certain threshold value, the local field at the center area in the gap between two particles reaches the critical value for superpolarized state, and hence, a thin pillar made of the polar molecules in the superpolarized state is formed between the particles. The pillar fixes the dispersed particles by its high yield stress [5.32].

As the applied field increases further, the region where the electric field is higher than the critical value expands, and the pillar becomes thicker.

To determine the yield stress, force $F = \partial\Delta U/\partial y$ is calculated as follows:

$$F = \pi W_{\text{neck}}^2 \ln(1/H) - \frac{(\varepsilon - 1)\varepsilon_0}{2}\pi^2 W^4 + 2\pi\gamma W, \quad (5.131)$$

where ΔU is the function of y, which is the vertical coordinate. The corresponding stress $\tau = F/S$ is

$$\tau = \ln(1/H) - \frac{(\varepsilon - 1)\varepsilon_0}{2}\pi W^2 + 2\gamma/W. \tag{5.132}$$

As the yield stress is the stress at one special spot, the cross section of the pillar is proportional to the yield stress. According to this picture, the shear stress between two dispersed particles, i.e., the yield stress of the ER fluids, is proportional to the pillar's cross section, which increases linearly when the applied field exceeds the threshold. It is predicted that the yield stress of such ER fluids would also reach a value as high as an order of MPa when a local electric field is as high as an order of 10^9 V/m [5.32, 5.64].

(4) Effect of many polar molecules in the gap

Xu, Tian *et al.* [5.65] take into account the effect of polar molecules between the particles (or equivalently, oriented within the field-directed gap between them) on the ER effect. According to the Green's function approach, the effects of charges on the potential at any position can be written as

$$\Phi(\boldsymbol{r}) = -E_0 z + \frac{1}{4\pi}\int q(\boldsymbol{r}')G(\boldsymbol{r} - \boldsymbol{r}')d\boldsymbol{r}', \tag{5.133}$$

where $q(\boldsymbol{r}')$ includes the surface polarized charges induced by an electric field and the polar molecules oriented within the gap between the particles. The effective dipole moment of all polar molecules is related to the Langevin function. The local field at the center of the gap between the two particles is derived as

$$E_{\text{loc}} = \frac{16(\varepsilon - \varepsilon_h)a^3 + (\varepsilon + 2\varepsilon_h)R^3}{3R^3\varepsilon_h}C_1, \tag{5.134}$$

where a is the radius of a particle, R the center-to-center distance between the two particles, ε (or ε_h) the dielectric constant of a bare particle (or host oil), and

$$C_1 = \frac{12p - 3E_0\pi R^3\varepsilon_h}{\pi[(\varepsilon - \varepsilon_h)a^3 + (\varepsilon + 2\varepsilon_h)R^3 - 3(\varepsilon - \varepsilon_h)a^3\cos^2\theta]}. \tag{5.135}$$

The change in electrostatic energy due to the introduction of particles and polar molecules into the host medium in the presence of an external electric field $\boldsymbol{E}_0$ is

$$W = \frac{1}{3}(\varepsilon - \varepsilon_h)a^3 C_1 E_0 + pE_{\text{loc}}. \tag{5.136}$$

This leads to the yield stress expressed as

$$\begin{aligned}\tau(\theta) &= -\frac{1}{V}\frac{\partial W}{\partial \theta} \\ &= -\frac{1}{V}\frac{6a^3(\varepsilon - \varepsilon_h)(\pi\varepsilon_h R^3 E_0 - 4p)\Omega\cos\theta\sin\theta}{\pi\varepsilon_h R^3[a^3(\varepsilon - \varepsilon_h) + R^3(\varepsilon + 2\varepsilon_h) - 3a^3(\varepsilon - \varepsilon_h)\cos^2\theta]^2},\end{aligned} \tag{5.137}$$

with $\Omega = a^3(\varepsilon - \varepsilon_h)(R^3\varepsilon_h E_0 + 16p) + R^3(\varepsilon + 2\varepsilon_h)p$, where V is the average volume occupied by a particle (modified by polar particles). This is a result of the first-order approximation, which is good enough to interpret the experiment result.

The effective induced dipole moment p consists of n polar molecules. The number of the polar molecules, n^c, is linked with the Langevin function $L(E_0)$:

$$n = \alpha + \beta L(E_0), \tag{5.138}$$

where

$$L(E_0) \approx \frac{1}{3}\frac{p_0 E(E_0)}{k_{\text{B}}T} \tag{5.139}$$

is the linear Langevin function when the local field is not extremely strong, where p_0 is a dipole moment of single polar molecules, and E is directing local field as a function of the external field. A theoretical fitting of the curve in Fig. 5.18(b) gives $\alpha = 9.2816 \times 10^4$ and $\beta = 2.0891 \times 10^8$. From the values of L for differing local field, one may find that n, the number of interparticle polar molecules that are aligned with the external electric field, is as much as 10^{-6}.

This reveals that n ($\sim 10^6$) polar molecules oriented within the field-directed gap between the colloidal particles can strengthen colloidal electrostatic interactions, thus yielding the unusual enhancement of yield stresses.

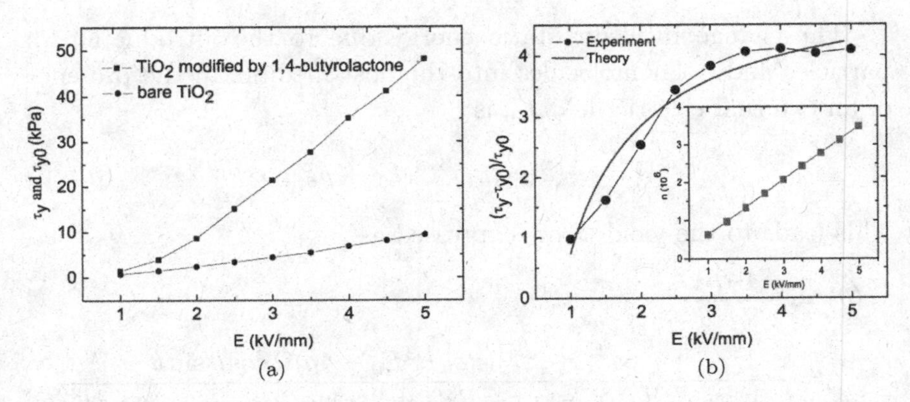

Fig. 5.18 (a) Experimental result for yield stresses y and $y_0 \cdot y$: TiO_2 modified by polar molecules (with liquid-solid ratio 0.45 ml/g or mol ratio 1.5); y_0: bare TiO_2. (b) Theoretical fitting experimental results. The inset shows the number n of polar molecules (dipoles) within the gap between the two particles as a linear function of external electric fields [5.65].

Tan *et al.* [5.66] proposed a model of polar molecule dominated ER effect, and the model explains the effect of particle size and the thickness and dipole moment of polar molecular layer. The model considers that only the outermost layers of the shell of coated polar molecules are dominant in the interaction between two spheres consisting of a dielectric core and polar molecular shell.

In this case the induced electronic dipole moment μ_i is the summation of electrostatic μ_e and orientational ones μ_o, or $\mu_i = \mu_e + \mu_o$, where

$$\mu_{\text{electronic}} = \frac{1}{2}\pi\varepsilon_0 d^3 E \tag{5.140}$$

and

$$\begin{aligned}\mu_{\text{orient}} &= \langle \mu \cos\Theta \exp(\mu E \cos\Theta / kT)\rangle \\ &= \mu\left(\frac{e^{\alpha}+e^{-\alpha}}{e^{\alpha}-e^{-\alpha}} - \frac{1}{\alpha}\right) \\ &\equiv \mu L_{\alpha}(E)\end{aligned} \tag{5.141}$$

where $\alpha = \mu E/kT$ and μ is the permanent dipole moment of a polar molecule. The equivalent dielectric constant of the inner shell can be readily estimated as

$$\varepsilon_{\text{innershell}} = 1 + \frac{n\mu_{\text{induced}}}{\varepsilon_0 E}, \tag{5.142}$$

where n is the number density of polar molecules. In this equation, we may assume $\mu_o \approx \mu 2E/3kT$ for μE, kT owing to the screening effect arising from the outer shell.

We may express the maximum local electric field E_{local} in a self-consistent equation like

$$E_{\text{local}} = E_1 + L_\alpha(E_{\text{local}})\frac{n\mu}{\varepsilon_f}. \tag{5.143}$$

The major contribution of electrostatic interaction between two nearby PMER particles is from the outer shell-outer shell interaction, while those between two dielectric cores and between polar molecules of one particle and core of the other are ignored. The polarization P of the outer shell is determined by the local field. By the definition of an electric field distribution function $f(\theta)$, $E_{\text{local}}(\theta) = E_{\text{local}}\, f(\theta)$ (here θ is an angle with respect to the external electric field), the electrostatic energy of the polarized outer shell (with a thickness of a single polar molecule) is given by

$$U = 2\pi R^2 d \int_0^{\pi/2} \mathbf{P} \cdot \mathbf{E}_{\text{local}}(\theta) \sin\theta \cos\theta \, d\theta. \tag{5.144}$$

If $f(\theta)$ is replaced with the Lorentz distribution, one has

$$U = 2\pi R^2 \int_0^{\pi/2} n\mu d L_\alpha(E_{\text{local}} f(\theta)) E_{\text{local}} f(\theta) \sin\theta \cos\theta \, d\theta. \tag{5.145}$$

Since the local field decays from the center point when θ increases, the interaction between two PMER particles can be thought to be equivalent to an interaction between two small virtual planes of polar molecules, each of which has an area S

$$S = 2\pi R^2 \int_0^{\pi/2} \frac{L_\alpha(E_{\text{local}} f(\theta)) f(\theta) \sin\theta \cos\theta}{L_\alpha(E_{\text{local}})} d\theta \tag{5.146}$$

and the distance between the two virtual planes is

$$h_1 = \frac{2S}{\pi R} + \delta. \tag{5.147}$$

The shear stress τ_s of a PMER fluid is given by

$$\tau_s = \frac{F_\gamma}{S_p} = \frac{\partial U}{S_p h_1 \partial_\gamma} \tag{5.148}$$

where $S_p = \pi(2R)^2/6\varphi$ is the average cross sectional area occupied by a chain of PMER particles, φ is the volume fraction of PMER particles, γ is the tilted angle of a chain. To obtain τ_s, the FEA was used to determine the relation between E_{local} [Eq. (5.143)] and γ, by assuming that the second term in the right hand side of Eq. (5.143) is independent of γ due to the saturation of the orientational polarization. It has been noted that the major contribution of Eq. (5.148) is from the integration of $\theta \in [0, 0.17]$. Namely, the PMER effect is mainly due to the interaction of the tail-head connected polar molecules within the two outer shells between the two PMER particles.

For general PMER fluids, the static yield stress is nearly proportional to R^{x-1}. When h/R, the ratio of shell thickness h and particle radius R, changes from 0.05 to 0.5, the index x changes accordingly from 0.64 to 0.51. The relation between x and h/R can be fitted with an exponential decay function, $x = x_0 + A_1 \exp(-h/Rt_1)$, where $x_0 = 0.50946$, $A_1 = 0.21025$ and $t_1 = 0.10459$. According to this model, when particle radius R is 10 nm, shell thickness h is 0.2 nm, the gap between two particles δ is 0.2 nm, the dielectric constants of core and shell are 1000 and 60, respectively, the yield stress of such polar molecular ER fluids can reach as high as 628 kPa at an external electric field E_0 of 5 kV/mm.

(5) Effect of wettability on colloidal ER effect

All ER fluids require that particles are well dispersed under zero field, and at this occasion, the particles exhibit a smooth and well mixed state with the oil media. The particles are usually considered to be wet with the oil. Many researchers [5.15–5.17] have found that particle wettability plays an important role.

The mixture of $(BaTiO(C_2O_4)_2$ particles and a hydrocarbon oil (PAO) is in a lumpy-like state — a partial-wetting state, and gives little ER effect; in contrast, after a very small amount of an oleic acid is added, the mixture becomes smooth and dilute, although the solid particle volume fraction has almost no change. The modified material gives a wettability-induced giant ER effect. It is believed that wet particles would be dispersed well in a liquid [5.15, 5.20, 5.67].

The disjoining pressure of liquid is

$$\prod = \frac{A}{6\pi h^3}, \tag{5.149}$$

where h is the local thickness of liquid between two particles, and $h = 2(R - R\cos\theta) + g$. Here R is the particle radius, and g is the surface-to-surface gap between the two particles. The wetting force (lyophilic repulsion force) along the joining line of two particles is

$$F_w = \int \prod dS = \int \frac{A \cdot 2\pi R\cos\theta}{6\pi[2(R - R\cos\theta) + g]^3} dR\cos\theta$$
$$= \int_0^{\frac{\pi}{2}} \frac{A \cdot 2\pi R^2 \cos\theta \sin\theta}{6\pi[2(R - R\cos\theta) + g]^3} d\theta = \frac{A(2R - g)}{24g^2} + \frac{A}{24(2R + g)} \tag{5.150}$$

where A is the Hamaker constant.

The van der Waals attractive force is expressed as

$$F_v = \frac{1}{6} A \left[\frac{64R^6}{(2R + g)^3 (4Rg + g^2)^2} \right]. \tag{5.151}$$

The difference of the wetting force and the van der Waals force is plotted in Fig. 5.19.

It has been proven both theoretically and experimentally that there exists a repulsion force between the wet particles in a liquid. It is this force that overcomes [5.16, 5.18] van der Waals attraction force [5.19] between colloidal ER particles when there is no electric field and makes the particles well dispersed. Good wettability, or good zero-field dispersity of ER particles is a prerequisite of forming the necessary strong particle structure after the electric field is applied. Aggregated non-wetting particles show much less yield stress [5.20].

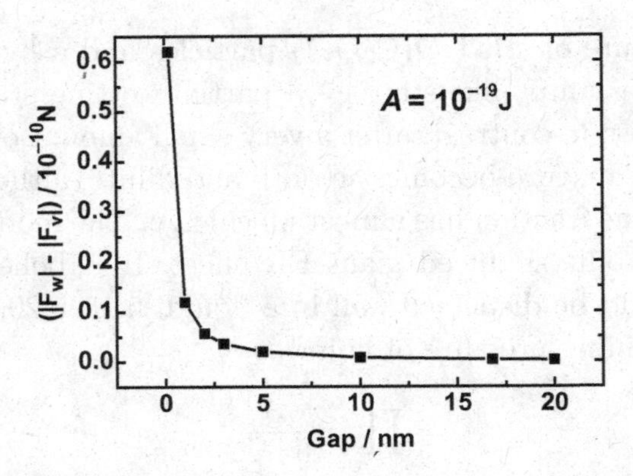

Fig. 5.19 The difference between wetting force $|F_w|$ and van der Waals force $|F_v|$ [5.20].

The interaction between the coated materials and base liquids is an important issue in the microscopic mechanism of the effects of core-shell colloidal ER fluids. Chen *et al.* [5.68] observed via their molecular dynamics simulation that when the electric field is turned on, the urea molecules diffuse into the silicone oil layer from both sides, with their molecular dipoles generally aligned along the field direction, forming filaments that bridge the two sides of the gap. When the applied electric field increases, the allowed gap size where the filaments still exist also increases. However, no formation of the filaments was ever observed in their simulations for gaps $>10\,\text{nm}$.

It has been proven by a molecular dynamics simulation [5.68] that the bridges consisting of polar molecule filaments across the coated dielectric particles are responsible for the polar molecules' dominant giant ER effect. The formation of polar molecule filaments may be attributed to the confinement effect exerted by the oil chains. Because of the hydrophobic nature of the oil chains, there are repulsive interactions between oil chains' methyl groups and atoms in the urea molecules that can act as either a donor or an acceptor when forming hydrogen bonds. Such repulsive interactions facilitate the formation of a single urea molecular file, consisting of multiple urea filaments confined to the "axes" of the hydrophobic pore space that is forced open by the urea dipolar file. The formation of the filament structure comes from the fact that urea is not wet with silicone oil.

Existence of silicone oil in the gap between particles is the premise of polar molecule filaments. This requires that the dielectric particle cores are wet with the oil. The fact that the silicone oil wets the GER particle [5.9] is attributed to the presence of the oxalate groups in the core nanoparticles, and the non-uniformity of the urea coating. The hydrophilicity of silicone oil with particle cores knits networks of silicone oil molecules in the gap between the two particles. The hydrophobicity of the oil with urea facilitates the formation of filaments of urea molecules that squeeze in oil networks. The authors of Ref. [5.68] conclude that it is the urea filament that results in GER effect. This would be instructive to the giant ER material design: the core dielectric material must be wet to the base liquid, but the non-uniformly coated material should, at the same time, be non-wet to the liquid.

5.4 Sedimentation of Particles

5.4.1 *Importance of sedimentation study*

The particle sedimentation problem in magnetorheological (MR) fluids most attracts the attention of those in the field, including material scientists and manufacturers, application engineers and, of course, final users. In vehicle MR damper applications, it is generally required that the MR fluids would be homogenized within five cycles of a piston. In building MR dampers, the MR fluids may sit in the damper chambers for years, even decades of years before they exhibit their potential ability of reducing the building vibration. Long term stability is a basic and, in many applications, a critical characteristic for an MR fluid. Many MR fluids are made of particles (usually spheres) of iron oxides, and their density is around $5\,\mathrm{g/cm^3}$. The particle size of MR fluids is on the order of magnitude of micrometers, and the bared particles do not show obvious Brownian motion as their nanometer counterparts do in magnetic fluids. The surface modification cannot do too much in reducing the particle density. However, material scientists try every effort to attach various chemicals to particle surfaces. The major function of such attachments on the particle surfaces of MR fluids is to increase the steric effect which prevents the particles from forming larger clusters. We will try to understand the science behind this in the next paragraphs.

5.4.2 *Derivation of Stokes' formula* [5.69]

At a low and constant velocity, the drag force F_d is proportional to the radius R of a falling sphere, the velocity of main strain u, and the dynamic viscosity η of the medium, while the proportionality constant under the condition of small Reynolds numbers is 6π.

The Reynolds number is defined as $Rn = \rho ul/\eta$, where ρ and η (in Pa · s) are the density and dynamic viscosity of the base liquid, respectively; u is the velocity of the main stream, and l is one linear dimension determining the geometrical properties of a body with a given shape. In the case of flow with small Rn and for steady flow of an incompressible fluid, the authors [5.69] start with the simplified Navier–Stokes equation

$$(\boldsymbol{v} \cdot \mathbf{grad})\boldsymbol{v} = -(l/\rho)\mathbf{grad}\, p + (\eta/\rho)\Delta\boldsymbol{v}.$$

The term $(\boldsymbol{v} \cdot \mathbf{grad})\boldsymbol{v}$ is on the order of magnitude of u^2/l. The quantity $(\eta/\rho)\Delta\boldsymbol{v}$ is on the order of magnitude of $\eta u/\rho l^2$. The ratio of the two is just the Reynolds number. Hence the term $(\boldsymbol{v} \cdot \mathbf{grad})\boldsymbol{v}$ may be neglected if the Reynolds number is small, and the equation of motion reduces to a linear equation

$$\eta\Delta\boldsymbol{v} - \mathbf{grad}\, p = 0. \tag{5.152}$$

Together with the equation of continuity

$$\mathbf{div}\, \boldsymbol{v} = 0, \tag{5.153}$$

it completely determines the motion. It is useful to note also the equation

$$\Delta\mathbf{curl}\, \boldsymbol{v} = 0, \tag{5.154}$$

which is obtained by taking the curl of Eq. (5.152) and noting the vector formula $\mathbf{curl}\, \mathbf{grad}\, p = 0$.

G. G. Stokes considered, in 1851, a rectilinear and uniform motion of a sphere in a viscous fluid with velocity $\boldsymbol{u}$ at infinity.

Since $\operatorname{div}(\mathbf{curl}\,\boldsymbol{A}) = 0$ satisfies for any vector $\boldsymbol{A}$, and we have here $\operatorname{div}(\boldsymbol{v} - \boldsymbol{u}) = \operatorname{div}\boldsymbol{v} = 0$, one must have

$$\boldsymbol{v} - \boldsymbol{u} = \mathbf{curl}\,\boldsymbol{A},$$

with curl $\boldsymbol{A} = 0$ at infinity. The vector $\boldsymbol{A}$ must be axial, in order for its curl to be polar, like the velocity. In flow past a sphere, a completely symmetrical body, there is no preferred direction other than that of $\boldsymbol{u}$. This parameter $\boldsymbol{u}$ must appear linearly in $\boldsymbol{A}$, because the equation of motion and its boundary conditions are linear. The general form of a vector function $\boldsymbol{A}(\boldsymbol{r})$ satisfying all these requirements is $\boldsymbol{A} = f'(r)\boldsymbol{n} \times \boldsymbol{u}$, where $\boldsymbol{n}$ is a unit vector parallel to the position vector $\boldsymbol{r}$ (the origin being taken at the center of the sphere), and $f'(r)$ is a scalar function of r. The product $f'(r)\boldsymbol{n}$ can be represented as the gradient of another function $f(r)$. Using vector formula $\mathbf{curl}\,(f\boldsymbol{u}) = \mathbf{grad}\,f \times \boldsymbol{u} + f\,\mathbf{curl}\,\boldsymbol{u}$, we shall thus look for the velocity in the form

$$\boldsymbol{v} = \boldsymbol{u} + \mathbf{curl}(\mathbf{grad}f \times \boldsymbol{u}) = \boldsymbol{u} + \mathbf{curl}\,\mathbf{curl}(f\boldsymbol{u}); \tag{5.155}$$

the last expression is obtained by noting that $\boldsymbol{u}$ is constant.

To determine the function f, we use Eqs. (5.155), (5.154) and a vector formula: $\mathbf{curl}\,\mathbf{curl}\,\boldsymbol{v} = \mathbf{grad}\operatorname{div}\boldsymbol{v} - \Delta\boldsymbol{v}$. Since

$$\begin{aligned}\mathbf{curl}\,\boldsymbol{v} &= \mathbf{curl}\,\mathbf{curl}\,\mathbf{curl}(f\boldsymbol{u}) = (\mathbf{grad}\operatorname{div} - \Delta)\mathbf{curl}(f\boldsymbol{u})\\ &= -\Delta\,\mathbf{curl}(f\boldsymbol{u})\end{aligned}$$

(as $\operatorname{div}(\mathbf{curl}\,f\boldsymbol{u}) = 0$), Eq. (5.154) takes the form $\Delta^2\,\mathbf{curl}(f\boldsymbol{u}) = \Delta^2$ $(\mathbf{grad}f \times \boldsymbol{u}) = (\Delta^2\,\mathbf{grad}\,f) \times \boldsymbol{u} = 0$. It follows from this that

$$\Delta^2\,\mathbf{grad}\,f = 0. \tag{5.156}$$

A first integration of Eq. (5.156) gives

$$\Delta^2 f = \text{constant}.$$

The constant must be zero, since the velocity difference $v - u$ must vanish at infinity, and so must its derivatives. The expression $\Delta^2 f$ contains fourth derivatives of f, whilst the velocity is given in terms of the second derivatives of f. Thus we have

$$\Delta^2 f \equiv \frac{1}{r^2}\frac{\mathrm{d}}{\mathrm{d}r}\left(r^2\frac{\mathrm{d}}{\mathrm{d}r}\right)\Delta f = 0.$$

Hence

$$\Delta f = 2a/r + c.$$

The constant c must be zero if the velocity difference $v - u$ vanishes at infinity. From $\Delta f = 2a/r$ one obtains

$$f = ar + b/r. \tag{5.157}$$

The additive constant is omitted, since it is immaterial (the velocity being given by derivatives of f).

Substituting Eq. (5.157) in Eq. (5.155), one finds

$$\boldsymbol{v} = \boldsymbol{u} - a\frac{\boldsymbol{u} + \boldsymbol{n}(\boldsymbol{u}\cdot\boldsymbol{n})}{r} + b\frac{3\boldsymbol{n}(\boldsymbol{u}\cdot\boldsymbol{n}) - \boldsymbol{u}}{r^3}. \tag{5.158}$$

The constants a and b have to be determined from the boundary conditions: at the surface of the sphere ($r = R$), $\boldsymbol{v} = 0$ with non-slipping hypothesis considering molecular attraction between the solid ball and the liquid, i.e.

$$-\boldsymbol{u}\left(\frac{a}{R} + \frac{b}{R^3} - 1\right) + \boldsymbol{n}(\boldsymbol{u}\cdot\boldsymbol{n})\left(-\frac{a}{R} + \frac{3b}{R^3}\right) = 0.$$

Since this equation must hold for all $\boldsymbol{n}$, each of the coefficients of $\boldsymbol{u}$ and $\boldsymbol{n}(\boldsymbol{u}\cdot\boldsymbol{n})$ must vanish. Hence $a = (3/4)R$, $b = (1/4)R^3$. Thus one has

$$f = \frac{3}{4}Rr + \frac{1}{4}R^3/r, \tag{5.159}$$

$$\boldsymbol{v} = -\frac{3}{4}R\frac{\boldsymbol{u} + \boldsymbol{n}(\boldsymbol{u}\cdot\boldsymbol{n})}{r} - \frac{1}{4}R^3\frac{\boldsymbol{u} - 3\boldsymbol{n}(\boldsymbol{u}\cdot\boldsymbol{n})}{r^3} + \boldsymbol{u}, \tag{5.160}$$

or, in spherical polar components with the axis parallel to $\boldsymbol{u}$,

$$\left.\begin{aligned} v_r &= u\cos\theta\left[1 - \frac{3R}{2r} + \frac{R^3}{2r^3}\right], \\ v_\theta &= -u\sin\theta\left[1 - \frac{3R}{4r} - \frac{R^3}{4r^3}\right]. \end{aligned}\right\} \tag{5.161}$$

This gives the velocity distribution about the moving sphere. To determine the pressure, one may substitute Eq. (5.155) in Eq. (5.152):

$$\begin{aligned}\mathbf{grad}\, p &= \eta\Delta \boldsymbol{v} = \eta\Delta\, \mathbf{curl}\, \mathbf{curl}(f\boldsymbol{u}) \\ &= \eta\Delta(\mathbf{grad}\, \mathrm{div}(f\boldsymbol{u}) - \boldsymbol{u}\Delta f).\end{aligned}$$

But $\Delta^2 f = 0$, and so

$$\mathbf{grad}\, p = \mathbf{grad}[\eta\Delta \mathrm{div}(f\boldsymbol{u})] = \mathbf{grad}(\eta\boldsymbol{u}\cdot\mathbf{grad}\,\Delta f).$$

Hence

$$p = \eta\boldsymbol{u}\cdot\mathbf{grad}\,\Delta f + p_0, \tag{5.162}$$

where p_0 is the fluid pressure at infinity. Substitution for f leads to the final expression

$$\begin{aligned}f &= \frac{3}{4}Rr + \frac{1}{4}R^3/r, \\ p &= p_0 - \frac{3}{2}\eta\frac{\boldsymbol{u}\cdot\boldsymbol{n}}{r^2}R.\end{aligned} \tag{5.163}$$

Using the above formulae, the authors calculate the drag force $\boldsymbol{F}$ exerted on the sphere as it moves through the fluid. To do so, they take spherical polar coordinates with the axis parallel to $\boldsymbol{u}$; by symmetry, all quantities are functions only of r and of the polar angle θ. The force $\boldsymbol{F}$ is evidently parallel to the velocity $\boldsymbol{u}$. The magnitude of this force can be determined from the force acting on unit surface area

$$P_i = -\sigma_{ik}n_k = pn_i - \sigma'_{ik}n_k. \tag{5.164}$$

The first term is the ordinary pressure of the fluid, while the second is the force of friction, due to the viscosity, acting on the surface. Note that $\boldsymbol{n}$ in Eq. (5.164) is a unit vector along the outward normal to the fluid, i.e. along the inward normal to the solid surface. Taking from this formula the components, normal and tangential to the surface, of the force on an element of the surface of the sphere, and projecting these components on the direction of $\boldsymbol{u}$, it is found that

$$F = \oint (-p\cos\theta + \sigma'_{rr}\cos\theta - \sigma'_{r\theta}\sin\theta)\mathrm{d}f, \tag{5.165}$$

where the integration is taken over the whole surface of the sphere.

Substituting the expressions Eq. (5.161) in the formulae of the rr- and $r\theta$-components of the stress tensor

$$\sigma'_{rr} = 2\eta\frac{\partial v_r}{\partial r}, \quad \sigma'_{r\theta} = \eta\left(\frac{1}{r}\frac{\partial v_r}{\partial \theta} + \frac{\partial v_\theta}{\partial r} - \frac{v_\theta}{r}\right),$$

one finds that at the surface of the sphere

$$\sigma'_{rr} = 0, \quad \sigma'_{r\theta} = -(3\eta/2R)u\sin\theta,$$

while the pressure Eq. (5.163) is $p = p_0 - (3\eta/2R)u\cos\theta$. Hence the integral Eq. (5.165) reduces to $F = (3\eta u/2R)\oint \mathrm{d}f$. In this way the authors finally arrive at Stokes' formula for the drag on a sphere moving slowly in a fluid:

$$F = 6\pi\eta R u. \tag{5.166}$$

5.4.3 *Derivation of Stokes' law* [5.70]

The gravitational force F_g acting on a sphere of radius R is the difference between the downward force due to sphere mass and, according to Archimedes' principle, the upward buoyancy force due to the mass of the liquid that is expeled by the submerged sphere:

$$F_g = (4/3)\pi R^3(\rho - \rho_0)g, \tag{5.167}$$

where ρ_0 is the density of the liquid medium, and g is the gravitational acceleration.

Under conditions of steady state

$$F_d = F_g, \tag{5.168}$$

one will obtain Stokes' law:

$$u = \frac{2R^2(\rho - \rho_0)g}{9\eta}. \tag{5.169}$$

If the particle is not a sphere, then R is the radius of the equivalent sphere, and the size distribution is given in terms of equivalent spherical radii.

One may find sedimentation time t from particle sedimentation velocity u due to Eq. (5.169) if the height h of the clear liquid is known ($t = h/u$). On the other hand, one may estimate the average particle radius R from Eq. (5.169) if the sedimentation time t and clear liquid height h are known. As the particle size distribution would be a good physical quantity for particle dispersity of an MR fluid, the average radius R can be compared with known radius of bared MR particle to judge the characteristics of particle dispersions. Particle dispersion can be better examined with an electromicroscope and it is crucial to provide an evidence of the effect of particle surface modification on repulsing from each other.

The robust performance, of the surface modification, namely the "tightness" of the coating, is important for industrial applications. In many applications, MR fluids are in flows of high shear rate which would strongly affect the coating chemicals. Only those materials whose coating chemicals are strongly attached to particle surface can continue to be used in applications.

References

[5.1] US DOE, Assessment on Electrorheological Fluids, May (1992).

[5.2] W. M. Winslow, *J. Appl. Phys.* **20** (1949), 1137.

[5.3] T. J. Chen, R. N. Zitter and R. Tao, *Phys. Rev. Lett.* **68** (1992), 2555.

[5.4] T. C. Halsey and W. Toor, *Phys. Rev. Lett.* **65** (1990), 2820.

[5.5] T. C. Halsey, *Science* **258** (1992), 761.

[5.6] R. Tao and J. M. Sun, *Phys. Rev. Lett.* **67** (1991), 398.

[5.7] R. Tao and Q. Jiang, *Phys. Rev. Lett.* **73** (1994), 205.

[5.8] R. Tao, Lecture in Fudan University (1993).

[5.9] W. J. Wen, X. X. Huang, S. H. Yang, K. Q. Lu and P. Sheng, *Nature Materials* **2** (2003), 1.

[5.10] H. R. Ma, W. J. Wen, W. Y. Tam and Ping Sheng, *Phys. Rev. Lett.* **77** (1996), 2499.

[5.11] H. R. Ma, W. J. Wen, W. Y. Tam and Ping Sheng, *Adv. Phys.* **52** (2003), 343.

[5.12] Ping Sheng and Weijia Wen, *Solid State Commun.* **150** (2010), 1023.

[5.13] Kunquan Lu, Rong Shen *et al.*, *Chin. Phys.* **15** (2006), 2476.

[5.14] Rong Shen, Xuezhao Wang *et al.*, *Adv. Mat.* **21** (2009), 4631.

[5.15] W. Wen, X. Huang and P. Sheng, *Appl. Phys. Lett.* **85** (2004), 299.

[5.16] R. Shen, X. Z. Wang, W. J. Wen and K. Q. Lu, *Intern. J. Mod. Phys. B* **19** (2005), 1104.

[5.17] B. X. Wang and X. P. Zhao, *Adv. Funct. Mat.* **15** (2005), 1815.

[5.18] H. T. Xue, Z. N. Fang, Y. Yang, J. P. Huang and L. W. Zhou, *Chem. Phys. Lett.* **432** (2006), 326.

[5.19] J. T. Woestman and A. Widom, *Phys. Rev. E* **48** (1993), 1063.

[5.20] Z. N. Fang, H. T. Xue, W. Bao, Y. Yang, L. W. Zhou and J. P. Huang, *Chem. Phys. Lett.* **441** (2007), 314.

[5.21] Manjun He, Weixiao Chen, Xixia Dong, *Polymer Physics*, Fudan University Press, Shanghai, (1990). 371–403 (in Chinese).

[5.22] Hanru Li, *Introduction to Dielectric Physics*, Chengdu University of Science and Technology Press, (1990) (in Chinese).

[5.23] C. J. F. Boettcher, *Theory of Electric Polarization*, Vol. **1**, Elsevier Scientific Publishing Co., (1973).

[5.24] X. Donald, L. Klass and Thomas W. Martinek, *J. Appl. Phys.* **38** (1967), 67.

[5.25] Yu Tian, Doctoral Thesis, Tsinghua University, (2000). 26–28 (in Chinese).

[5.26] C. G. Wei, *Electrorheological Technology — Theory, Material and Engineering Applications*, Beijing University of Science and Technology Press, Beijing, (2000), 106.

[5.27] Wenjian Weng, Doctoral Thesis of Zhejiang University, (1993) (in Chinese); see also R. Tao, Q. Jiang and H. K. Sim, *Phys. Rev. E* **52** (1995), 2727.

[5.28] P. A. Arp and S. G. Mason, *Coll. Polym. Sci.* **255** (1977), 566.

[5.29] R. Friedberg, *Phys. Rev. B* **46** (1992), 6582.

[5.30] M. H. Davis, Quart. *J. Mech. Appl. Mech.* **17** (1964), 499.

[5.31] B. J. Cox, N. Thamwattana and J. M. Hill, Electric field-induced force between two identical uncharged spheres, *App. Phys. Lett.* **88** (2006), 152903.

[5.32] Mingchun Jiao, Gang Sun, Qiang Wang and Kunquan Lu, Electrorheological effect induced by liquid to superpolarized state transitions, 12th International Conference of ERMR Suspensions, Philadelphia, (2010).

[5.33] Zhiyong Wang, Zheng Peng, Kunquan Lu and Weijia Wen, *App. Phys. Lett.* **82** (2003), 1796.

[5.34] J. N. Foulc, P. Atten and N. Felici, *J. Electrostat.* **33** (1994), 103.

[5.35] Y. L. Siu, J. T. K. Wan and K. W. Yu, *Phys. Rev. E* **64** (2001), 051506.

[5.36] J. P. Huang and K. W. Yu, *Phys. Rev. E* **70** (2004), 061401.

[5.37] M. Brunner, J. Dobnikar and H. H. von Gruenberg, *Phys. Rev. Lett.* **92** (2004), 078301.

[5.38] P. Sheng, *Introduction to Wave Scattering, Localization and Mesoscopic Phenomena*, 2nd ed., Springer, Berlin, (2006), 89.

[5.39] B. Khusid and A. Acrivos, *Phys. Rev. E* **52** (1995), 1669.

[5.40] R. A. Anderson, in R. Tao (ed.) Electrorheological Fluids, Mechanisms, Properties, Structure, Technology, and Applications, *Proc. 5th Int. Conf. Electrorheo. Fluids*, Carbondale, Illinois, USA, World Scientific, Singapore, (1992), 81.

[5.41] L. C. Davis, *Appl. Phys. Lett.* **60** (1992), 319.
[5.42] M. Parthasarathy and D. Klingenberg, *Mat. Sci. and Engine.* R**17** (1996), 57 (and references cited there).
[5.43] M. J. Chrzon *et al.*, in R. Tao (ed.) *Proc. 3rd Int. Conf. Electrorheo. Fluids*, World Scientific, (1992), 175.
[5.44] L. C. Davis and J. M. Ginder, in K. O. Havelka and F. E. Filisko (eds.) *Progress of Electrorheology*, (Plenum Press, NY, 1995) 107.
[5.45] D. J. Klingenburg, C. F. Zukoski and J. C. Hill, *J. App. Phys.* **73** (1993), 4644.
[5.46] D. J. Klingenberg and C. F. Zukoski, *Langmuir* **6** (1990), 15.
[5.47] H. Block and J. P. Kelly, *J. Phys. D. Appl. Phys.* **21** (1998), 1661.
[5.48] K. D. Weiss, J. D. Carlson and J. P. Coulter, *J. Intell. Mat. Sys. Struct.* **4** (1993), 13.
[5.49] Y. D. Kim and D. J. Klingenberg, *Polym. Prep.* **35** (1994), 389.
[5.50] C. F. Zukoski, *Ann. Rev. Mater. Sci.* **23** (1993), 57.
[5.51] C. W. Wu and H. Conrad, *J. Phys. D* **29** (1996), 3147.
[5.52] N. Felici, J. N. Foulc and P. Atten, in R. Tao and G. D. Roy (eds.) Electrorheological Fluids, Mechanisms, Properties, Technology, and Applications, Proc. of the Fourth Int. Conf. on Electrorheological Fluids, World Scientific, Singapore, (1994), 139.
[5.53] H. Conrad, *J. Phys. D:* **30** (1997) 2634.
[5.54] H. Jie, M. Shen, J. Xu, W. X. Chen, Y. Jin, W. N. Peng, X. B. Fu and L. W. Zhou, *Appl. Phys. Lett.* **85** (2004), 2646.
[5.55] P. Gonon, J.-N. Foulc, P. Atten and C. Boissy, *J. Appl. Phys.* **86** (1999), 7160.
[5.56] J. N. Israelachvili, Intermolecule and Surface Force, (London: Academic), (2000).
[5.57] C. J. F. Boettcher, *Theory of Electric Polarization*, Vol. **1**, Elsevier Scientific Publishing Co., (1973), 163.
[5.58] F. Frenkel, *Phys. Rev.* **54** (1938), 647.
[5.59] F. Borghese, P. Denti, R. Saija, G. Toscano and O. I. Sindoni, Multiple Electromagnetic Scattering from a Cluster of Spheres. I. Theory, *Aerosol Sci. Tech.* **3** (1984), 227.
[5.60] M. Washizu, Precise calculation of dielectrophoretic force in arbitrary field, *J. Electrost.* **29** (1992), 177.
[5.61] T. Matsuyama, H. Yamamoto and M. Washizu, Potential distribution around a partially charged dielectric particle located near a conducting plane, *J. Electrost.* **36** (1995), 195.
[5.62] Sacha Gómez-Moñivas and José Sáenz, Montserrat Calleja and Ricardo Garcı'a, Field-induced formation of nanometer-sized water bridges, *Phys. Rev. Lett.* **91** (2003), 056101.
[5.63] Tobias Cramer, Francesco Zerbetto and Ricardo GarcÍa, Molecular mechanism of water bridge buildup: Field-induced formation of nanoscale menisci, *Langmuir* **24** (2008), 6116.
[5.64] H. J. Xu, Superpolarization, Soft Matter Physics Lecture paper, Fudan University, (2011).

[5.65] L. Xu, W. J. Tian, X. F. Wu, J. G. Cao, L. W. Zhou, J. P. Huang and G. Q. Gu, *J. Mater. Res.* **23** (2008), 409.

[5.66] P. Tang, X. F. Wu, W. J. Tian, L. W. Zhou and J. P. Huang, Saturated Orientational Polarization of Polar Molecules in Giant Electrorheological Fluids, *J. Phys. Chem. B* **113** (2009), 9092.

[5.67] Cai Shen, Weijia Wen, Shihe Yang and Ping Sheng, Wetting-induced electrorheological effect, *J. Appl. Phys.* **99** (2006), 106104.

[5.68] S. Y. Chen, X. X. Huang *et al.*, *Phys. Rev. Lett.* **105** (2010), 046001.

[5.69] L. D. Landau and E. M. Lifshitz, *Fluid Mechanics* (Course of Theoretical Physics, v. 6), 2nd Eng. Ed., Trans. J. B. Sykes and W. H. Reid, Pergamon Press, (1987), 58–61.

[5.70] I. D. Morrison and S. Ross, *Colloidal Dispersions — Suspensions, Emulsions and Foams*, John Wiley & Sons, New York, (2002), 70.

Chapter 6

Dynamic Electrorheological Effects

6.1 Dynamic Behaviors of ER Fluids

6.1.1 *Dynamic phenomena*

(1) Relaxation time of rotating and stationary spheres

When there exists a speed gradient of shear flow, electrorheological (ER) particles will spin [6.1, 6.2]. Wan *et al.* [6.3] have examined the effect of an oscillatory rotation of a polarized dielectric particle. The rotational motion leads to a redistribution of the polarization charge on the surface of the particle. On the basis of Maxwell–Wagner relaxation time

$$t_{mw} = \varepsilon_0 \frac{\varepsilon_p + 2\varepsilon_c}{\sigma_p + 2\sigma_c}, \tag{6.1}$$

they derived the relaxation time of spheres that experience oscillatory rotation

$$\tau = \tau_\infty + \frac{\tau_0 - \tau_\infty}{\sqrt{1 + \theta_0^2 k^2 \tau^2}}, \tag{6.2}$$

where k is the oscillation frequency, θ_0 is the oscillation amplitude,

$$\tau_\infty = \epsilon_0 \left(\frac{1 + 2\epsilon_m}{\sigma_1 + 2\sigma_m} \right) \quad \text{and} \quad \tau_0 = \epsilon_0 \left(\frac{\epsilon_1 + 2\epsilon_m}{\sigma_1 + 2\sigma_m} \right). \tag{6.3}$$

They found that $\tau = \tau_0$ for $k\theta_0 = 0$ and $\tau \to \tau_\infty$ for $k\theta_0 \to \infty$. They show that the time-averaged steady-state dipole moment is along the field direction, but its magnitude is reduced by a reduction factor R

that depends on the angular velocity of rotation:

$$R = \frac{\langle p \rangle}{p_0} = \frac{1}{\sqrt{1 + \theta_0^2 k^2 \tau^2}}. \tag{6.4}$$

As a result, the rotational motion of the particle reduces the electrorheological effect.

In many models it is assumed that the polarization is instantaneous. This is certainly reasonable for electronic, atomic and dipolar polarization mechanisms. Employing the Maxwell–Wagner model, the polarization time constant describing migration polarization is $t_{\mathrm{MW}} = 0.2\,\mathrm{ms}$ when $\varepsilon_p = 20$, $\varepsilon_f = 2.5$, $\sigma_p = 10^{-6}$ mho m^{-1}, and $\sigma_f = 10^{-8}$ mho m^{-1}. Under these conditions and for low to moderate field strengths, the assumption that polarization is much faster than structure formation is valid. As the formation rate is predicted to increase with the square of the electric field strength, the assumption is not valid at sufficiently large field strengths. This regime is likely to be further complicated by nonlinear dielectric phenomena, and has not been considered in microscopic models. The influence of finite polarization rates on the early stages of aggregation is discussed in Ref. [6.4].

(2) Dynamic nonlinearity

Huang and Yu [6.5] found that under the application of electric fields, the structure of ER solids can be changed from body-centered tetragonal (bct) lattice (ground state) to other lattices. For a particle in the lattice, they derived its dipole factor by taking the following two effects into account: the local-field effect arising from all the other particles and the multipolar interaction between two touching particles. It was found that the electrorotation spectrum of ER solids can be affected significantly by the structure transformation. It was proposed that the structure of ER fluids can be monitored by detecting the electrorotation spectrum.

Based on the Ewald–Kornfeld formulation, Huang [6.6] investigated the nonlinear responses of an ER solid with nonlinear spherical particles embedded in a linear host fluid under a sinusoidal AC electric field. He found that the dielectric response of the composite will generally consist of AC fields at frequencies of higher-order harmonics, and the fundamental and third-order harmonic AC responses are sensitive to the degree of anisotropy within the ER solid. He proposed that one could perform a real-time monitoring of structure transformations by measuring the nonlinear AC responses of ER solids.

Martin *et al.* [6.7] studied light-scattering of ER fluids in the quiescent state, steady shear, and oscillatory shear. Their studies of an ER fluid in steady shear show that the structure reaches a steady state wherein droplets are rotated in the direction of fluid vorticity at some angle relative to the applied electric field. This angle increases with the cube root of the shear rate, in agreement with their model for the shear thinning viscosity. In this model the equation of motion of elliptical droplets is found by balancing the hydrodynamic and electrostatic torques.

Their studies of ER fluids in oscillatory shear demonstrate that the chain dynamics, and thus the electrorheology, is nonlinear. They have described a simple kinetic chain model of the dynamics that describes the approach of a chain to its maximum stable size by a kinetic equation. Much of their experimental data can be described by taking the instantaneous approach to stability; however, at low fields strong nonlinearities suggest that the approach to stability is slow compared to the shear period. This model is then used to compute the nonlinear rheology of an ER fluid and it is concluded that light scattering is an indirect probe of stress. At high voltages, they observed a phase bifurcation in the scattering pattern that they attribute to the onset of a shear slip zone, and they concluded that the shear slip instability may be the cause of phase bifurcation.

6.1.2 *Lorentz local field*

ER particles would rotate under a shear flow due to a speed gradient of liquid. Do such rotations affect the *Lorentz local field* (LLF), the local field in the vicinity of particles?

(1) Observation of LLF reductions

The systematic experimental study of Tao and Lan [6.8] shows the reduction of the interaction force between a rotating polarized sphere and stationary one in an electric field. The experiments show that the interaction between two dielectric spheres reduces by 22% (from 1.8 to 1.4 dynes) when one of the spheres rotates at 2750 rpm; that between one rotating metallic sphere and a dielectric sphere reduces by 30% (from 4.3 to 3 dynes) when the metallic sphere rotates at the above speed; that between two metallic spheres reduces by 26% (from 75.6 to 55.8 dynes) when one of the spheres rotates at the above speed. The authors proposed that the interaction reduction is due to the local field reduction in the interparticle gap. Their experiment actually implies the reduction of the LLF.

Tan and Tian *et al.* [6.9] directly detected the LLF between a rotating sphere and a stationary one (Fig. 6.1(a)) according to the electro-optical Pockels effect and the computerized tomography principle. The experimental results give a reduction of LLF, E_L,

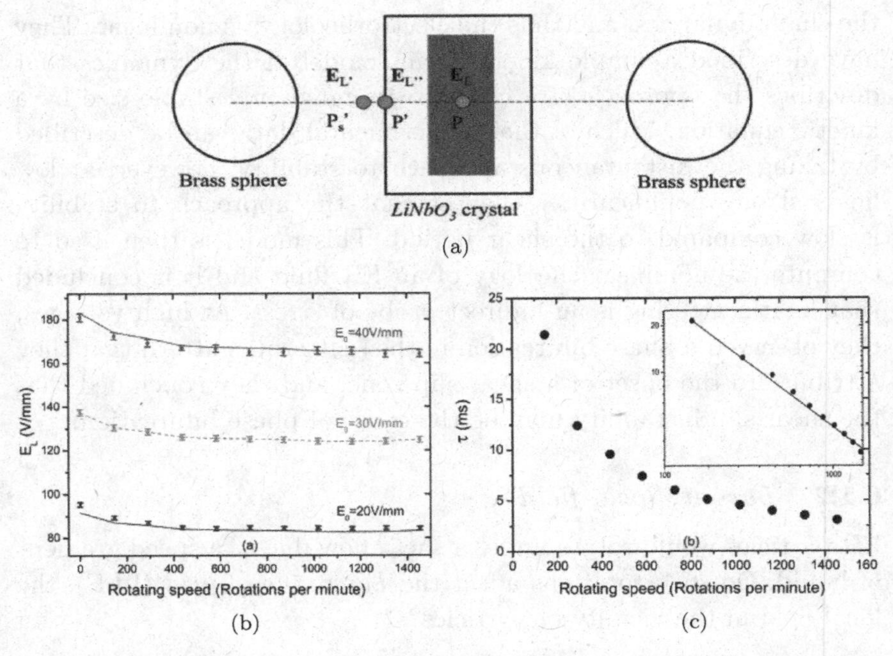

Fig. 6.1 (a) Arrangement of two metallic spheres and $LiNbO_3$ (LN) crystal. (Not to scale.) (b) The experimental data (symbols) and the theoretical fitting (lines) of the LLF E_L at the center P. (c) Relaxation times τ vs. rotating speed, also in the inset in a logarithmic scale [6.9].

at the center P between the two metallic spheres with increasing rotating speed of one sphere for different applied DC electric fields E_0 (Fig. 6.1(b)). The experimental result of relaxation times τ is extracted when $E_0 = 30\,\text{V/mm}$ (Fig. 6.1(c)). The other two solid lines for $E_0 = 20\,\text{V/mm}$ and $E_0 = 40\,\text{V/mm}$ are corresponding theoretical results by using the obtained relaxation times. The logarithmic scale of τ versus rotating speeds (inset of Fig. 6.1(c)) approximately displays a power-law relation.

Rotation-driven reduction of a LLF has significant consequences for electric-field-responsive soft-matter systems. For instance, it brings corrections to the reconstruction or revisiting of the fundamental theories for *electrorotation* of biological cells due to a rotating electric field (which is related to nonlinear responses of an electric torque to an LLF in biophysics), to the *Quincke* rotation due to a DC electric field (which is related to an LLF-threshold phenomena in colloidal physics), and to *nonlinear harmonics* in suspensions in a shear flow (which is related to nonlinear responses of harmonics to an LLF in nonlinear physics).

(2) Theoretical calculation of LLF using the multiple image method of dipoles

For the system displayed in Fig. 6.1(a), the theoretical calculation [6.9] for the LLF, E_L, at point P_s' is derived by considering the continuity condition of the normal electric displacement at the point P$''$ at the left (or right) surface of the $LiNbO_3$ (LN) crystal

$$E_{L'}\varepsilon_1 = E_{L''}\varepsilon_2, \tag{6.5}$$

where $E_{L''}$ denotes the local electric field at the position P_s' (which is very close to P') in silicone oil and $E_{L''}$ the corresponding local field at P' in the LN crystal. For an approximation $E_L = E_{L''}$, and $E_{L'}$ is expressed as

$$\begin{aligned} E_{L'} = E_0 &+ (E_\alpha + E_{\alpha 1} + E_{\alpha'} + E_{\alpha 2} - E_{\alpha 0}) \\ &+ (E_\beta + E_{\beta 1} + E_{\beta'} + E_{\beta 2} - E_{\beta 0}). \end{aligned} \tag{6.6}$$

By using the multiple image method of dipoles [6.10], the multipolar interactions between each sphere and its corresponding images to the nearest electrode and LN crystal can be captured, and the total dipole moments in each sphere can be expressed analytically. This would yield the electric fields E_α and $E_{\alpha 1}$ at the position P_s', respectively. In the mean time, the interaction between the sphere α and the LN crystal can also be seen as that between the sphere α and its image sphere α_2. Similarly, the multiple image method of dipoles helps in determining the total dipole moment in α or α_2, which yields $E_{\alpha'}$ and $E_{\alpha 2}$ at P_s', respectively. In Eq. (6.6), $E_{\alpha 0}$ denotes the electric field at P_s' produced by the dipole moment of the single static sphere α in the presence of E_0. On the same footing, E_β, $E_{\beta 1}$, $E_{\beta'}$, $E_{\beta 2}$, and $E_{\beta 0}$ in Eq. (6.6) arise from the rotating metallic sphere β on the right.

The multiple image method of dipoles for two unequal spheres [6.10] can be generally performed to solve each of the following four pairs of electric fields: E_α and $E_{\alpha 1}$, $E_{\alpha'}$ and $E_{\alpha 2}$, E_β and $E_{\beta 1}$, and $E_{\beta'}$ and $E_{\beta 2}$, respectively. Therefore, in the following, let us consider two general unequal metallic spheres, A and B, with radius a and b, respectively. Then we apply a uniform electric field to the system. This induces a dipole moment into each of the spheres. The dipole moments of spheres A and B are given by $p_{\mathrm{A}1}$ and $p_{\mathrm{B}1}$, respectively. Next, we include the image effects. The dipole $p_{\mathrm{A}1}$ induces an image dipole $p_{\mathrm{A}2}$ into sphere B, while $p_{\mathrm{A}2}$ induces another image dipole in sphere A. As a result, multiple images are formed. Similarly, $p_{\mathrm{B}1}$ induces an image $p_{\mathrm{B}2}$ into sphere A. The formation of multiple images leads to an infinite series of image dipoles. The sum of the dipole moments inside each sphere is first calculated according to Eqs. 38–41 of Ref. [6.10] and then used to produce the local field at P_s'.

To better understand the underlying physics, a review is briefly given for the formulae derivation from the multiple image method of dipoles for two unequal spheres A and B. The authors of Ref. [6.9] use r_0 to denote the center-to-center distance between the two spheres A and B. The original dipole moments parallel with the line joining the centers of the two spheres (longitudinal case) are $p_{1,\mathrm{A}}$ (Eq. (6.17)) for sphere A and $p_{1,\mathrm{B}}$ (Eq. (6.17)) for sphere B. Then, the local field at P_s' contributed by the sum of the longitudinal dipole moments in spheres A and B [6.10] can be, respectively, given by

$$E_{\rm A}^{\rm L}(R)=\frac{1}{4\pi\varepsilon_1}\left(\sum_{n=1}^{\infty}\frac{2p_{n,{\rm A}}^{\rm L}}{(R/2-x_{n,{\rm A}})^3}+\sum_{n=1}^{\infty}\frac{q_{n,{\rm A}}^{\rm L}}{(R/2-x_{n,{\rm A}})^2}\right.$$
$$\left.-\sum_{n=1}^{\infty}\frac{q_{c{\rm A}}^{\rm L}\lambda_{n,{\rm A}}^{o}+q_{c{\rm B}}^{\rm L}\lambda_{n,{\rm A}}^{e}}{(R/2-x_{n,{\rm A}})^2}\right), \tag{6.7}$$

$$E_{\rm B}^{\rm L}(R')=\frac{1}{4\pi\varepsilon_1}\left(\sum_{n=1}^{\infty}\frac{2p_{n,{\rm B}}^{\rm L}}{(R'/2-x_{n,{\rm B}})^3}+\sum_{n=1}^{\infty}\frac{q_{n,{\rm B}}^{\rm L}}{(R'/2-x_{n,{\rm B}})^2}\right.$$
$$\left.-\sum_{n=1}^{\infty}\frac{q_{c{\rm B}}^{\rm L}\lambda_{n,{\rm B}}^{o}+q_{c{\rm A}}^{\rm L}\lambda_{n,{\rm B}}^{e}}{(R'/2-x_{n,{\rm B}})^2}\right). \tag{6.8}$$

Here R (or R') denotes the distance between the center of sphere A (or B) and P'_s. Since an infinite series of image dipole moments is formed in sphere A, the nth-order result $p_{n,{\rm A}}^{\rm L}$ is given in odd-order ($p_{n,{\rm A}}^{o}$) and even-order ($p_{n,{\rm A}}^{e}$) terms as

$$p_{n,{\rm A}}^{o}=p_{1,{\rm A}}\left(\frac{b\sinh 2\vartheta}{a\sinh(n-1)\vartheta+b\sinh(n+1)\vartheta}\right)^3, \tag{6.9}$$

$$p_{n,{\rm A}}^{e}=p_{1,{\rm B}}\left(\frac{a\sinh 2\vartheta}{r_0\sinh n\vartheta}\right)^3. \tag{6.10}$$

Both $p_{1,{\rm A}}$ and $p_{1,{\rm B}}$ are given by Eq. (6.17). Similarly, $x_{n,{\rm A}}$ (the position of the nth-order image dipole moment $p_{n,{\rm A}}^{\rm L}$ and image charge $q_{n,{\rm A}}^{\rm L}$) can be expressed as

$$x_{n,{\rm A}}^{o}=\frac{ar_0\sinh(n-1)\vartheta}{a\sinh(n-1)\vartheta+b\sinh(n+1)\vartheta}, \tag{6.11}$$

$$x_{n,{\rm A}}^{e}=\frac{a^2\sinh n\vartheta+ab\sinh(n-2)\vartheta}{r_0\sinh n\vartheta}. \tag{6.12}$$

And $q_{n,{\rm A}}^{\rm L}$ admits

$$q_{n,{\rm A}}^{o}=\frac{p_{1,{\rm A}}}{a}\frac{br_0\sinh 2\vartheta\sinh(n-1)\vartheta}{(a\sinh(n-1)\vartheta+b\sinh(n+1)\vartheta)^2}, \tag{6.13}$$

$$q_{n,{\rm A}}^{e}=\frac{p_{1,{\rm B}}}{b}\frac{a\sinh 2\vartheta(a\sinh(n-2)\vartheta+b\sinh n\vartheta)}{(r_0\sinh n\vartheta)}. \tag{6.14}$$

In addition, since the two spheres are isolated, two charges q_{cA}^{L} and q_{cB}^{L} [6.10] should be put at the centers of the spheres A and B, respectively, in order to keep charge neutralization. Here, q_{cA}^{L} is given by

$$q_{cA}^{L} = \frac{\sum_{n=1}^{\infty} q_{n,B}^{L} \sum_{n=2}^{\infty} \lambda_{n,A}^{e} - \sum_{n=1}^{\infty} q_{n,A}^{L} \sum_{n=1}^{\infty} \lambda_{n,B}^{o}}{\sum_{n=2}^{\infty} \lambda_{n,A}^{e} \sum_{n=2}^{\infty} \lambda_{n,1B}^{e} - \sum_{n=1}^{\infty} \lambda_{n,A}^{o} \sum_{n=1}^{\infty} \lambda_{n,B}^{o}}. \tag{6.15}$$

Further, the two point charges q_{cA}^{L} and q_{cB}^{L} also induce a series of image charges, whose normalized magnitudes in sphere A, $\lambda_{n,A}^{o}$ and $\lambda_{n,A}^{e}$ can be found in Ref. [6.10] are, respectively, represented as

$$\lambda_{n,A}^{o} = \frac{b \sinh 2\vartheta}{a \sinh(n-1)\vartheta + b \sinh(n+1)\vartheta}, \tag{6.16}$$

$$\lambda_{n,A}^{e} = \frac{a \sinh 2\vartheta}{r_0 \sinh n\vartheta}. \tag{6.17}$$

Here, $\lambda_{n,A}^{o}$ (or $\lambda_{n,iA}^{e}$) was normalized to q_{cA}^{L} (or q_{cB}^{L}).

If the sphere is rotating, the x-directed component of the first-order dipole moment is given by

$$p_1 = \frac{p^{(0)}}{1 + (\omega\tau)^2} \tag{6.18}$$

where $p^{(0)} = 4\pi\varepsilon_2 a^3 E_0$, and ω is the angular velocity. The relaxation time can be determined by fitting experimental data using Eq. (6.1).

In Fig. 6.1(b), the dashed line (corresponding to $E_0 = 30\,\text{V/mm}$) is used to extract the fitting relaxation times τ's, and the solid lines are theoretical calculations based on the extracted relaxation times. The relaxation times shown in Fig. 6.1(c) are $\sim 10^{-3}$ s, while it is known that the relaxation time of brass is on the order of 10^{-17} s. It is concluded that the LLF's reduction is due to the rotation-driven relaxation of the dipole moment, namely it is due to the dipolar molecules adsorbed to the spheres.

6.1.3 *Shear stress under static shear flow and transient electric field*

The ER effect is also affected by the applied field frequency. Three different phenomena have been recognized as contributing to the frequency dependence of the steady shear response. Particle polarization can be frequency-dependent, due either to the frequency dependence of the intrinsic dielectric properties or to Maxwell–Wagner-type dispersion, resulting in frequency-dependent electrostatic forces and shear stresses. Particle rotation at large shear rates can interfere with particle polarization, producing shear rate, as well as frequency-dependent behavior. Finally, transient electric field will produce time-dependent structures and potentially frequency-dependent rheological properties, even if particle polarization is frequency-independent.

See and Doi [6.11] examined the influence of electric field frequency and shear rate on structure dynamics and rheological properties in steady shear flow, assuming that particle polarization is independent of frequency and shear rate. Dynamic simulations with point dipoles were employed, with sinusoidal and unipolar square-wave electric fields. The authors focused on pulsed fields, but observed similar behavior in sinusoidal fields.

The time-averaged shear stress and the mean cluster size are calculated as a function of field frequency at fixed shear rate. The shear stress passed through a minimum at a field frequency $\omega_e = \omega_{\min}$. $\omega_{\min}$ is independent of concentration, but proportional to $\dot{\gamma}(\omega_{\min} \approx 1.6\dot{\gamma})$. This behavior can be understood in terms of phenomena that control the microstructure and their time scales-hydrodynamic forces, which destroy field-induced structures on a time scale $\dot{\gamma}^{-1}$, and electrostatic polarization forces, which control aggregation on the two time scales $t_s \alpha \eta_c / \varepsilon_0 \varepsilon_c \beta^2 E_0^2$, and ω_e^{-1}.

For $\omega_e \ll \dot{\gamma}$, the time scale for shear-induced rupture and reformation ($\dot{\gamma}^{-1}$) is much smaller than the period for which the electric field is on, and thus the mean cluster size simply follows the oscillating field strength. The instantaneous shear stress also follows the field frequency, and the time-averaged stress is independent of ω_e. For $\omega_e \gg \dot{\gamma}$, the mean cluster size no longer changes with frequency; the period for which the field is off is too short to allow clusters to disintegrate. The time-averaged stress is again independent of frequency. At intermediate frequencies, and for the range of $\dot{\gamma}$ investigated, the period for which the field is applied is too short for clusters to develop completely before the field is subsequently removed. Incomplete structure formation results in reduced stress, giving rise to the observed minimum.

What people who study ER fluids want to explain is why some ER materials may have very high static yield stress, but their shear stress decreases very fast with increasing shear rate, while some ER materials may not have very high static yield stress, but at the same high shear rate, their shear stress may not be low, and can even be very high.

Ginder *et al.* [6.12, 6.13] examined the role of the time dependence of particle polarization on the shear stress of ER suspensions in unipolar and bipolar pulsed electric fields. Experiments were performed on three types of suspended materials chosen for their polarization mechanisms. The polarization of amorphous and crystalline aluminosilicates was ascribed to ion transport and that of barium titanate was attributed to conventional dielectric polarization.

Transient shear stresses within the period of the pulsed field were examined; very different responses were observed for the different systems. When the electric field was applied, the shear stress in aluminosilicate systems increased monotonically with time to a plateau stress. When the field was removed, the stress decreased uniformly to the zero field value. When the field was applied to the barium titanate system, the stress increased rapidly to a maximum, then decreased slowly. When the field was removed, the stress again increased rapidly before decaying to the zero field value.

The authors employed the Maxwell–Wagner model (Section 5.2.2) to describe these transients, considering the polarization of an isolated sphere in pulsed unipolar and bipolar electric fields. For a step increase in electric field strength form 0 to E_0 at $t = 0$, the time-dependent dipole magnitude is

$$p(t) = \begin{cases} 0 & t < 0 \\ p_o^c\left[1 + \left(\dfrac{\beta_d}{\beta_c} - 1\right)e^{-t/t_{\mathrm{MW}}}\right] & t \geq 0 \end{cases}, \tag{6.19}$$

where $p_o^c = (\pi/2)\varepsilon_o\varepsilon_c\sigma^3\beta_c E_o$, and the parameters β_c, and t_{mw} are defined in Eqs. (5.39) and (5.40), respectively. When the field is removed from a steady-state value of E_0 at $t = 0$, the dipole strength is

$$p(t) = \begin{cases} p_o^c & t < 0 \\ p_o^c\left(1 - \dfrac{\beta_d}{\beta_c}\right)e^{-t/t_{\mathrm{MW}}} & t \geq 0. \end{cases} \tag{6.20}$$

And when the field is reversed from a steady-state magnitude of $-E_0$ to $+E_0$ at $t = 0$, the dipole magnitude in that direction is

$$p(t) = \begin{cases} -p_o^c & t < 0 \\ p_o^c\left[1 - 2\left(1 - \dfrac{\beta_d}{\beta_c}\right)e^{-t/t_{\mathrm{MW}}}\right] & t \geq 0. \end{cases} \tag{6.21}$$

Ginder *et al.* attributed the stress transients entirely to the time-dependent polarization, representing the shear stress as

$$\tau_{xz} = c_1 + c_2 p^2(t), \tag{6.22}$$

where C_1 and C_2 are constants for each system and shear rate. This model reproduced the experimental stress transients extremely well, suggesting that the Maxwell–Wagner model was an accurate description of their systems over the range of parameters investigated. The character of the response is therefore determined by the relative magnitudes of β_c and β_d, which describe the polarization due to conductivity and dielectric constant differences, respectively.

For the aluminosilicates, $\beta_c/\beta_d > 1$, and polarization is dominated by conductivity differences. When the field strength was increased from 0 to E_0, the polarization, and thus the shear stress, increased monotonically over the period t_{MW}. For barium titanate, $\beta_c/\beta_d < 1$, so polarization was controlled initially by the large dielectric constant differences resulting in a rapid rise in the stress. However, over the time scale t_{MW}, polarization by conductivity begins to build, screening the field within the particle, and reducing the shear stress. Over long times, dielectric particle polarization will always be screened, regardless of the value of β_d.

This description also explains qualitatively why suspensions of particles with large ε_p show little ER activity in DC fields, but show an increase in the stress with increasing electric field frequency. At low frequencies, the potential for large polarization in the large ε_p material is screened by the polarization arising from conductivity differences; at large frequencies, the mobile charges cannot respond and the large permittivity differences dominate polarization and stress transfer.

6.2 Lamellar Structure

6.2.1 *Lamellar structure stability under shearing*

Tang *et al.* [6.14] studied the lamellar structure of ER fluid of Na-PMA particle (w 20%) and silicon oil when both electric and shear fields are applied. Two parallel and concentric transparent disks were used as electrodes, the top one of which could rotate and the distance between the two was 1.25 mm. The process of particle structure formation can be recorded through the bottom electrode with a video camera. Figures 6.2(a)–(f) shows the temporal evolution of shear-induced lamellar structure at 1.0, 2.0, 3.0, 4.0, 5.0 and 9.0 s after an electric field of 1.6 kV/mm was applied. When $t = 1$ s, polarized particles aggregate, forming a particle structure that spans the two electrodes. Shear strain increases with time, and the particle structure experiences a process of destruction and reformation. Under electrostatic force, lamellar structure forms and reaches an equilibrium state.

The shear-induced lamellar structure under equilibrium state is related to electric field, and also to shear rate. Figure 6.3 gives a group of pictures of particle structure at equilibrium state under six different conditions. The stronger the electric field is, the more lamellar rings appear. It was also found that the torque exerted by the top rotating electrode increases first during the formation of lamellar structure, and then reaches an equilibrium value after a steady structure is formed.

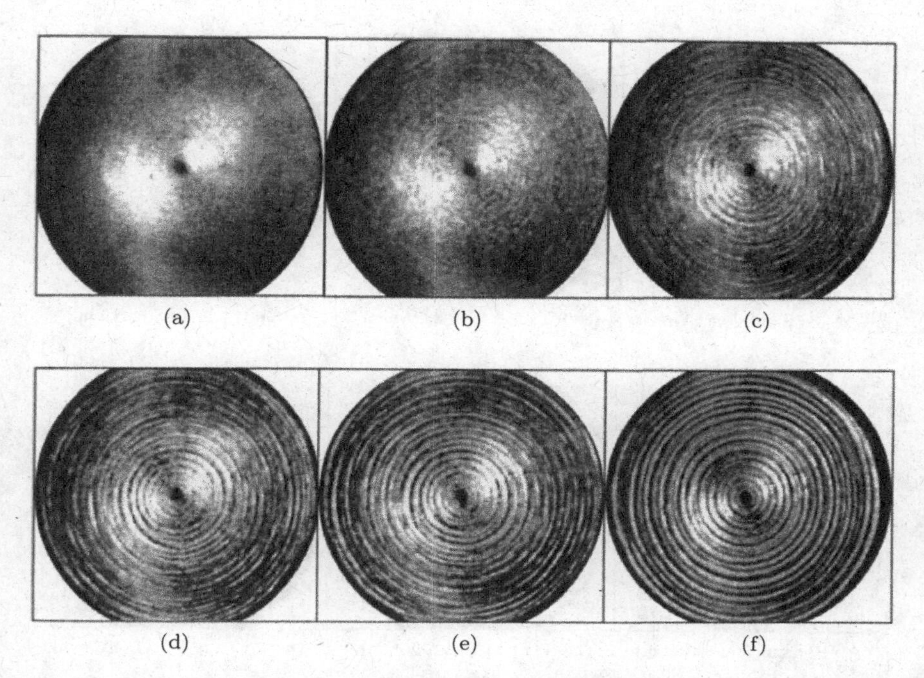

Fig. 6.2 The diameter of the above circular diagrams is 35 mm. (a)–(f) are the top views at 1.0, 2.0, 3.0, 4.0, 5.0 and 9.0 s after an electric field of 1.6 kV/mm is applied in the experiment. The dark area is the polarized particle aggregation area [6.14].

The authors of [6.12] gave a possible phase diagram of their ER fluid on the plane of external electric field and shear rate (Fig. 6.4). Under the solid line, particle structure does not exist 5 min after the electric field is applied (showing homogeneous fluid phase). Above the dashed line, there exists a shear-induced lamellar structure 15s after electric field is applied (LSII). LSI implies that it takes at least 15 s after the electric field is applied to form a clear lamellar structure, and it is called weak lamellar structure (LSI).

6.2.2 *Criterion of ER activity*

In introducing electrorheological (ER) or magnetorheological (MR) fluids, materials' ability to form chain or column structures is usually

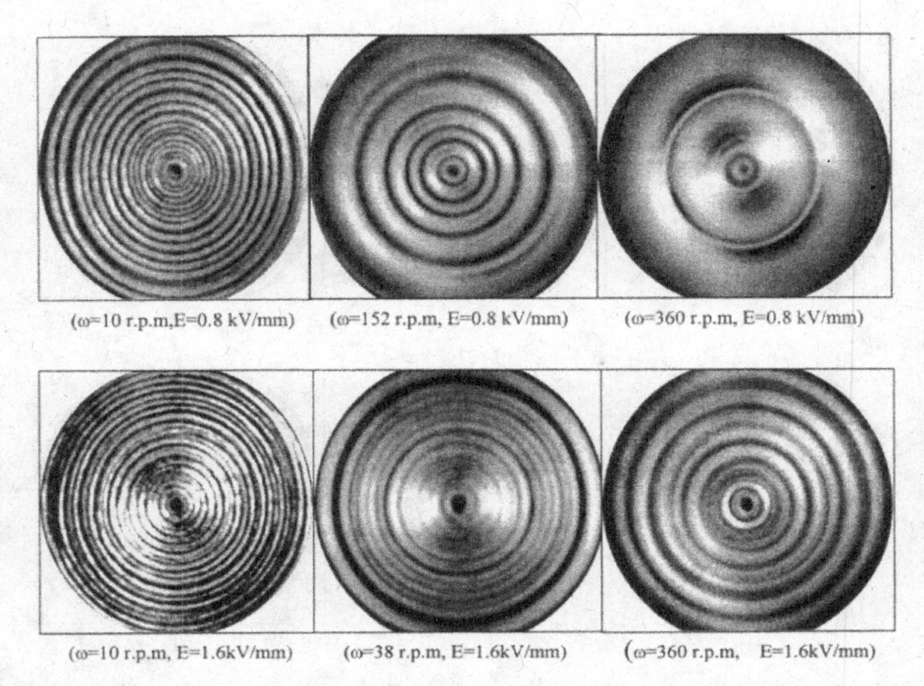

Fig. 6.3 The lamellar structure of an ER fluid at equilibrium state formed between two relatively rotating discs at different electric and shear fields [6.14].

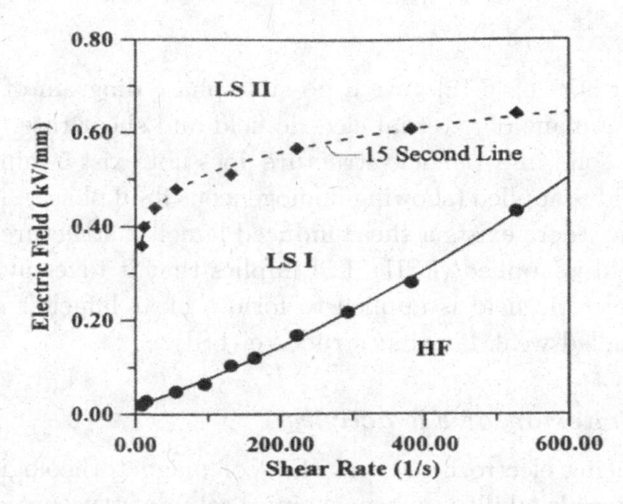

Fig. 6.4 A possible phase diagram of the ER fluid on the plane of electric field and shear rate [6.12].

emphasized. However, not all the materials that are good at chaining are ER active, or those not good at chaining are not ER active. Many materials can form chain structure under an electric field, and they are strong in ER activity. These materials form lamellar structure under both electric and shear fields. Some materials can form chain structure under an electric field, but they are not strong in ER activity. These materials cannot form lamellar structure under both electric and shear fields. Some materials that do not show clear chain or column structure under electric field, however, show pretty strong ER activity. These materials do form clear particle lamellar structure under both electric and shear fields. It seems that the formation of particle lamellar structure under both electric and shear fields is a good criterion for strong ER active materials. The length of such lamellar structures is along the direction of shear field, while the thickness is along the external electric field. The lamellar structure between electrodes of parallel plates is similar to wall structure; the lamellar structure between rotating parallel discs is similar to the structure of a cylindrical ring (Fig. 6.5).

It was proved that a column structure transmits very little shear stress between electrodes; while lamellar structure can transmit a large stress. In other words, lamellar structures are necessary for excellent ER fluids.

It is necessary to understand why a column structure transmits very little shear stress between electrodes, and why a lamellar structure can transmit a large stress.

Fidisko and Henley [6.15] gave a model simulation showing the flow of ER fluids under both electric and shear fields for a Couette and parallel disc flow geometrics (Fig. 6.5). They also show how the moving direction of rotating walls between two discs changes alternatively.

Henley and Fidisko [6.16] made an ER fluid using wet $BaTiO_3$ mixed with paraffin oil and compared it with an ER fluid using dry $BaTiO_3$ mixed with paraffin oil. The former has some ER activity (shear stress of 140 Pa at an electric field of 6 kV/mm), while the latter has almost no ER activity. Figure 6.6 shows a particle structure between two parallel rotating and transparent disc electrodes.

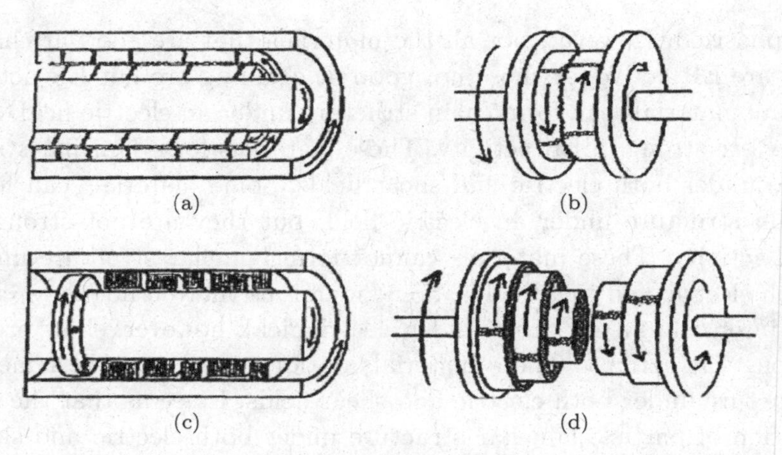

Fig. 6.5 Simulation of models for flow of ER fluids simultaneously under electric and shear fields for a couette and parallel disc flow geometrics. (a) and (b) show how a chain breaking model might appear; (c) and (d) show how the alternative model presented here would appear [6.15].

Two minutes after an electric field of 2000V/mm is on, and there is no relative rotation between the two electrodes, it can be seen from Fig. 6.6(a) that there are some non-uniformly distributed particle columns of irregular shapes in both ER fluids. In the ER fluid of dry $BaTiO_3$, the average diameter of the columns is 0.61 mm (Fig. 6.6(a)), while in the ER fluid of wet $BaTiO_3$, the average diameter of the columns is 0.83 mm.

When an electric field is applied to the two $BaTiO_3$ ER fluids, the two electrodes rotate relatively (with a relative rotating speed of 10 rmp). No particle structure can be identified in the dry $BaTiO_3$ ER fluid, but the homogeneous gray color indicates that particles are distributed homogeneously. On the other hand in the wet $BaTiO_3$ ER fluid, there is obvious particle aggregation of lamellar ring structure.

Filisko and Henley [6.15] also made such a comparison study using ER fluids made of paraffin oil with polystyrene or sulfonated polystyrene particles. After an electric field of 6 kV/mm is applied, the former shows almost no ER activity, while the latter gives a shear stress 400 times higher than that at zero electric field.

When there is only electric field (2000 V/mm) but no shear field is applied, Fig. 6.7(a) shows that there exist black columns of irregular

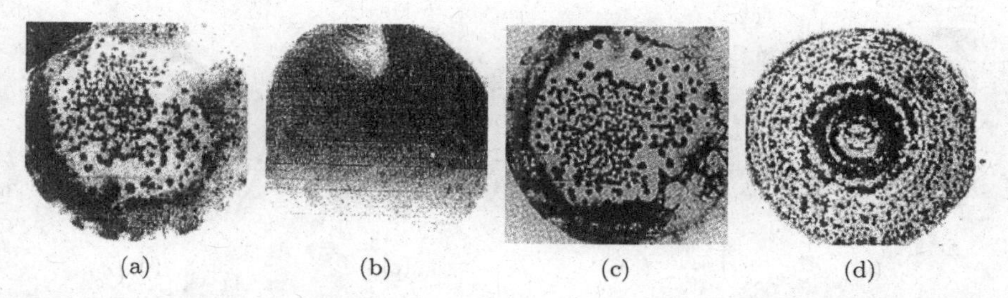

Fig. 6.6 Dry $BaTiO_3$ ER fluid (a, b) and wet $BaTiO_3$ ER fluid (c, d). Particle structure formation when only electric field is applied (a, c) or when both electric and shear fields are applied (b, d). A thick column may be a sign of active ER effect in an static case provided that it also shows lamellar structure under both electric and shear fields [6.15].

shapes in polystyrene ER fluids with the average diameter of columns being 0.63 mm. The author of Ref. [6.15] noted that in a material with no ER activity, columns of particle aggregation may still exist when only an electric field is applied. However, when electric and shear fields are applied simultaneously, a polystyrene ER fluid with poor ER activity shows only gray color and no identical particle structure. (Fig. 6.7(b)).

When only an electric field is applied in sulfonated polystyrene ER fluid, there are also columns of particle aggregation, but the average diameter of the columns is only 0.45 mm, while when an electric field (2 kV/mm) and shear field (rotating speed of 10 rmp) are applied simultaneously, there is an apparent cylindrical ring shaped lamellar structure of particle aggregation in the sulfonated polystyrene suspension. The length of the lamellar structure is along the shear direction, and the height is the same as the distance between the two electrodes. People are still not clear whether the columns are not destroyed, but transformed into the lamellar structure via movement, or the columns are destroyed at the initial period of shearing, and re-aggregated into lamellar structure.

However, the materials that show lamellar structure under both electric and shear field must be ER active.

Henley and Filisko [6.16] observed that several-hundred kinds of ER fluids with strong ER activity possess the above phenomena. They pointed out that a suspension with ER activity possesses the

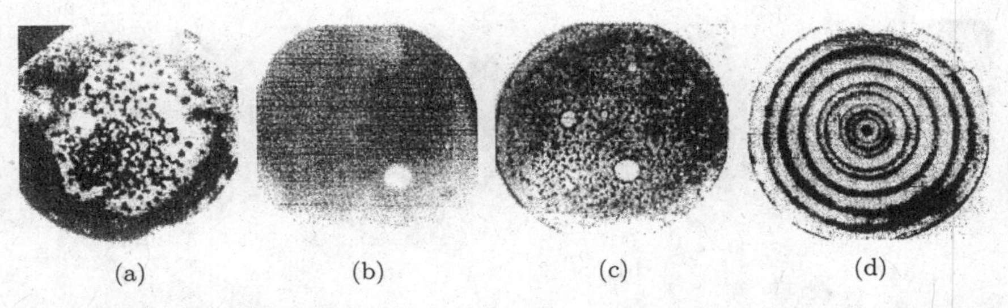

Fig. 6.7 The formation of particle structure of polystyrene ER fluid (a, b) and sulfonated polystyrene ER fluid (c, d) when only an electric field is applied (a, c) or both electric and shear fields are simultaneously applied (b, d) [6.15].

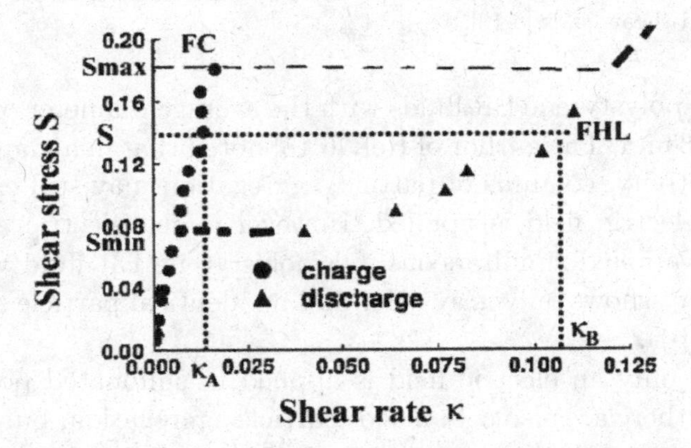

Fig. 6.8 Schematic construction of two-phase shear banded flow. The units of shear stress and shear rate are k_BT/a^3 and $\sqrt{k_BT/ma^2}$, respectively [6.17].

characteristics that particles aggregate into lamellar structure under a joint application of electric and shear fields with directions not parallel to each other. They also found whether a material is ER-active or not depends not on whether it can form chains or columns, but on whether it can form a lamellar structure.

Tao [6.17] examines Fig. 6.8 and concludes that their result of phase separation may lead to shear-banded flow [6.18, 6.19]. There are two distinct shear bands: the flowing-chain (FC) state has the maximum shear stress, $S_{\max}$, higher than the minimum shear stress, $S_{\min}$, of the flowing-hexagonal-layer (FHL) state (Fig. 6.9). Figure 6.8 shows a shear-banded flow. In the charging process, shear stress climbs the curve to reach $S_{\max}$. Further increase of shear rate

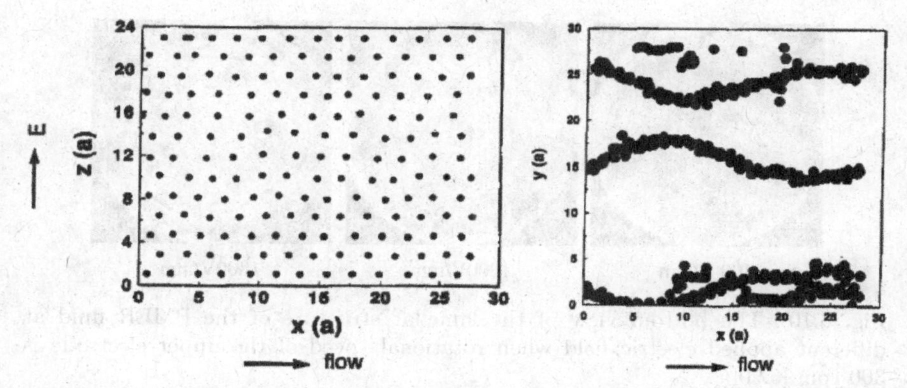

Fig. 6.9 A flowing-hexagonal layer (FHL) flow state on the x-z plane at left and x-y plane at right. The applied electric field is along the z-axis and the flow is along x [6.17].

will not change the shear stress, but change the flow structure from the FC state to the FHL state. Once the FHL state is fully developed shear stress will climb along the curve associated with FHL as shear rate increases. If the course of the shear rate is reversed and begins to reduce, the shear stress goes down along the curve associated to the FHL state until $S_{\min}$ is reached. Then the shear stress will remain as a constant until the shear rate is down to the value where the fluid is fully in the FC state. Afterwards, the shear stress will continue to go down along the curve associated with the FC state. If the shear stress S is between $S_{\min}$ and $S_{\max}$, the shear rate can be any value between κ_A and κ_B since the flow may be shear banded flow. Tao writes the shear rate $\kappa = x_A\kappa_A + x_B\kappa_B$ with $x_A + x_B = 1$. The values of x_A and x_B determine the percentage of the fluid in the FC state and FHL state, respectively.

Liu *et al.* [6.20] developed an inverted electrorheoscope which can recorded the images of particle structure patterns at the same time measuring the shear stress and viscosity of ER fluids. The images viewed from the lower electrode for differing electric field at the rotational speed of 300 rpm are shown in Fig. 6.10. Its shear stress versus applied electric field is shown in Fig. 6.8 when the rotating speed of one of the disc electrodes is 300 rpm and the other is zero. In comparison between Figs. 6.10 and 6.11, it is clear that the thinner the

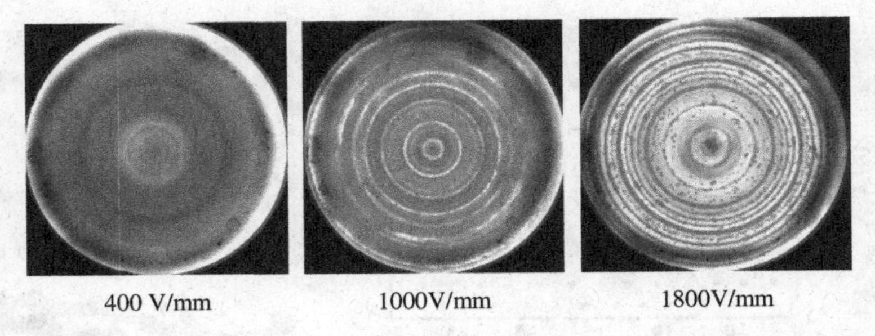

Fig. 6.10 The bottom view of the lamellar structure of the PMER fluid at different applied electric field when rotational speed of the upper electrode is 300 rpm [6.20].

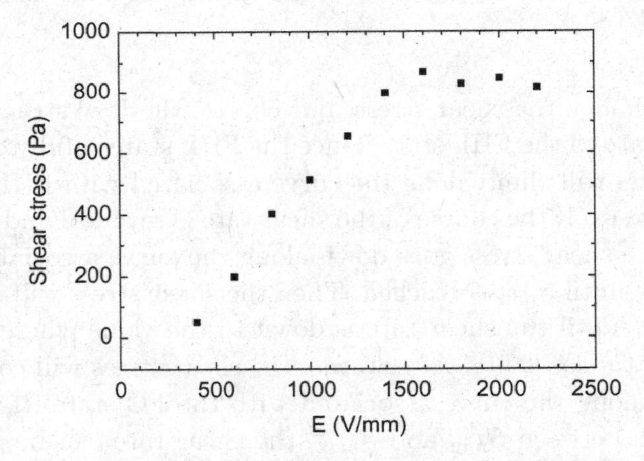

Fig. 6.11 The shear stress of polar-molecules dominant ER fluids at different fields when the rotating speeds of one disc electrode is 300 rpm and the other, zero. Part of its lamellar structures at certain electric fields are shown in Fig. 6.10 [6.20].

particle rings (the more the number of rings), the stronger the ER effect is.

Particle ring position, ring width, particle concentration and particle speed v against different applied electric field E are determined from the pattern videos and collectively plotted in Fig. 6.12. The following characteristics can be drawn from the plot: (a) the particle bands with zero linear speed (white boxes) can be cached at low

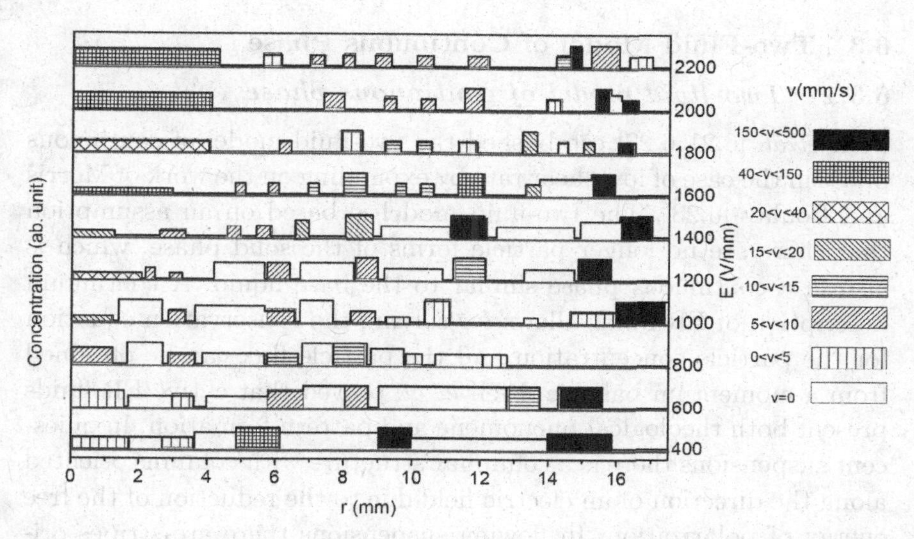

Fig. 6.12 Particle concentration and linear speed v of particle rings in the bottom image of lamellar structure of the PMER fluids at different fields while the rotational speed is kept at 330 rpm [6.20].

fields, while they are eliminated when the field is increased and even disappear at fields as high as 2200 V/mm. (b) At low fields, almost the entire area of the patterns is covered with particle bands, while when the field increases, more and more empty bands where the particle concentration is zero can be detected. It seems that these empty bands become unstable, and particles that happen to be there are thrown away and kept on nearby steady bands.

When compared with Fig. 6.11, the above two characteristics imply that thin and highly contracted rings are responsible for the high shear stress. The reason why smaller walls give larger force can be drawn from the Stokes' equation of viscous force for spherical particles, $F = 6\pi\eta rv$, where η is the viscosity of base liquid, r is particle radius, and v, the velocity difference between the particle and base liquid. If a large particle of radius R and volume V, is broken into three small particles of radius r while the total volume of the three particles remains V, $3F(r)/F(R) \sim 2.08$; namely, the three small particles will have shear stress twice as high as the large one.

6.3 Two-Fluid Model of Continuous Phase

6.3.1 *Two-fluid model of continuous phase*

Pfeil *et al.* [6.21, 6.22] established the two-fluid model of continuous phase in the case of low shear rate by expanding on the work of Morris and Boulay [6.23]. The two-fluid model is based on an assumption that there are no longer particle forms of the solid phase, which is also in a continuous phase similar to the base liquid. A continuum description of ER fluids allows for solving the conservation equation for the particle concentration and the particle flux can be obtained from a momentum balance. Pfeil *et al.* proved that active ER fluids present both rheological phenomena and pattern formation. In quiescent suspensions there is a columnar structure with columns oriented along the direction of an electric field due to the reduction of the free energy of polarization. In flowing suspensions there are stripes oriented in flow because the shear flow stabilizes the fluctuation in the shear plane. Reference [6.22] and references cited there have detailed derivations, so, in this section, there is only a rough description of the derivations with an explanation of the solution.

(1) The mass conservation equation

The two-fluid model starts from the mass conservation equation

$$\frac{\partial p}{\partial t} + \nabla \cdot (pu) = 0 \tag{6.23}$$

where $\rho(\boldsymbol{x}, t)$ is the mass density and $u(\boldsymbol{x}, t)$ is the velocity. This equation is valid at every point in both components of the mixture.

Equation (6.23) can be transformed into a particle-phase conservation equation by an average procedure [6.21, 6.22]. This involves multiplying Eq. (6.22) by the phase indicator function χ ($\chi = 1$ in particles and $\chi = 0$ in fluid), multiplying by the N-particle configurational probability $P_N(\boldsymbol{x}_N)$, integrating over configurations $\boldsymbol{x}_N$, and dividing out the constant particle density ρ_p from the equation. The resulting particle phase conservation equation is

$$\frac{\partial \varphi}{\partial t} + \nabla \cdot (U\phi) = 0, \tag{6.24}$$

where $\boldsymbol{U}$ is the local average velocity of the particle phase, and ϕ is the particle volume fraction,

The flow model of ER and MR fluids is established in Ref. [6.22] and the references cited therein. The flow model yields the general mass conservation equation.

$$\frac{\partial \varphi}{\partial t} + \langle \boldsymbol{u} \rangle \cdot \nabla \phi = \nabla \cdot \boldsymbol{j}, \tag{6.25}$$

where $\langle \boldsymbol{u} \rangle$ is the suspension average velocity, $\boldsymbol{j} \equiv \phi(\boldsymbol{U} - \langle \boldsymbol{u} \rangle)$ is the particle flux relative to the mean suspension motion,

$$\boldsymbol{j} = \frac{2a^2}{9\eta_c} f(\phi) \nabla \cdot \boldsymbol{\sigma}^{(p)}, \tag{6.26}$$

with a being the particle radius, η_c, the viscosity of the suspending fluid, $f(\phi) = (1-\phi)^4$, hindered mobility (or sedimentation function), and $\boldsymbol{\sigma}^{(p)}$, the particle contribution to stress.

(2) Electrostatic contribution of the stress

With the definition of the electrostatic contribution of the Maxwell stress tensor [6.24]

$$4\pi \boldsymbol{\sigma}^{(E)} = \varepsilon_0 \varepsilon \boldsymbol{E}\boldsymbol{E} - \frac{1}{2} \varepsilon_0 \varepsilon E^2 \boldsymbol{I}. \tag{6.27}$$

The state of a slightly deformed body is described by the strain tensor [6.25]

$$u_{ij} = \frac{1}{2} \left(\frac{\partial u_i}{\partial u_j} + \frac{\partial u_j}{\partial x_i} \right) \tag{6.28}$$

where $\boldsymbol{u}(x, y, z)$ is the displacement vector for points in the body. The dielectric tensor of the deformed material can be represented as [6.25]

$$\varepsilon_{ij} = \varepsilon_0 \delta_{ij} + a_1 u_{ij} + a_2 u_{kk} \delta_{ij} \tag{6.29}$$

where ε_0 is the permittivity of the undeformed body, and the other two terms, which contain the scalar constants a_1, a_2, form the most general tensor of rank two which can be constructed linearly from the components. u_{ik} yields the following formula for the stress tensor.

Suppose that the dielectric spheres are immersed in a dielectric base liquid, and for such a suspension, inserting Eq. (6.29) into Eq. (6.27) yields the electrostatic contribution of the stress [6.26],

$$4\pi\boldsymbol{\sigma}^{(E)} = \epsilon_0\left[\left(\epsilon - \frac{1}{2}a_1\right)\boldsymbol{E}\boldsymbol{E} - \frac{1}{2}(\epsilon + a_2)\boldsymbol{E}^2\boldsymbol{\delta}\right], \tag{6.30}$$

where $\epsilon(\phi)$ is the suspension dielectric constant.

$$\epsilon(\phi) = \epsilon_c\frac{1+2\beta\phi}{1-\beta\phi}, \tag{6.31}$$

and the electrostriction coefficients of the suspension, $a_1(\phi) = 0$, and

$$a_2(\phi) = -\frac{(\epsilon(\phi)-\epsilon_c)(\epsilon(\phi)+2\epsilon_c)}{3\epsilon_c}. \tag{6.32}$$

Taking the divergence of Eq. (6.26) and utilizing Mawell's equation $\boldsymbol{\nabla}\cdot\boldsymbol{D} = \rho^{(e)}$, ($\rho^{(e)} = 0$, is the free charge density) and $\boldsymbol{\nabla}\times\boldsymbol{E} = \boldsymbol{0}$ gives

$$\nabla\cdot\boldsymbol{\sigma}^{(E)} = \rho^{(e)}\boldsymbol{E} - \frac{1}{2}\epsilon_0 E^2\nabla\epsilon - \frac{1}{2}\epsilon_0\nabla(a_2E^2). \tag{6.33}$$

The terms on the right side are body forces.

If the average speed of the flow is $\langle\boldsymbol{u}\rangle = (\dot{\gamma}z, 0, 0)$, and the external electric field is $\boldsymbol{E} = (0, 0, E_0)$, the conservation equation becomes

$$\frac{\partial\phi'}{\partial t} + \dot{\gamma}z\frac{\partial\phi'}{\partial x} = -M\left(\frac{\partial^2\phi'}{\partial x^2} + \frac{\partial^2\phi'}{\partial y^2} - \kappa\frac{\partial^2\phi'}{\partial z^2}\right) \tag{6.34}$$

where $\phi'(x,t) \equiv \phi(x,t) - \phi_0$,

$$\begin{aligned} M &= -\frac{a^2\epsilon_0E_0^2f(\phi_0)}{9\eta_c}\left(\frac{d\epsilon}{d\phi} + \frac{da_2}{d\phi}\right)_{\phi_0} \\ &= \frac{2a^2\epsilon_0\epsilon_c\beta^2E_0^2f(\phi_0)\phi_0}{3\eta_c(1-\beta\phi_0)^3}, \end{aligned} \tag{6.35}$$

and $\kappa = 2(1-\beta\phi_0)/(1+2\beta\phi_0) > 0$.

6.3.2 *Electric field to a quiescent suspension*

(1) Maximum in free energy

Consider the electrostatic contribution to the free energy, $F^{(E)} = -\varepsilon_0\varepsilon(\phi)E^2/2$, where E is a local electric field. From Eq. (6.7), $\partial^2\varepsilon/\partial\phi^2 > 0$, one has $\partial^2 F^{(E)}/\partial\phi^2 < 0$, so there is a maximum in $F^{(E)}$, which corresponds to the homogeneous distribution, which implies that the free energy can be reduced by a phase separation.

(2) Solution of stress equation

The electrodes are located at $z = 0$ and L. The solution of the stress equation, Eq. (6.30), has the form

$$\phi'(x,t) = f(z)e^{i(k_x x + k_y y)}e^{st}, \tag{6.36}$$

which indicates fluctuations with sinusoidal variation in x and y directions. The z dependence is yet to be determined, and s is the growth rate of fluctuations. The full solution to Eq. (6.34) is a summation of Eq. (6.36). When shear rate is zero, substituting Eq. (6.36) into Eq. (6.34) yields

$$\kappa f''(z) + (k_x^2 + k_y^2 - s/M)f(z) = 0. \tag{6.37}$$

Using the no-flux boundary conditions $f'(z) = 0$ at $z = 0$ and $z = L$ gives

$$f(z) = \cos n\pi z/L \quad n = 0, 1, \ldots, \tag{6.38}$$

and

$$\frac{sL^2}{M} = (k_x L)^2 + (k_y L)^2 - \kappa n^2\pi^2. \tag{6.39}$$

(3) Fluctuation in concentration

Fluctuations will grow for all $s > 0$. The largest s implies $n = 0$ since other positive integers of n lead to smaller s (Eq. (6.39)) and smaller fluctuation (Eq. (6.36)). The largest s means the largest fluctuation with the fluctuation rate of $s_{\max}/M = k_x^2 + k_y^2 > 0$, or largest k_x and k_y, which implies the smallest wavelength of the pattern, or thinnest column structure. The growth rate s is symmetric with respect to k_x and k_y, and if the initial fluctuations are random and isotropic,

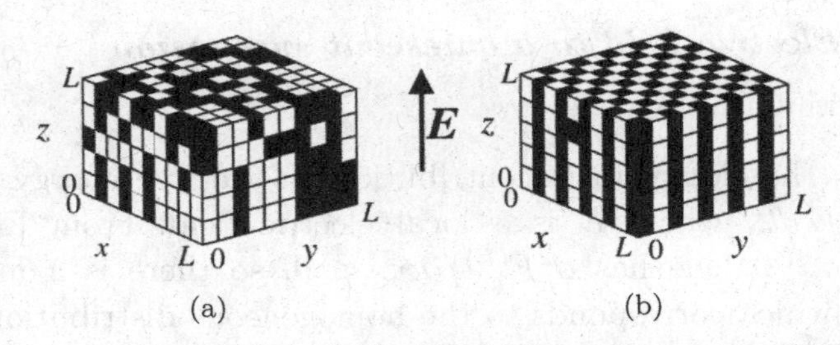

Fig. 6.13 (a) Initial structure and (b) structure after the application of an electric field to a quiescent suspension ($tM/L^2 = 0.05$). Dark and light parts represent positive and negative concentration fluctuations, respectively [6.21, 6.22].

the resulting structure is cylindrical columns along z as shown in Fig. 6.13.

It is worth learning that the variation of the free energy would also lead to a column formation in quiescent suspension under field.

The electrostatic contribution to the free energy

$$\mathcal{F}^E = -\epsilon_0 \epsilon(\phi) E^2/2 \tag{6.40}$$

where $\epsilon(\phi) = \epsilon_c \frac{1+2\beta\phi}{1-\beta\phi}$. Since $\partial^2\epsilon/\partial\phi^2 > 0$, $\partial\mathcal{F}^E/\partial\phi = 0$, and $\partial^2\mathcal{F}^E/\partial\phi^2 < 0$. This means that when an electric field is applied, the homogeneous distribution of ER fluid changes to a phase separated state to reduce the free energy. The authors of Refs. [6.21] and [6.22] proved that it is true no matter in a quiescent or a flow state.

The variation of the free energy is

$$\delta\mathcal{F}^E = -\epsilon_0 \int_V \delta\epsilon(\phi) E \cdot E_0 dV/2. \tag{6.41}$$

Supposing concentration fluctuation is parallel to the direction of the applied field, $\phi(x) = \phi_0 + A_\parallel \cos\beta_\parallel x$, $(A_{//} \ll 1)$, and the change in the electrostatic free energy is $\delta\mathcal{F}^E = -\epsilon_0 E_0^2 (d^2\epsilon/d\phi^2) A_\parallel^2 V/8 < 0$, implying that such fluctuations parallel to the field are unstable and will increase, forming a phase separated state; namely, there would be a concentration change along the x and y directions. Supposing concentration fluctuation is perpendicular to E, $\phi(z) = \phi_0 + A_\perp \cos\beta_\perp z$, prove that $\delta\mathcal{F}^E = +\epsilon_0 E_0^2[(d\epsilon/d\phi)^2/\epsilon - (d^2\epsilon\, d\phi^2)/2] A_\perp^2 V/4 > 0$, implying that such fluctuations are stable and will decay, namely, concentration along z axis is homogeneous. From the above two considerations, there should be column formation as shown in the right panel of Fig. 6.14.

6.3.3 *Electric field to a flowing suspension*

(1) Concentration fluctuation for flowing suspension

For non-zero shear rates, substituting Eq. (6.36) into Eq. (6.34) yields

$$\kappa f''(z) + (k_x^2 + k_y^2 - s/M - i\dot{\gamma}k_x z/M)f = 0. \quad (6.42)$$

The numerical solution to Eq. (6.42) determines $s(k_x, k_y)$. Figure 6.14 illustrates the results when $\dot{\gamma}L^2/M = 10^3$ and 10^4 [6.21, 6.22], showing the contour plots of sL^2/M as a function of k_x and k_y. The curve of $s = 0$ between S and U represents the boundary between the stable (S, taking negative s in Eq. (6.37)) and unstable (U, positive s) areas. Since the fluctuations ϕ' is proportional to e^{st}, the stabilized fluctuations mean ϕ' with $s \leq 0$, which locates in the area on or below the contour $s = 0$ in Fig. 6.15. When shear rate $\dot{\gamma}$ increases, or $\dot{\gamma}L^2/M$ increases, the area below the contour $s = 0$ increases. The larger the area under the zero sL^2/M contour implies the larger stable zone, which means that taking the values of k_xL and k_yL under the contour guarantees a stable fluctuation of particle concentration. In other words, a pattern with a wavelength of $2\pi/k_x$ and f $2\pi/k_y$ would be stable. One may conclude that the shear rate tends to stabilize the spatial fluctuations of particle concentration, and then more particles have their concentration fluctuation stabilized.

Figure 6.14(b) implies that when the shear rate is large enough, under the cutoff of large $|\boldsymbol{k}|$, an unstable fluctuation appears only when k_x is very small while k_y can be almost any number. It can then be predicted that an electric field applied to a flowing suspension would cause a dominant structure in which particles are cumulated in the area along the flow direction of the x axis, while there is an apparent wavelength along the y axis. This is the lamellar structure depicted in Fig. 6.15 and also proven by many experimental results [6.15–6.20, 6.27].

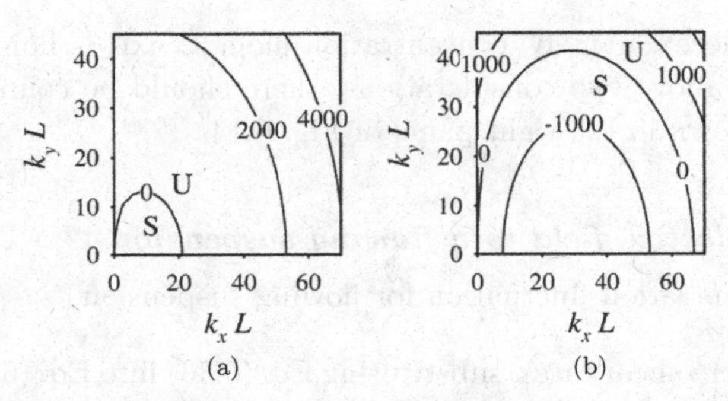

Fig. 6.14 Contour plots of sL^2/M as a function of k_xL and k_yL for (a) $\dot{\gamma}L^2/M = 10^3$ and (b) $\dot{\gamma}L^2/M = 10^4$. Stable and unstable regions are denoted by S and U, respectively [6.21, 6.22].

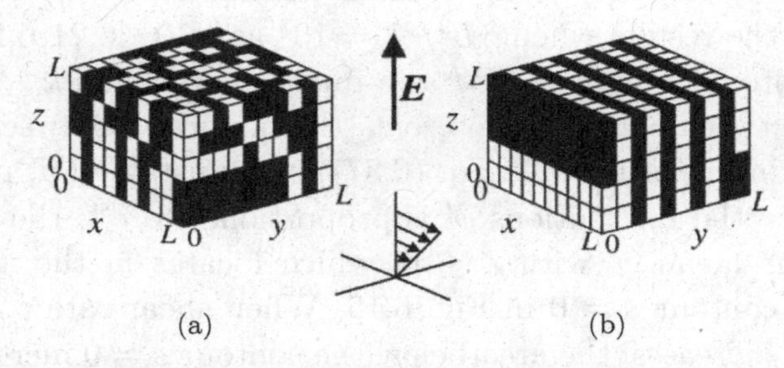

Fig. 6.15 (a) Initial structure and (b) structure after the application of an electric field to a flowing suspension ($\dot{\gamma}L^2/M = 10^4$, $tM/L^2 = 0.05$). Dark and light parts represent positive and negative concentration fluctuations, respectively [6.21, 6.22].

(2) Fluctuation in concentration

The mechanism of stripe formation in shear flow is related to the mechanism of column formation in quiescent suspensions. In quiescent suspensions, particles form columns parallel to the electric field, so the major direction of spatial concentration fluctuations is parallel to the electric field. Namely, fluctuations grow when the fluctuations are parallel to the electric field and decay when they are perpendicular to the field. In shear flow such as that shown in the left panel of Fig. 6.16, column-like fluctuations parallel to the applied field (the dashed curve in Fig. 6.16) will be rotated toward the flow direction (the solid curve in Fig. 6.16) becoming a fluctuation perpendicular

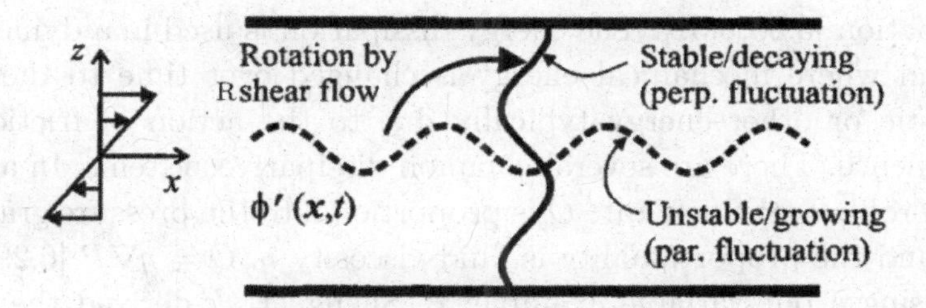

Fig. 6.16 Unstable fluctuations parallel to the applied electric field (along z-axis) would rotate by simple shear flow and become stable fluctuations perpendicular to the electric field [6.21, 6.22].

to the field. As we have already seen in the last section, any fluctuations perpendicular to the field will be decayed (this is also true in the shear flow state). However, since fluctuations in the vorticity direction (i.e., $\boldsymbol{k} = (0, k_y, 0)$) are unaffected by shear and will continue to grow, the resulting structure will be sheets of higher concentration in the plane of shear, as illustrated in Fig. 6.15(b).

In the two-fluid continuum model for mass transport in ER suspensions, the particle flux is related to the divergence of the particle contribution to the stress, which in turn is related to the suspension dielectric and electrostrictive properties. Solutions of the resulting particle conservation equation capture common observations: column formation in quiescent suspensions and stripe formation in sheared suspensions. Column formation arises because only these structures produce a decrease in the free energy of polarization. Stripe formation in shear flow arises because the flow stabilizes fluctuations in the shear plane.

6.4 Onsager Principle of Least Energy Dissipation

6.4.1 *Derivation of the Onsager principle*

The dynamic behavior of ER fluid is especially important since almost all their applications are under high shear. Starting from the Onsager principle, Jianwei Zhang *et al.* [6.28, 6.29] calculated the shear stress of ER.

The Onsager principle states that the dynamics of dissipative systems is governed by the principle of minimum energy dissipation [6.30,6.31]. The energy dissipation is used in a dynamical system where mechanical energy is changed over time to thermal, acoustic or other energy typically due to the action of friction or turbulence. There are several common dissipative systems. In a fluid flow problem, the flow rate Q is proportional to the pressure gradient ∇P and the proportionality is fluid viscosity η, $Q = \eta \nabla P$ [6.29].

Using a one-variable example, P. Sheng *et al.* derived the principle of minimum energy dissipation [6.30, 6.31]. In an overdamped dissipative system, the dynamics may be described by the Langevin equation

$$\eta \dot{\alpha} = -\frac{\partial F(\alpha)}{\partial \alpha} + \xi(t), \tag{6.43}$$

where α is the displacement from equilibrium, $\dot{\alpha}$ denotes the change of α, $F(\alpha)$ is the relevant free energy, and $\xi(t)$ is the white noise with a zero mean, satisfying the correlation $\langle \xi(t)\xi(t') \rangle = 2\eta k_B T \delta(t - t')$, where k_B denotes the Boltzmann constant and T, the temperature. The left hand side of Eq. (6.43) is the dissipative force, which is balanced by the conservative force plus a stochastic force on the right. With the white noise term $\xi(t)$, the dynamics of α is no longer deterministic and its trajectory is best described by a probability density $P(\alpha, t)$ that is governed by the Fokker–Planck equation

$$\frac{\partial P}{\partial t} = D\left[\frac{\partial^2 P}{\partial \alpha^2} + \frac{1}{k_B T}\frac{\partial}{\partial \alpha}\left(\frac{\partial F}{\partial \alpha}P\right)\right], \tag{6.44}$$

where the diffusion constant D satisfies the Einstein relation $\zeta D = k_B T$, where drag coefficient $\zeta = 6\pi r \eta_m$, and η_m is the viscosity of media where spherical particles of radius r are located. It is simple to verify that the *stationary* solution of Eq. (6.44) is given by the Boltzmann distribution $P_{eq} \propto \exp[-F(\alpha)/k_B T]$. The *dynamic* transition probability for α at t to α' at $t + \Delta t$ is given by

$$\begin{aligned} &P(\alpha', t + \Delta t | \alpha, t) \\ &\quad = \frac{1}{\sqrt{4\pi D \Delta t}} \exp\left[-\frac{(\alpha' - \alpha)^2}{4D\Delta t}\right] \exp\left[-\frac{F(\alpha') - F(\alpha)}{2k_B T}\right] \end{aligned} \tag{6.45}$$

for α' in the vicinity of α and small Δt. By using the Einstein relation, the two exponents can be combined:

$$P(\alpha', t + \Delta t | \alpha, t) = \frac{1}{\sqrt{4\pi D \Delta t}} \exp\left[-\frac{A}{2k_B T}\right], \tag{6.46}$$

where

$$A = \frac{\zeta(\alpha' - \alpha)^2}{2\Delta t} + [F(\alpha') - F(\alpha)] \approx \left[\frac{\zeta}{2}\dot{\alpha}^2 + \frac{\partial F(\alpha)}{\partial \alpha}\dot{\alpha}\right] \Delta t \tag{6.47}$$

is the quantity to be minimized if we want to maximize the probability of transition with respect to α'. For small Δt, it is seen that instead of minimizing A with respect to the target state α', the same is achieved by minimizing with respect to the *rate* $\dot{\alpha}$. Indeed, if we carry out the simple minimization on the right hand side of Eq. (6.47), we obtain the force balance equation

$$\zeta\dot{\alpha} = -\frac{\partial F(\alpha)}{\partial \alpha}, \tag{6.48}$$

i.e., the Langevin equation without the stochastic force term. This is reasonable, since the stochastic force has a zero mean, so Eq. (6.48) is true on average.

Thus we learn from the above that (a) there can be a variational functional, of which the quantity A is the one-variable version, which should be minimized with respect to the rates; (b) the result of such minimization would guarantee the force balance on average; and (c) the minimization would also yield the equations of motion and the related boundary conditions, which represent the *most probable course of a dissipative process.*

The last statement essentially guarantees that in the statistical sense, the most probable course will be the only course of action observed macroscopically.

For the general case of multivariables, the variational functional can be simply generalized from Eq. (6.47) as

$$A = \frac{1}{2}\sum_{i,j} \zeta_{ij}\dot{\alpha}_i\dot{\alpha}_j + \sum_{i=1}^{n} \frac{\partial F(\alpha_1, \ldots, \alpha_n)}{\partial \alpha_i}\dot{\alpha}_i \tag{6.49}$$

where, in the case of α_i's being field variables, the summation should be replaced by integrals, and partial derivatives by functional derivatives. In Eq. (6.49) the dissipation coefficient matrix elements ζ_{ij} must be symmetric with respect to the interchange of the two indices, as shown by Onsager [6.32, 6.33, from 6.31] based on microscopic reversibility.

In equilibrium, we can use functions of states — free energies, thermodynamic potentials — to determine the most probable state. Out of equilibrium, we can use a functional of a sequence of states to determine the probability of trajectories. In the case of small, linear deviations from equilibrium, the Onsager "action" gives us such a functional [6.34, 6.35].

6.4.2 *Establishment of the Navier–Stokes equations*

The *Onsager action functional* A is introduced as follows [6.28]

$$A = \Phi + \dot{F}, \tag{6.50}$$

where the dissipation function Φ is expressed as

$$\Phi = \frac{1}{2}\varsigma\dot{\alpha}^2, \tag{6.51}$$

while $\dot{\alpha}$ is the time rate of the change of fluid displacement from equilibrium, and $\dot{F}$ is the time rate of change of functional free energy density

$$\dot{F} = \int \mu \frac{\partial n}{\partial t} d\boldsymbol{x}, \tag{6.52}$$

where $\frac{\partial n}{\partial t} = \dot{n} - \boldsymbol{V}_s \cdot \nabla n$ and $\dot{n} = -\nabla \cdot \boldsymbol{J}$ with n being the ER sphere density and $\boldsymbol{V}_s$, the velocity of ER spheres.

Viscous dissipation of fluid flow is given by [6.30, 6.31]

$$\Phi = \frac{1}{2}\int \left\{ \frac{1}{2}\eta_s[\partial_n(\boldsymbol{V}_s)_\tau]^2 + \frac{\gamma}{n}J^2 + K(\boldsymbol{V}_f - \boldsymbol{V}_s)^2 \right\} d\boldsymbol{x}, \tag{6.53}$$

where subscripts n and τ indicate the normal and tangential directions on the liquid-solid surface, respectively. The first term of the right hand side of the above equation is the viscous dissipation, with

η_s denoting the colloidal viscosity. The second term is the energy dissipation rate per unit volume with γ denoting a frictional coefficient related to the convective-diffusive current's dissipation, and can be obtained as follows. The convective-diffusive current density is $\boldsymbol{J}= n\boldsymbol{V}_d$, where $\boldsymbol{V}_d$ denotes the drift velocity. The dissipative force acting on a single microsphere is $\gamma\boldsymbol{V}_d$. Hence the force density is given by $n\gamma V_d$, and the energy dissipation rate per unit volume is $n\gamma V_d^2 = \gamma J^2/n$. Taking into account the factor of $\frac{1}{2}$ leads to the expression above. The third term is the dissipation caused by the friction between the solid and fluid components, characterized by a constant $K = \gamma n$.

Variational minimization of the Onsager action functional $A = \dot{F} + \Phi$ with respect to $(\boldsymbol{J}, \boldsymbol{V}_s)$ using the Lagrange multiplier method with $\lambda = -2p_s$ leads to

$$\boldsymbol{J} = -\frac{1}{\gamma} n \nabla \mu, \tag{6.54}$$

and the Stokes equation for the solid component:

$$0 = -\nabla p_s + \eta_s \nabla^2 \boldsymbol{V}_s + n\nabla\mu + K(\boldsymbol{V}_f - \boldsymbol{V}_s), \tag{6.55}$$

where the condition of incompressibility flow $\nabla \cdot \mathbf{V}_s = 0$ is adopted. This condition comes from jointly solving the following two equations. The continuity equation leads to $\partial n/\partial t + n(\nabla \cdot \boldsymbol{V}_s) + \nabla n \cdot \boldsymbol{V}_s = 0$. Incompressibility flow means that density of the fluid (so is its solid phase) is constant when following the fluid motion, or $\dot{n}$, the material derivative of the density of the solid phase of the fluid, is zero: $\dot{n} \equiv \frac{\partial n}{\partial t} + (\nabla n) \cdot \boldsymbol{V}_s = -n(\nabla \boldsymbol{V}_s) = 0$, which results in $\nabla \cdot \boldsymbol{V}_s = 0$.

For the fluid component, the Stokes equation is

$$0 = -\vec{\Delta} p_f + \eta_f \nabla^2 V_f + K(V_s - V_f). \tag{6.56}$$

When the inertial effect is not ignored, the momentum balance requires [6.28] Eqs. (6.51) and (6.52) to be respectively replaced with

$$\rho_s\left(\frac{\partial \vec{V}_s}{\partial t} + \vec{V}_s \cdot \nabla \vec{V}_s\right) = -\nabla p_s + \nabla \cdot \tau^s_{visc} + \nabla \cdot \tau_s + K(\vec{V}_f - \vec{V}_s) \tag{6.57}$$

and

$$\rho_f \left(\frac{\partial \vec{V}_f}{\partial t} + \vec{V}_f \cdot \nabla \vec{V}_f \right) = -\nabla p_f + \nabla \cdot \tau^f_{visc} + K(\vec{V}_s - \vec{V}_f) \qquad (6.58)$$

with the supplementary incompressibility conditions $\nabla \cdot \vec{V}_{s,f} = 0$. $\rho_s = mn(\vec{x}) + (1 - f_s)\rho_f$ is the local mass density of the "s" phase, p_s and p_f are the pressures in the two phases. $\nabla \cdot \tau_s$ is the force density, and $\tau^s_{visc} = \eta_s(\nabla \vec{V}_s + \nabla^T \vec{V}_s)/2$ and $\tau^f_{visc} = \eta_f(\nabla \vec{V}_f + \nabla^T \vec{V}_f)/2$.

6.4.3 *Numerical calculation*

(1) Linear sample cell

Numerical calculations of the above equations lead to an excellent agreement with the experiments. The authors of Ref. [5.25] have shown that whereas the usual configuration of applied electric field being perpendicular to the shearing direction can lead to shear thinning at high shear rates and thus the loss of ER effect, the inter-digitated, alternating electrodes configuration can eliminate the shear thinning effect.

The electric field configuration is obtained by solving the Laplace equation $\nabla \cdot \varepsilon(x) \nabla \varphi$, where φ denotes the electrical potential, and the effective dielectric constant $\varepsilon(\boldsymbol{x})$ is locally updated by using the Maxwell–Garnett relation $[\varepsilon(\boldsymbol{x}) - \varepsilon_f]/[\varepsilon(\boldsymbol{x}) + 2\varepsilon_f] = f_s(\boldsymbol{x})(\varepsilon_s - \varepsilon_f)/(\varepsilon_s + 2\varepsilon_f)$. Consistency between the local field and $\boldsymbol{p}$ is obtained iteratively. In the numerical calculations, n is taken to be 2D, with periodic condition along x (the shearing direction) and the thickness along the y direction taken to be 2a, i.e., one layer of ER fluid.

Their numerical result of shear stress in shown in Fig. 6.17. When the top plate is moved at a constant speed relative to the bottom plate to generate a Couette flow, the resulting shear stress experienced on the top plate is plotted as a function of time in the inset to Fig. 6.17. Fluctuations are seen which reflect the breaking and re-attachment of the columns. The time-averaged stress is plotted as a function of shear rate in Fig. 6.17. The behavior is very similar to the Bingham fluid at low shear rates, with an extrapolated dynamic yield stress that is 30% lower than the static yield stress.

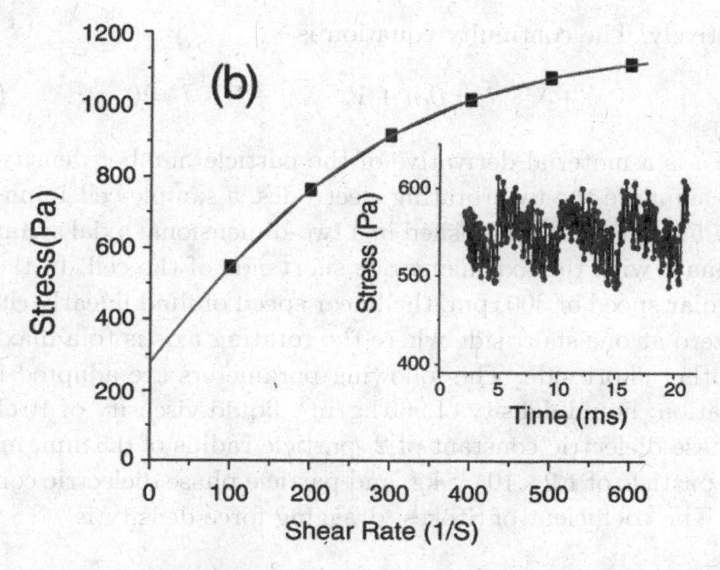

Fig. 6.17 The shear stress of an ER fluid changed with shear rate numerically calculated on the basis of the Onsager principle of least energy dissipation [6.25].

(2) Circular sample cell

Li *et al.* [6.36] implemented a numerical calculation utilizing finite element analysis with COMSOL to solve the Navier–Stokes equation for colloidal ER fluids.

In the simulation, a Couette flow is generated by rotating the upper circular electrode while keeping the lower one steady. In this geometry, the Navier–Stokes equations of particle and liquid phases are [6.28]

$$\rho_s\left(\frac{\partial \vec{V}_s}{\partial t}+\vec{V}_s\cdot\nabla\vec{V}_s\right)=-\nabla p_s+\nabla\cdot\tau^s_{visc}+\nabla\cdot\tau_s+K(\vec{V}_f-\vec{V}_s) \tag{6.59}$$

and

$$\rho_f\left(\frac{\partial \vec{V}_f}{\partial t}+\vec{V}_f\cdot\nabla\vec{V}_f\right)=-\nabla p_f+\nabla\cdot\tau^f_{visc}+K(\vec{V}_s-\vec{V}_f), \tag{6.60}$$

respectively. The continuity equation is

$$\dot{n} + \vec{\nabla} \cdot \vec{J} = \partial_t n + V_s \cdot \vec{\nabla}_n + \vec{\nabla} \cdot \vec{J} = 0 \tag{6.61}$$

where $\dot{n}$ is a material derivative of the particle number density n.

To simulate the two rotating electrodes, a sample cell 1 mm thick and 17.5 mm long is established in a two-dimensional axial symmetric coordinate with the axis along one short side of the cell. In the case of angular speed of 300 rpm, the linear speed of fluid linearly changes from zero at one short side where the rotating axis is to a maximum at another short side. The following parameters are adapted in the calculation: liquid density of 960 kg/m^3, liquid viscosity of 10 cP, liquid phase dielectric constant of 2, particle radius of 0.5 mm, mass of single particle of 1.2×10^{-12} kg, and particle phase dielectric constant of 40. The coefficient of Stokes' dragging force density is

$$K = 9 f_s \eta_f / 2a^2. \tag{6.62}$$

The local electric field between particles is obtained from the following equation

$$[\vec{E}_1(\vec{x})]_i - [\vec{E}_{ext}(\vec{x})]_i + \int G_{ij}(\vec{x}, \vec{y}) p_j(\vec{y}) n(\vec{y}) d\vec{y}. \tag{6.63}$$

The repulsive potential function is defined as

$$\varepsilon_0 \int \left(\frac{a}{|\vec{x} - \vec{y}|} \right)^{12} n(\vec{y}) d\vec{y}. \tag{6.64}$$

Equations (6.59)–(6.61) are solved with finite element approach (FEA) by implementing COMSOL Multiphysics using parameters mentioned above, and the lower and upper views of the particle structure patterns are obtained as shown in Fig. 6.18 for the applied electric field of 1000V/mm.

The density of the pattern is proportional to the particle concentration. These figures qualitatively display the major characteristics of lamellar structure. Further calculation also gives the speed of particle phase ring, time evolution of pattern as well as the shear stress of the model colloidal ER fluid.

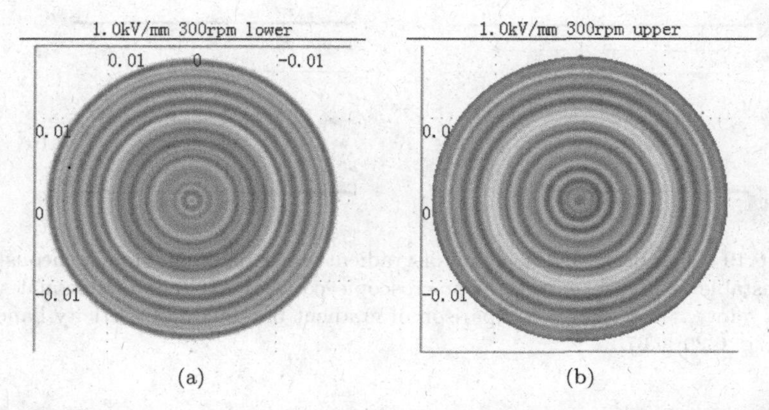

Fig. 6.18 The result of numerical calculation depicts the lower and upper lamellar structure of a colloidal ER fluid when applied electric field is 1000 V/mm and rotating speed of one electrode is 300 rpm. The gap between the electrodes is 1 mm, and the radius of the electrodes is 17.5 mm (the unit of the vertical axes is m) [6.36].

6.5 Shear Banding

6.5.1 *Experimental phenomena of shear banding*

In Chap. 1 we saw a sketch of Couette flow (Fig. 1.1 left, also Fig. 6.19(a)), and this is true when the flow structure is homogeneous, and the flow is in its quiescent state. However, when an applied shear rate exceeds a characteristic structural relaxation rate, the fluid structure would be re-organized into an inhomogeneous one, and the fluid is in its non-equilibrium state. As we saw in Section 6.2, the lamellar structures of ER fluids under certain high electric field and shear rate are such inhomogeneous ones, and ER fluids are in their non-equilibrium states.

One of the simple scenarios of inhomogeneous structure is the coexisting two bands of unequal shear rates shown in Fig. 6.19(b). [6.37]

There are two types of shear bandings (see Fig. 6.20). One is called *gradient* (*shear*) *banding*, where the resulting inhomogeneous fluid structure has different apparent viscosities which are separated along the flow speed gradient direction (y direction in Fig. 6.19(b)) and have the same shear stress. Another type of shear banding is

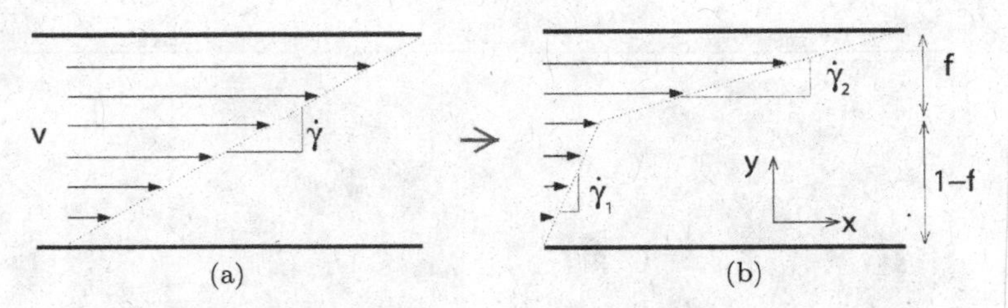

Fig. 6.19 The basic phenomenon of gradient shear banding. Homogeneous flow is unstable to the formation of macroscopic coexisting bands of unequal shear rates γdot$_1$, γdot$_2$. For a comparison of gradient banding and vorticity banding, see Fig. 6.20 [6.37].

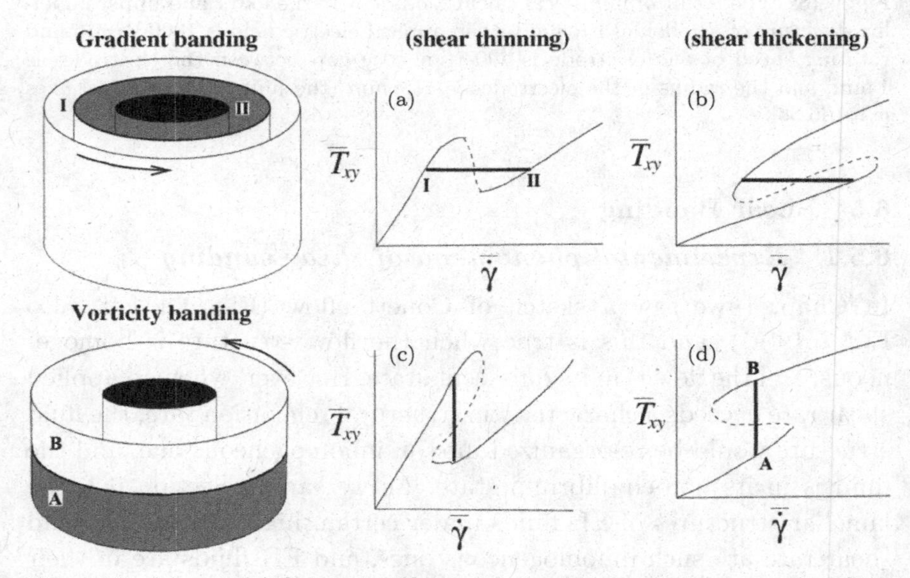

Fig. 6.20 (a) and (b) [(c) and (d)] are the candidate constitutive curves (*thin lines*) of average total stress $\bar{T}_{xy}$ vs. average shear rate $\bar{\dot{\gamma}}$ ← for shear thinning and shear thickening flow, respectively, in case of gradient [vorticity] shear banded flows. The *thick lines* could correspond to shear banding under either gradient (in (a) and (b)) or vorticity (in (c) and (d)) shear banding conditions. In all cases, homogeneous states could be observed for imposed conditions corresponding to the *thin solid lines*, while inhomogeneous shear-banding states could be observed for imposed conditions corresponding to the *thick solid lines*, while the *dashed lines* are putative unstable steady flows that must be resolved into inhomogeneous shear banding states. The upper (lower) left diagram shows schematically the arrangement of the gradient (vorticity) shear bands in curved Couette flow [6.38].

called *vorticity* (*shear*) *banding*, in which case, for an applied shear rate, the inhomogeneous fluid structure is separated into bands of different shear stress along the vorticity direction.

Shear banding occurs in many different soft matters. To list a few examples, rheological signatures consistent with that of Fig. 6.20(a) have been seen in wormlike micelle solutions, lyotropic lamellar surfactants, side-chain liquid crystal polymers, and many other systems. Figure 6.20(b) has been seen in wormlike micelle solutions. Figure 6.20(c) has been seen in colloidal. suspensions, and Fig. 6.20(d) has been seen in lyotropic lamellar surfactant solutions [6.37, 6.38].

6.5.2 *Constitutive models of shear banding [6.37–6.39]*

It is important to identify the microscopic mechanisms of flow-structure coupling: possibilities include entanglement effects, micellar breakage or enhanced length, liquid–crystalline effects, changes in charge and association, or other changes in micellar topology. Unfortunately, there remains little known about these effects and observation of these issues is an outstanding experimental challenge.

The origin of shear banding lies in the existence of the multiple branches in the constitutive "shear stress-shear rate" ($T - \dot{\gamma}$) curve in homogeneous flow. One of such curves is shown in Fig. 6.21(a) (see the dashed curve). In the regime of negative slope where $dT_{xy}/d\dot{\gamma} < 0$, homogeneous flow is unstable. This triggers the formation of bands of unequal shear rates $\dot{\gamma}_1$ and $\dot{\gamma}_2$, which coexist at a common shear stress $T_{xy} = T^*$. The bands formed lie in the direction perpendicular to the flow gradient direction. In this way, the system bypasses the unstable part of the constitutive curve, forming the shear bands which can stay stable for days to months.

(1) Phenomenological constitutive model

The nonlinearity of the shear stress-shear rate curve mentioned before can be analyzed with several constitutive models. These models can be divided into two classes, according to whether they are explicitly derived by considering the mesoscopic dynamics of the

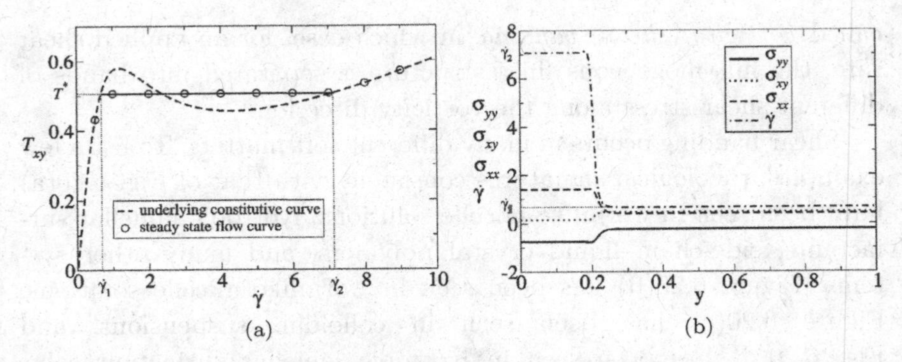

Fig. 6.21 (a) Homogeneous constitutive curve and steady state flow curve of the non local Johnson–Segalman model. (b) Corresponding steady state shear banded profile predicted by one-dimensional calculations. Shown are the shear rate and the various components of the viscoelastic stress. All quantities here are reduced ones and are dimensionless, and y is a normalized distance between the two plates (see Fig. 6.19(b)) [6.37].

solutes themselves, or just motivated phenomenologically. The former models are often too complicated to be used in studying shear banding, particularly when complex dynamics arises. So, here we are going to briefly introduce only the simple phenomenological models.

For a small cell of a constitutive polymer fluid, if the system has the density ρ and the pressure P, the velocity field is marked as $V(R)$, and using $\Sigma(R)$ denotes the viscoelastic stress carried by the polymeric molecules, the momentum balance provides

$$\rho(\partial_t + V \cdot \nabla)V = \nabla \cdot (\Sigma + \eta\nabla V - P\mathrm{I}). \tag{6.65}$$

For homogeneous planar shear, $V = y\dot{\gamma}\hat{x}$, and the total shear stress $T_{xy} = \Sigma_{xy}(\dot{\gamma}) + \eta\dot{\gamma}$.

Besides, incompressibility condition means zero divergence of flow velocity,

$$\nabla \cdot V = 0 \tag{6.66}$$

which gives the following continuity equation

$$\partial_t\rho + \nabla \cdot (\rho V) = 0. \tag{6.67}$$

Equations (6.64) and (6.65) are two basic equations of a constitutive fluid.

Next, suppose that the polymeric stress obeys diffusive Johnson–Segalman (DJS) model of fluid dynamics,

$$\overset{\diamond}{\Sigma} = 2GD - \frac{\Sigma}{\tau} + \frac{l^2}{\tau}\nabla^2\Sigma, \tag{6.68}$$

where the Gordon–Schowalter time derivative of viscoelastic stress Σ is

$$\overset{\diamond}{\Sigma} = (\partial_t + V \cdot \nabla)\Sigma - a(\mathrm{D} \cdot \Sigma + \Sigma : \mathrm{D}) - (\Sigma \cdot \Omega - \Omega \cdot \Sigma). \tag{6.69}$$

D and Ω are the symmetric and anti-symmetric parts of the velocity gradient tensor, respectively, $(\nabla V)_{\alpha\beta} = \partial_\alpha v_\beta$, G is the elastic modulus which depends on the concentration and here it is assumed to be a constant, and τ is the relaxation time.

If a =1 and D=0 are chosen, this model reduces to the Oldroyd B model. However, in this limit, the constitutive curve is $T_{xy} = G\dot{\gamma}\tau + \eta\dot{\gamma}$, which is monotonic, and so fails to capture the dramatic shear thinning needed to trigger shear banding.

To capture shear thinning, the DJS model invokes a "slip parameter" a with $|a| < 1$, so that the curve of $T_{xy}(\dot{\gamma})$ is then capable of non-monotonicity. The result is [6.40]

$$T_{xy}(\dot{\gamma}) = \frac{G\dot{\gamma}\tau}{1 + (1 - a^2)\tau^2\dot{\gamma}^2} + \eta\dot{\gamma}. \tag{6.70}$$

First, obviously, if a $\to 1$, $T_{xy}(\dot{\gamma}) = G\dot{\gamma}\tau + \eta\dot{\gamma}$ returns to a monotonic curve; second, if $\eta \geq G\tau/8$, the curve also returns to the monotonic curve, as Fig. 6.22 shows. So, to get the non-monotonic curve for shear banding, we need the viscosity $\eta < G\tau/8$ and the slip parameter $|a| < 1$, which means that the slip between the polymer and fluid extension is kept in $(-1, 1)$ [6.37].

(2) Mechanism of formation of two shear bands

The non-monotonic curve of shear stress-shear rate relation has been achieved by a simple DJS model in previous part. It predicts a region of negative slope which is instable for the fluid.

For an imposed shear rate $\bar{\dot{\gamma}}$ in this negatively sloping regime, this instability triggers the formation of shear bands. Force balance requires the shear stress T_{xy} to be uniform across the gap of the

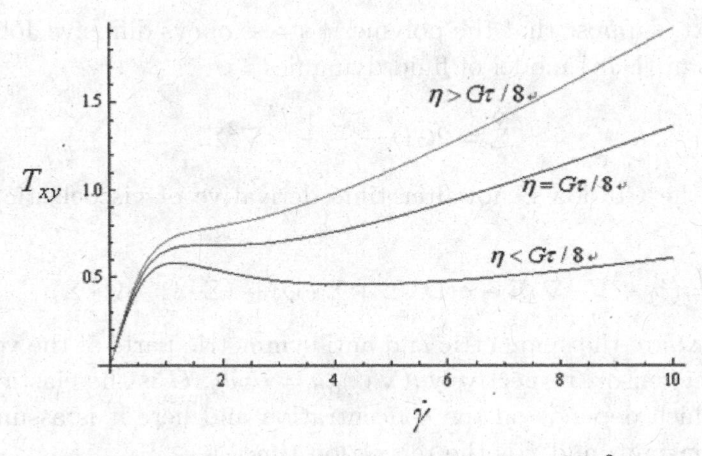

Fig. 6.22 Sketch map of Eq. (6.70) with different value of η. a^2 is set to be 0.1, and other parameters are set to 1 [6.39].

bands, and therefore common to all of the bands. However, while the imposed shear rate $\bar{\dot{\gamma}}$ remains constant, the local shear rate varies across the banding regime. Assume that the distance y between the two plates has been normalized to 1, and one of the bands in Fig. 6.19(b) has the thickness of f, and the other, $1 - f$. Then, the spatially averaged shear rate across the plates is

$$\bar{\dot{\gamma}} = (1 - f)\dot{\gamma}_1 + f\dot{\gamma}_2. \tag{6.71}$$

The system can adjust the thickness of the bands (by choosing a proper value of f), so that the shear rates varying across shear bands will have an effect as an average shear rate $\bar{\dot{\gamma}}$ of the whole system.

Therefore, the mechanism of shear banding can be drawn as follows: when the shear stress is promoted to an instable regime in the non-monotonic curve in a local area, the flow is separated into two bands, which have the same shear stress T^* but different shear rate $\dot{\gamma}_1$ and $\dot{\gamma}_2$. The bands would adjust their thickness to keep the spatially averaged shear rate consistent with the average shear rate $\bar{\dot{\gamma}}$, which is imposed by a rheometer.

While this model gives a brief mechanism of shear banding, it is too simple to take the interface of shear bands into account.

The interface here is just considered infinitely sharp, which is unphysical. Also, the shear banded flow of this model is instable and needs insightful thought [6.41].

References

[6.1] L. Lobry, E. Lamaire, *J. Elect.* **47** (1999), 61.

[6.2] J. G. Cao, J. P. Huang, L. W. Zhou, *Chem. Phys. Lett.* **419** (2006), 149.

[6.3] Jones T. K. Wan, K. W. Yu, and G. Q. Gu, *Phys. Rev. E* **64** (2001), 061501.

[6.4] B. Khusid, A. Acrivos, *Phys. Rev. E.*,**52** (1995), 1669.

[6.5] J. P. Huang and K. W. Yu, *Phys. Lett. A***333** (2004), 347.

[6.6] J. P. Huang, *J. Phys. Chem. B* **109** (2005), 4824.

[6.7] J. E. Martin, J. Odinek, T. C. Halsey, R. Kamien, Structure and dynamics of ERF, *Phys. Rev. E* **57** (1998), 756.

[6.8] R. Tao, Y. C. Lan, *Phys. Rev. E* **72** (2005), 041508.

[6.9] P. Tang, W.J. Tian, J.P. Huang, L.W. Zhou, On the Lorentz Local electric field in soft-matter systems, *J. Phys. Chem. B* **113** (2009), 5412–5417, and the references cited there.

[6.10] Z. Jiang, *J. Electrostat.* **58** (2003), 247; also Y. Ju, J.P. Huang, Dynamic Effects on Colloidal Electric Interactions, *J. Phys. Chem. B*, **112** (2008), 7865.

[6.11] H. See and M. Doi, *J. Phys. Soc. Jpn.* **61** (1992) 3984.

[6.12] J. M. Ginder. L. C. Davis and S. L. Ceccio, in F. E. Filisko and K. O. Havelka (eds). Proc. of the Electrorheological Materials and Fluids Symposium. Washington, DC. USA. 1994, Plenum, New York, (1995). J. M. Ginder and S. L. Ceccio. *J. Rheol.* **39** (1995), 1322.

[6.13] J. M. Ginder and L. C. Davis, in R. Tao and G. D. Roy (eds.), *Electrorheological Fluids: Mechanisms, Properties. Technology and Applications*, Proc. of the Fourth Int. Conf. on Electrorheological Fluids, World Scientific, Singapore (1994), 267

[6.14] X. Tang, W. H. Li, X. J. Wang, P. Q. Zhang, Structure Evolution of Electrorheological fluids under flow conditions, *Inter. J. Mod. Phys. B.* **13** (1999), 1806.

[6.15] F. E. Filisco and S. Henley, *Inter. J. Mod. Phys. B.* **15** (2001), 686.

[6.16] S. Henley and F. E. Filisco, *Inter. J. Mod. Phys. B* **16** (2002), 2286.

[6.17] R. Tao, Structure and dynamics of dipolar fluids under strong shear, *Chem. Engin. Sci.* **61** (2006), 2186.

[6.18] M. E. Cates, T. C. B. McLeish, G. Marrucci, The rheology of en tangled polymers at very high shear rates, *Europhys. Lett.* **21** (1993), 451.

[6.19] X. F. Yuan, Dynamics of a mechanical interface in shear-banded flow, *Europhys. Lett.* **46** (1999), 542–548.

[6.20] D. K. Liu, C. Li, J. Yao, L.W. Zhou, J.P. Huang, Relation of lamellar structure and shear stress of dynamic PM-ER fluids. Preprint.

[6.21] K. von Pfeil, M. D. Graham, D. J. Klingenberg, and J. F. Morris, Pattern Formation in Flowing Electrorheological Fluids, *Phys. Rev. Lett.* **88** (2002), 188301.

[6.22] K. von Pfeil, M. D. Graham, D. J. Klingenberg, and J. F. Morris, Structure evolution in electrorheological and magnetorheological suspensions from a continuum perspective, *J. App. Phys.* **93** (2003) 5769.

[6.23] J. F. Morris and F. Boulay, *J. Rheol.* **43** (1999), 1213.

[6.24] J. D. Jackson, *Classical Electrodynamics* (3rd. ed.) Wiley (1999) p. 261.

[6.25] L. D. Landau and E. M. Lifshitz. Vol. 8. *Electrodynamics of Continuous Media* (2nd ed.), Pergamon (1984) 64.

[6.26] Y. M. Shkel and D. J. Klingenberg,*J. Appl. Phys.* **83** (1998), 7834.

[6.27] T. C. Haley, R. T. Lahey, D.A. Drew, A characteristic analysis of void waves using two-fluid models, *Nucl. Engin. Design* **139** (1993), 45–57.

[6.28] Jianwei Zhang, Xiuqing Gong, Chun Liu, Weijia Wen and Ping Sheng, Electrorheological Fluid Dynamics, *Phys. Rev. Lett.* **101** (2008), 194503.

[6.29] Ping Sheng, Lectures in Kaliv Institute of Theoretical Physics, May-June 2008.

[6.30] Ping Sheng, Tiezheng Qian and Xiaoping Wang, Continuum Modelling of Nanoscale Hydrodynamics, in Zikang Tang and Ping Sheng eds., *Nanoscale Phenomena Basic Science to Device Applications, Lecture Notes in Nanoscale Science and Technology 2*, Chap. 9, Springer, (2008).

[6.31] Jianwei Zhang, Ping Sheng, Chun Liu, Onsager Principle and the Electrorheological Fluid Dynamics, *Prof. Theo. Phys. Supp.* **175** (2008), 131–143.

[6.32] L. Onsager, Reciprocal relations in irreversible process I, *Phys. Rev.* **37** (1931), 405–426.

[6.33] L. Onsager, Reciprocal relations in irreversible process II, *Phys. Rev.* **38** (1931), 2265–2279.

[6.34] L. Onsager and S. Machlup, Fluctuations and Irreversible Processes, *Phys. Rev.* **91** (1953), 1505.

[6.35] http://www.cscs.umich.edu/~crshalizi/notabene/noneq-sm.html

[6.36] C. Li, J. Zheng, J.W. Zhang, J.P. Huang and L.W. Zhou, Dynamics of PMER fluids, Preprint.

[6.37] S. M. Fielding, Complex dynamics of shear banded flows, *Soft Matter* **3** (2007), 1262, and references cited there.

[6.38] P. D. Olmsted, Perspectives on shear banding in complex fluids, *Rheol. Acta* **47** (2008), 283, and references cited there.

[6.39] Y. Zhang and X. Y. Zhou, A brief introduction to shear banding, Soft Matter Physics Lecture Midterm paper, Fudan University, (2010).

[6.40] M. W. Johnson, D. Segalman, *J. Non-Newton. Fluid* **2** (1977), 255.

[6.41] C. Y. D. Lu, P. D. Olmsted, R. C. Ball, *Phys. Rev. Lett.* **84** (2000), 642.

Chapter 7

Granular Systems

7.1 Introduction

One of the familiar granular systems is sand, with which you may play on a beach. Are sands soft or hard matters? They are so hard that it needs quite a lot of energy to grind them into fine powders. It does look like soft matter: if you hold a cup of sand, the shape of the sand pile looks exactly like the interior of the cup — just like water. You surely have seen sand waves in movies or pictures. Are sands somewhat like an ocean wave? Much more peculiar properties of granular systems that we will study in this chapter would show us that granular materials are, in many senses, like liquid. However, they are a very special type of soft matters.

Granular materials are usually considered to be particles of sizes ranging from 10 μm to 3 mm. These particles can be classified as in Table 7.1. The order of magnitude of the ratio of the potential energy of a sand grain of mass m at the height of its own diameter d over its thermal energy, mgd/k_BT, ranges from 10^2 to 10^{11} at room temperature when the particle sizes d range from 10 μm to 3 mm, so the temperature effect can be neglected in the study of granular materials. Interaction between granular materials can be hard or soft sphere interaction when particle weight, friction and collisions are considered. Granular materials can be treated as disperse media of solid, liquid and gas states. Granular systems are nonlinear and energy dissipative, or the granular material is of a stable and open system

Table 7.1. Particle size classification [7.1].

Particle size range	0.1–1 μm	1–10 μm	10–100 μm	0.1–3 mm	3–10 mm
Classification	Ultra-fine powder	Superfine powder	Granular powder	Granular solid	Broken solid
	⟵ Powders ⟶				
			⟵ Granular materials ⟶		
			⟵ Usual working range ⟶		

which is operating far from thermodynamic equilibrium within an environment that exchanges energy and entropy.

Dry granular materials consist of small discrete solid constituents, e.g., sand, rocks, snow, salt, grains, pills, milk powder, styrofoam, etc. Granular materials are important in agricultural, pharmaceutical, food industries and other industries for the storage, transport and manipulation of grains, seeds, tablets, ore, chemical powder, and so forth. The mechanics of individual components is known but their collective behavior is not fully understood.

The study of granular materials has a long history. In 1773, Coulomb's law of friction between solids was proposed by C.-A. de Coulomb in studying the stability of stone architecture. The law involves the following issues: friction between solids is proportional to the normal force between them, or $f \propto N$; f is independent of the contact area between the solids; and the static friction coefficient is larger than the dynamic one, or $\mu_s > \mu_d$. In 1780, E. Chladni discovered that light and smooth particles obey rules different from those for heavy and rough particles in the movement of mixed particles. In 1831, M. Faraday found that vibration of particles may lead to convection, and further, a heaping of particles. In 1885, O. Reynolds proposed that if particles are closely packed in an elastic bag, any deformation would expand their volume, and this is one of the basic rules of granular materials. In 1895, H.A. Janssen explained the silo effect by pointing out that the vertical force in a silo is decomposed, and the silo wall supports a large amount of weight due to friction. P.G. de Gennes pointed out in 1999 that "Granular matter is a new type of condensed matter, as fundamental as a liquid or a solid and showing in fact two states: one fluid-like, one solid-like. However, there is as yet no consensus on the description of these two states. Granular matter, in 1998, is at the level of solid-state physics in 1930" [7.2].

The *Janssen model* of the *silo problem* is a good example to explore the difference between granular materials and liquid.

Granular matter usually refers to a collection of many particles. Though individual particles seem to just obey Newtonian mechanics, the collective behavior is quite complicated, highly nonlinear and hard to solve [7.3, 7.4].

Vanel and Duran [7.3] defined apparent mass. Figure 7.1 experimentally simulates a silo with cylinder radius 20 mm and bead diameter 1.5, 2 and 3 mm. The apparent mass of the grains inside the container increases linearly at first, and soon deviates from the linearity and saturates at a certain value.

The microscopic definition of deformation for discrete hard particles is problematic. Usually, contact force laws are used to calculate contact forces between individual grains when they are considered as rigid bodies.

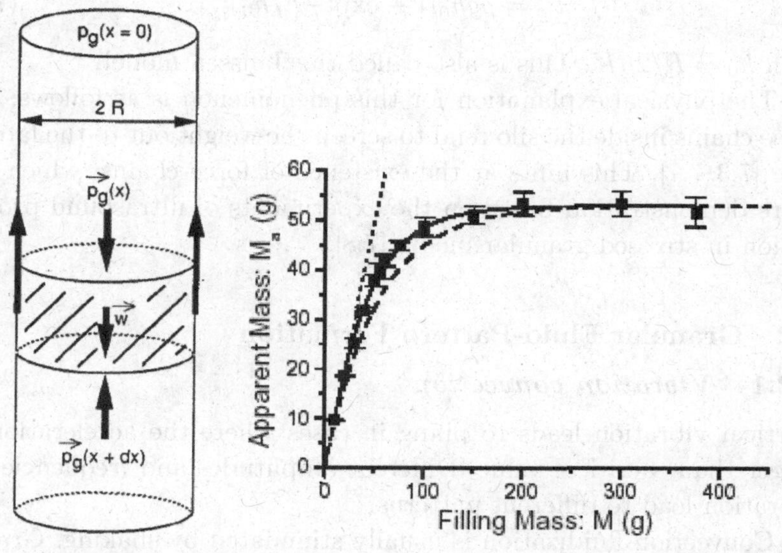

Fig. 7.1 The apparent mass vs. filling mass, the straight line indicates ρgh [7.6].

The bottom pressure P in a silo is quite different from what one sees in a liquid, $P = \rho gh(\rho =$ liquid density, $g =$ gravitational acceleration, and $h =$ liquid height.). In a silo, $P = \rho gh[1 - \exp(h/h_0)]$ (symbols to be explained later).

Janssen *et al.* [7.5] found the well-known "silo problem" that the pressure measured at the bottom of a silo filled with grains is much smaller than the hydrostatic pressure ρgh in a liquid and tends to be saturated when the silo height is larger than a critical h_0.

The Janssen model supposes that the weight is screened out, which indicates the existence of force chains pressure within packing. In [7.3], the weight of small slice dx supported by pressure gradient and friction at the wall is

$$\rho gAdx = [P_g(x + dx) - P_g(x)]A + \text{friction} \times 2\pi Rdx, \tag{7.1}$$

assuming normal force N at the wall is proportional to the pressure, the normal force $N = KP_g$, and then friction $= \mu N = \mu KP_g$ ($K =$ friction coefficient). One has

$$\rho g = dP_g/dx + 2\mu KP_g/R, \tag{7.2}$$

$$P_g = \rho gh_0[1 - \exp(-h/h_0)], \tag{7.3}$$

with $h_0 = R/2\mu K$. This is also called the Janssen model.

The physical explanation for this phenomenon is as follows: the force chains inside the silo tend to screen the weight out to the lateral side [7.3, 7.7]. This hints at the existence of force chains, which get more demonstration based on the experiments of ultrasound propagation in stressed granular media [7.8].

7.2 Granular Fluid-Pattern Formation

7.2.1 *Vibration convection*

Vertical vibration leads to piling in cases where the acceleration is larger than a critical value. Different amplitudes and frequencies of vibration lead to different patterns.

Convection fluidization is usually stimulated by shaking. Granular materials start to flow when shaking acceleration $A\omega^2$ (A is the

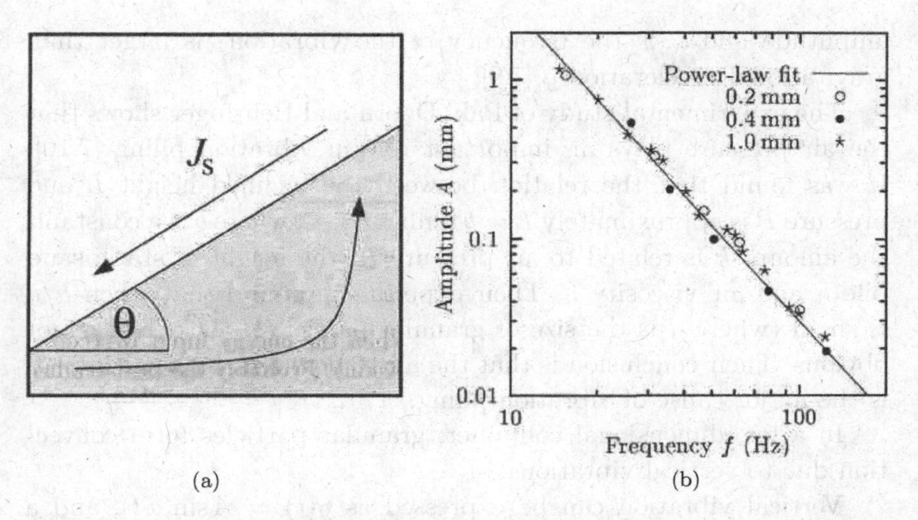

Fig. 7.2 The relation of amplitude and frequency shows a power law [7.9].

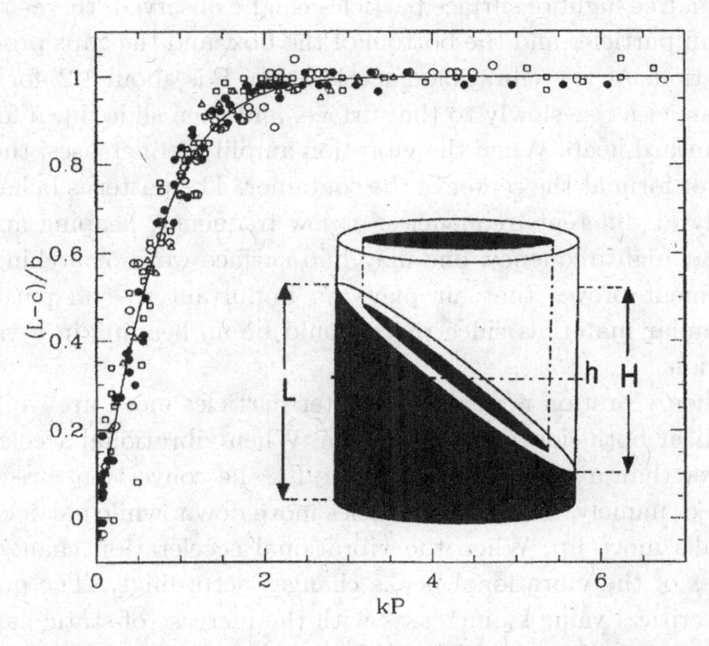

Fig. 7.3 Vibration is effected by the air pressure P in the container [7.10].

amplitude and ω is the frequency of the vibration) is larger than gravitational acceleration g [7.9].

The experimental study of Pak, Doorn and Behringer shows that the air pressure plays an important role in vibration piling [7.10]. It was found that the relation between the inclined height L and pressure P is approximately $L = b\tanh(kP)+\mathrm{c}$, where b is a constant, the amount k is related to air pressure P, the height of static sand pile h and air viscosity η. Their experiments found that when h/d is small (where d is the size of granular particles), the piling is not obvious. Their conclusion is that the air in between the particle gaps is the major cause of vibration piling.

In a two-dimensional container, granular particles form convection due to vertical vibrations.

Vertical vibration can be expressed as $y(t) = A\sin(\omega t)$, and a dimensionless vibration acceleration can be expressed as $\Gamma = A\omega^2/g$. When $\Gamma \leq 1$, there is no particle displacement, nor fluidization; when $\Gamma > 1$, a free flight of surface particles can be observed, there are gaps between particles and the bottom of the box, and the gaps produce a convection. At low vibrational acceleration, Γ is about 1.2, for example, particles rise slowly to the surface, and then slide down forming a downward heap. When the vibration amplitude increases, the heap does not form at the center of the container. The patterns behave differently at different frequencies: at low frequency, heaping appears, while at high frequency, one may find surface wave or arching. The experiment proves that air plays an important role in patterning of granular materials since there would be no heaping in a vacuum condition.

When vibration is weak, the center particles move up, while particles near both sidewalls move down. When vibrational acceleration is larger than a certain critical value Γ_c, the convection direction is reversed, namely, the center particles move down, while particles near sidewalls move up. When the vibrational acceleration changes, the number of the vibrational peaks changes accordingly. The quantity of the critical value Γ_c increases with the increase of static height h, and it is also affected with particle size d.

Through an experiment using magnetic resonance images, Knight found that when the outward inclining angle is above a certain value, the direction of particle convection is also reversed [7.11]. It was found that a horizontal spin-tagging pattern forms before vibration, and the deformation of a horizontal spin-tagging pattern appears after a single shake of $\Gamma = 5$ and $f = 20$ Hz. This experimental result is consistent with the result of computer simulation by Grossman [7.12], namely that the vertical vibration convection reverses its direction when the outward inclining angle is above 10° [7.13].

A simulation of convection [7.14] shows that the mechanism and the configuration of convection due to horizontal vibration are different from those due to vertical vibration. Konn did experiments on sand piling when both horizontal and vertical vibration are applied, and it was found that the characteristic behaviors of vibration piling and particle convection depend not only on acceleration, but also on the phase difference of the two vibrations. The horizontal vibration causes gaps existing between particles and the container wall, while the collapse of surface particles along the gaps is the source of the two convections.

7.2.2 *2D pattern formation*

(1) Comparisons between experiments and simulations

We will first introduce the experiment and simulation of two-dimensional quasi-pattern formation under vertical vibration.

Kim and Pak [7.16] studied the process of pattern formation in granular layers through both experiments and simulations. Ten layers of granular materials inside a vacuum container were placed under a vertical vibration of $A \sin 2\pi f t$. Control parameters were the dimensionless acceleration $\Gamma = A(2\pi f)^2/g$ and vibration frequency f (50 Hz). By increasing Γ in 50 μs, the system is quenched from a flat pattern state ($\Gamma = 2.4$) to a striped pattern state ($\Gamma = 2.8$), and more than 10^4 cycles were experienced before a full steady striped pattern appeared. This non-equilibrium and non-steady process showed dynamic scaling behavior. The growth exponent of the

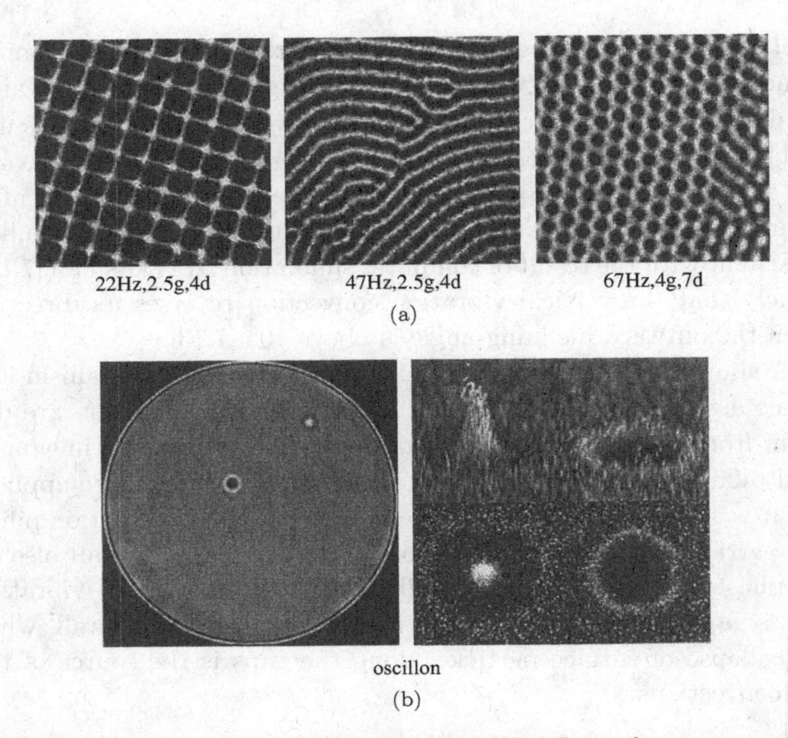

Fig. 7.4 (a) Patterns formed from vibrations with different frequency, acceleration and particle size; (b) Oscillons can be formed under certain condition [7.15].

characteristic length scale of the ordered domain was 0.25, which agrees with that of the Swift–Hohenberg system. Their experimental results (Fig. 7.5(a)–(d)) are compared with the simulation results (Fig. 7.5(e)–(h)). Bright parts in the recorded images corresponded to the high level area of the free surface, and dark parts corresponded to the low level area of the free surface.

(2) Simulation based on Swift–Hohenberg model

(i) Swift–Hohenberg equation

The Swift–Hohenberg model is based on three phenomenological rules. First, the system undergoes a type I instability (an instability associated with an inflection point in the profile of the normal

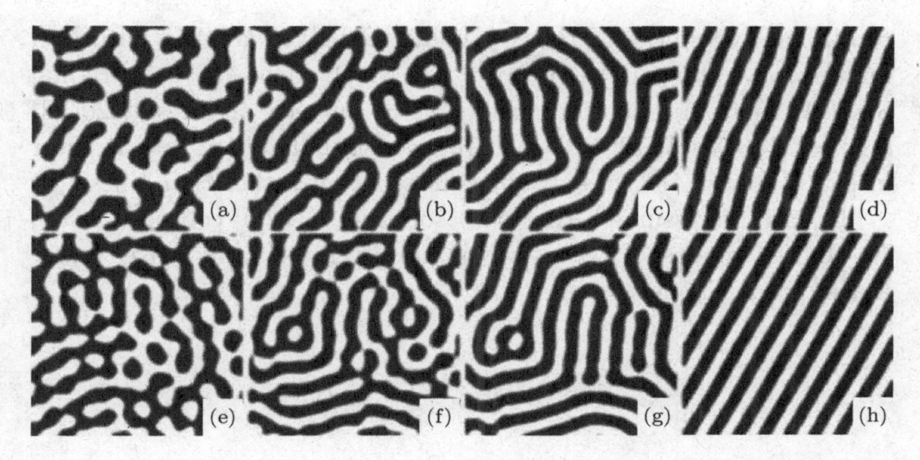

Fig. 7.5 Evolution of patterns in time. (a)–(d): The real images of the free surface of granular layers for $\Gamma_i = 2.4$ and $\Gamma_f = 2.8$ at t = 2 (a), $t = 10$ (b), $t = 200$ (c), and $t = 1000$ periods after quench (d). The bright parts correspond to the crests of the free surface, and the dark parts correspond to the troughs of the free surface. (e)–(h): The images of the numerical results of the 2D Swift–Hohenberg equation in time for $\varepsilon = 0.2$ [7.16].

velocity with respect to the disk plane [7.17]) which generates a pattern with a given wave vector around a critical value k_c. Second, the equation is invariant under the $u \to -u$ transformation, where u is a real scalar field. In this experiment, the transformation corresponds to the exchange of the crests and the troughs in the striped pattern. Third, due to the nonlinear interaction term, u^3.

Mueller [7.18] uses a two-dimensional simulation to show that the nonlinear mechanisms responsible for the pattern selection might give an insight into the physics prevailing in experiments. His starting point is a two-dimensional Swift–Hohenberg (SH)-type equation [7.19] for a real scalar field $u(x, y, t)$

$$\partial_1 u = \varepsilon u - (\nabla^2 + 1)^2 u - c(u). \tag{7.4}$$

The quantity $u(x, y, t)$ models the order-parameter fields in the system, ε measures the distance to the stability thresholds, and $\nabla = (\partial/\partial x, \partial/\partial y)$. The linear operator in Eq. (7.4) gives a stationary instability at a finite wave number $k_c = 1$. The direction of $k = (k_x, k_y)$ is infinitely degenerate, reflecting the continuous rotational symmetry in the x-y plane. c(u) changes the nature of nonlinearity. If c(u) $=$ $(1/3)u^3$, Eq. (7.4) comes to the original Swift–Hohenberg equation producing a pattern of lines [7.20]. If c(u) $= (1/3)\nabla \cdot [\nabla u(\nabla u)^2]$ as proposed by Chapman and Proctor in [7.20], Eq. (7.4) drives a square tessellation rather than lines.

It can be assumed that the solution u is a composition of N Fourier modes of waves with amplitude A_n and wave vector $\mathbf{k}_n(|\mathbf{k}_n| = 1)$, namely,

$$\bar{u} = \sum_{n=1}^{\bar{N}} A_n e^{ik_n \cdot r} + c \cdot c, \tag{7.5}$$

where each $\mathbf{k}_n$ lies on the critical unit circle, and the temporal evolution of the mode amplitudes $\{A_n\}$ is governed by a set of coupled Landau equations [7.21].

$$\partial_t A_n = \varepsilon A_n - \sum_{m=1}^{N} \beta(\theta_{nm})|A_m|^2 A_n, \quad n = 1, \ldots, N. \tag{7.6}$$

(ii) Free energy minimization

These Landau equations of amplitudes are gradient equations with

$$\partial_1 A_n = -\partial F / \partial A_n^*, \tag{7.7}$$

where the free energy is

$$F = -\varepsilon \sum_{i=1}^{N} |A_i|^2 + \frac{1}{2} \sum_{i,j=1}^{N} \beta(\theta_{i,j})|A_i|^2|A_j|^2. \tag{7.8}$$

The coupling function $\beta(\theta)$, evaluated at the angles θ_{ij} between $\mathbf{k}_i$ and $\mathbf{k}_j$, $\theta_{ij} = (\mathrm{n} - \mathrm{m})\pi/N$, is characteristic for the c(u) under consideration. It determines which pattern is to be selected. Reflection and rotational symmetry imposed on c(u) implies $\beta\,(\theta) = \beta\,(\pi - \theta) = \beta\,(-\theta)$.

As F is a function of $(A_1^*, A_2^*, A_3^*, \ldots)$, the minimization of F requires that Eq. (7.7) = 0. Thus, [7.22]

$$\varepsilon A_n - \sum_{m=1}^{N} \beta(\theta_{nm}) |A_m|^2 A_n = 0, n = 1, 2, \ldots, N \tag{7.9}$$

$$\varepsilon = \sum_{m=1}^{N} \beta(\theta_{nm}) |A_m|^2, n = 1, 2, \ldots, N \tag{7.10}$$

with $\beta(0) = 1$. There are N equations of ε and $|A_n| = A$, so $\beta(i\pi/N) = \beta[(N-i)\pi/N]$, except for $\beta(\theta = 0) = (1/2)\beta(\theta \to 0)$, ($\beta$ is discontinuous at $\theta = 0$ and π). Also,

$$\varepsilon = \sum_{0}^{N-1} \beta\left(\frac{\mathrm{n}\pi}{\mathrm{N}}\right) A^2. \tag{7.11}$$

Then

$$F = -\varepsilon \sum_{i=1}^{N} |A_i|^2 + \frac{1}{2} \sum_{i,j=1}^{N} \beta(\theta_{ij}) |A_i|^2 |A_j|^2$$
$$= -\mathrm{N}\varepsilon A^2 + \frac{1}{2}\mathrm{N}\varepsilon A^2. \tag{7.12}$$

Inserting Eq. (7.11) into Eq. (7.12) leads to

$$F_N = -\frac{1}{2}\varepsilon^2 \left[\frac{1}{N} \sum_{n=0}^{N-1} \beta(n\pi/N)\right]^{-1}. \tag{7.13}$$

From this expression the leading role of the coupling functions $\beta(\theta)$ becomes evident.

(iii) Two-dimensional simulation

Equation (7.6) allows equal amplitude fixed point solutions of the form $|A_n| = A$, $|k_n| = k_c = 1$, and the angle of $(k_n, k_m) = |n-m|\pi/N$, with $n, m = 1, 2, \ldots, N$. They describe regular periodic ($N \leq 3$) or quasiperiodic ($N > 3$) patterns of $2N$-fold orientational order. For $N = 1, F_1 = -(1/2)\varepsilon^2$, one has a pattern of lines; for $N = 2$, $F_2 = -(1/3)\varepsilon^2$, a pattern of squares; for $N = 3$, $F_3 = -(3/10)\varepsilon^2$, a hexagonal or triangle pattern; and $N = 4$, an octagonal quasipattern depending on the relative phases of the A_n. They are shown in Figs. 7.6–7.9 [7.18, 7.22, 7.23].

As mentioned before, we can work out the expression of amplitude $A_n(\varepsilon, \beta)$ in stable status by substitution of $\beta(\theta)$ for various n, N, and image the pattern in MATLAB by calculating out a matrix of the value U(x, y) for different N.

During the simulation, $\varepsilon = 0.04$ is chosen. The values of $\beta(\theta)$ are given below [7.18].

$$\begin{gathered} N = 1, \beta(0) = 1; \\ N = 2, \beta(0) = 1, \beta(\pi/2) = 2/3; \\ N = 3, \beta(0) = 1, \beta(\pi/3) = \beta(2\pi/3) = 2/3; \\ N = 4, \beta(0) = 1, \beta(\pi/4) = \beta(\pi/2) = \beta(3\pi/4) = 2/3. \end{gathered} \tag{7.14}$$

a. $N = 1$, Lines

Once amplitude A_n is solved, the expression of u will then be determined. For stable status, with the minimization of free energy, the amplitude for $N = 1$ (where $c(u) = (1/3)u^3$, line pattern formation) is [7.21]

$$|A_1|^2 = [\beta(0)]^{-1}\varepsilon A_1 = A_1^* = \sqrt{\varepsilon/\beta(0)} = \sqrt{\varepsilon}. \tag{7.15}$$

After substitution of the above expression into the expansion of u and definition of the direction of the only wavevector $\boldsymbol{k}_1 = (1, 0)$, then we reach

$$u_1 = A_1 e^{ik_1 \cdot r} + A_1^* e^{-ik_1 \cdot r} = 2\sqrt{\varepsilon}\cos(x). \tag{7.16}$$

The amplitude A_1 and solution u_1 are solved and only one Faraday wave with wavevector (1, 0) and the pattern is lines or rolls (Fig. 7.6).

b. $N = 2$, Squares [7.22, 7.23]

Since

$$u = \sum_{n=1}^{N} A_n e^{ik_n \cdot r} + c.c, \tag{7.17}$$

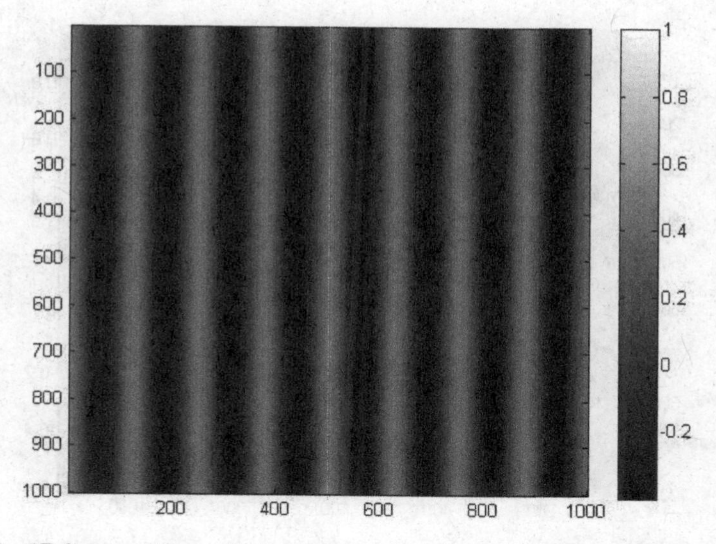

Fig. 7.6 2D linear pattern of granules under vertical vibration when $N = 1$ [7.22].

$$u = \sum_{n=1}^{N} A_n e^{i(\sin(\frac{n\pi}{N})x+\cos(\frac{n\pi}{N})y)} + A_n^* e^{-i(\sin(\frac{n\pi}{N})x+\cos(\frac{n\pi}{N})y)}$$

$$\begin{aligned}
&when\ N = 2\\
u &= A_1 e^{i(\sin(\frac{1}{2}\pi)x+\cos(\frac{\pi}{2})y)} + A_1^* e^{-i(\sin(\frac{\pi}{2})x+\cos(\frac{\pi}{2})y)}\\
&\quad + A_2 e^{i(\sin(\pi)x+\cos(\pi)y)} + A_2^* e^{-i(\sin(\pi)x+\cos(\pi)y)}\\
&= A_1 e^{ix} + A_1^* e^{-ix} + A_2 e^{-iy} + A_2^* e^{iy}\\
&= |A| \left(e^{i(x+\varphi)} + e^{-i(x+\varphi)} + e^{-i(y+\phi)} + e^{i(y+\phi)}\right)\\
&= |A| \left(2\cos(x+\varphi) + 2\cos(y+\phi)\right),
\end{aligned} \tag{7.18}$$

$$A_2 = A_2{}^* = \sqrt{\varepsilon / \left[\beta(0) + \beta\left(\frac{\pi}{2}\right)\right]} = \sqrt{\varepsilon/(1+2/3)}. \tag{7.19}$$

Since ϕ and φ have only impacts on phases, not on the patterns, they can be chosen as $\phi,\varphi = 0$.

$$u_2 = 2\sqrt{\varepsilon/(1+2/3)}(\cos(x) + \cos(y)). \tag{7.20}$$

The pattern for $N = 2$ is plotted in Fig. 7.7.

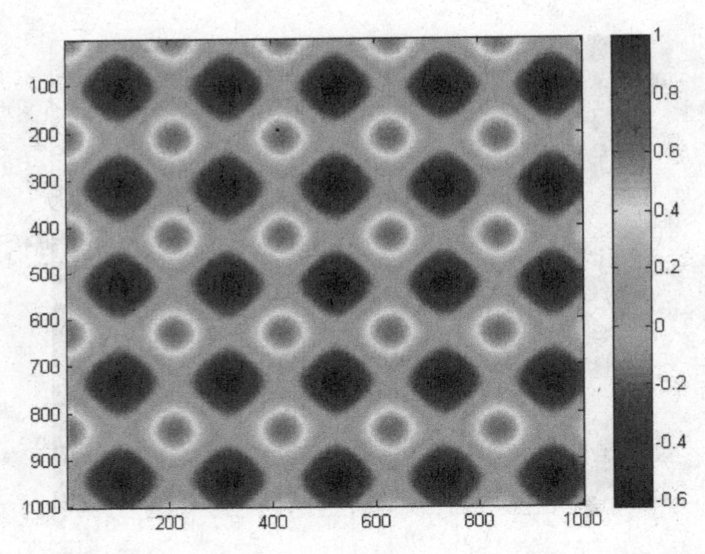

Fig. 7.7 Two orthogonal waves are superimposed to result in a square pattern. Red means the maximum value and blue, the minimum [7.22].

c. $N = 3$, Hexagons or Triangles [7.22]

Similarly, when $N = 3$,

$$\mathrm{A}_3 = \mathrm{A}_3{}^* = \sqrt{\varepsilon/\left[\beta(0) + \beta\left(\frac{\pi}{3}\right) + \beta\left(\frac{2\pi}{3}\right)\right]} = \sqrt{\varepsilon/(1 + 2 \times 2/3)} \tag{7.21}$$

$$u_3 = 2\sqrt{\varepsilon/(1+4/3)}\left[\cos(x) + \left(\frac{x+\sqrt{3}y}{2}\right) + \cos\left(\frac{x-\sqrt{3}y}{2}\right)\right]. \tag{7.22}$$

The corresponding patterns are hexagons (or triangles) as shown in Fig. 7.8.

d. $N = 4$, octagonal quasi-patterns [7.22]

When N =4,

$$\begin{aligned}\mathrm{A}_4 = \mathrm{A}_4{}^* &= \sqrt{\frac{\varepsilon}{\beta(0) + \beta\left(\frac{\pi}{4}\right) + \beta\left(\frac{\pi}{2}\right) + \beta\left(\frac{3\pi}{4}\right)}} \\ &= \sqrt{\varepsilon/(1 + 3 \times 2/3)},\end{aligned} \tag{7.23}$$

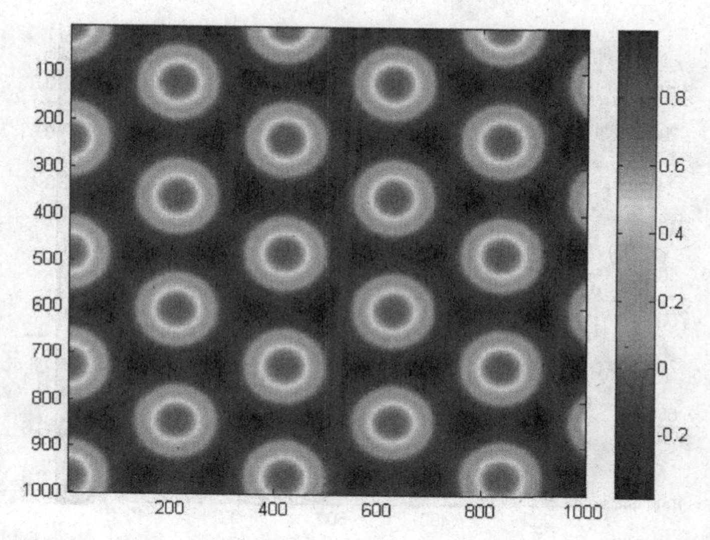

Fig. 7.8 Clearly, hexagons or triangles, depending on the relative phases of the amplitude A, can be found in the pattern [7.22].

and

$$u_4 = 2\sqrt{\varepsilon/3}\left(\cos(x) + \cos(y) + \cos\left(\frac{x+y}{\sqrt{2}}\right) + \cos\left(\frac{x-y}{\sqrt{2}}\right)\right). \tag{7.24}$$

The corresponding quasi-periodic pattern can be seen in Fig. 7.9.

For $N = 5, 6\ldots$, patterns can also be drawn in the MATLAB program with the calculation of the angles between waves.

7.2.3 *3D pattern formation*

In this section, the experimentation and simulation of 3D pattern formation under vertical vibration will be compared. Bizon *et al.* [7.24] did experiments and simulations of three-dimensional problem of granular particle vibration.

The simulation method used in Ref. [7.24] is an "event driven algorithm" [7.25–7.28]. In this simulation, particles interact only through instantaneous binary collisions and move only under the

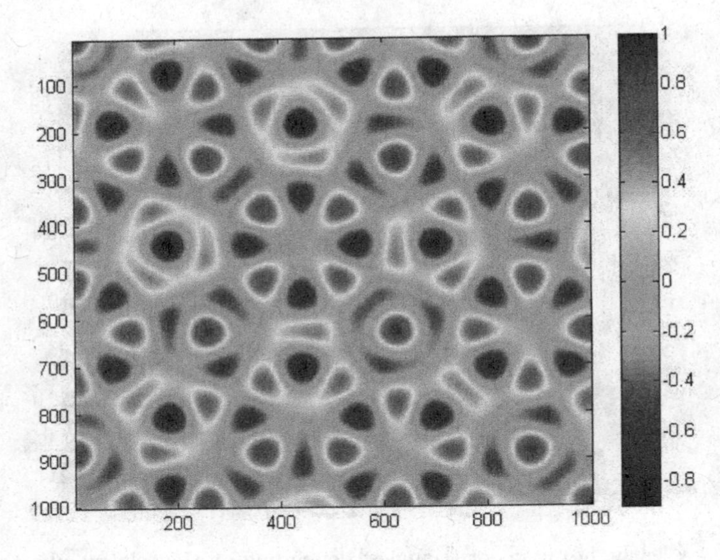

Fig. 7.9 Quasi-periodic pattern when $N = 4$ [7.22].

influence of gravity between collisions. Linear and angular momentum are conserved in collisions, while energy is dissipated. Thus the collision rule can be obtained as follows.

The value of the relative normal velocity v'_n of colliding particles after a collision is

$$v'_n = -v_n e(v_n), \tag{7.25}$$

where v_n is the pre-collision velocity and e is the coefficient of restitution. In their simulation it is assumed that

$$e(v_n) = 1 - B{v_n}^{\alpha} \quad \text{for } v_n < v_0, \tag{7.26}$$

$$e(v_n) = \varepsilon \quad \text{for } v_n > v_0, \tag{7.27}$$

where $v_0 = (gD)^{1/2}$ is a crossover velocity, $B = (1-\varepsilon)(v_0)^{-\alpha}$, ε is a constant, and index $\alpha = 0.75$.

The initial condition for all simulations is a uniform distribution of particles in physical space and a Maxwellian distribution in velocity space.

Particle numbers P are chosen to be either 30000 or 60000, side length L of the square container is $100D$ where D is the average diameter of particles. The layer depth $H = P(\pi/6)(D^3/L^2)/\varphi$, where the packing fraction φ was experimentally measured to be 0.58. Dimensionless frequency f^* is defined as $f^* = f\sqrt{H/g}, 0.1 < f^* < 2$, where the vibration frequency f, and g is gravitational acceleration. Dimensionless acceleration Γ is defined as $\Gamma = 4\pi^2 f^2 A/g$ for $2 \leq \Gamma \leq \tau$, where A is the amplitude of the vertically vibrating force $A \sin 2\pi ft$.

The particle velocities obtained from the collision rules using an event-driven algorithm will give the patterns of the granules.

The result of the computer simulation of the pattern formation under vertical vibration is shown in Fig. 7.10 [7.24]. Their results from simulation and experiment are remarkably similar over the

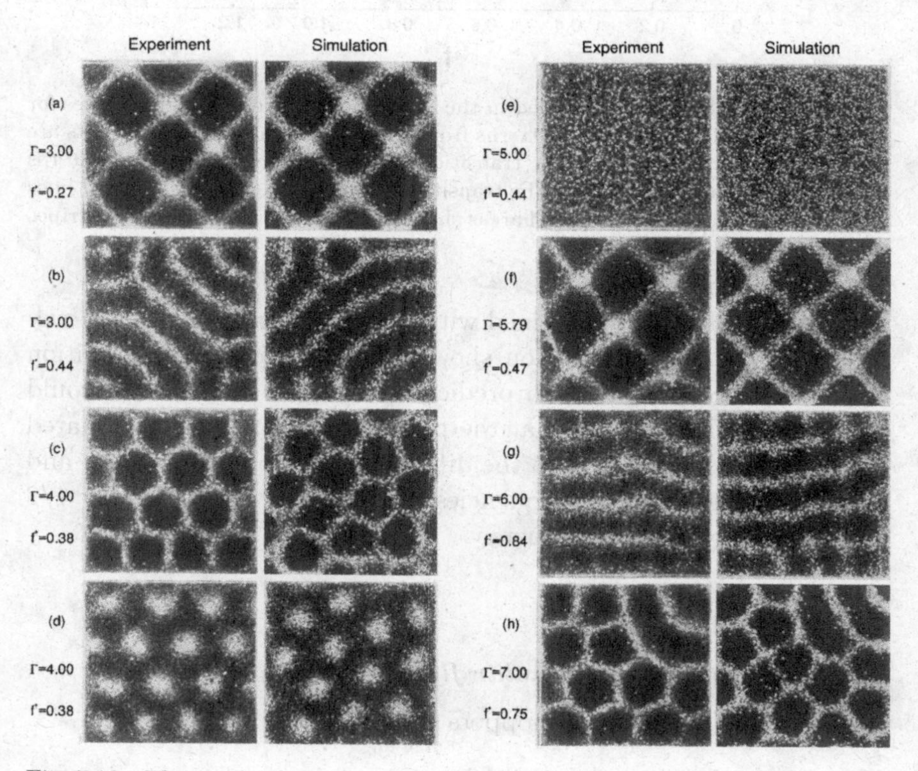

Fig. 7.10 The result of the computer simulation of the pattern formation under vertical vibration [7.24].

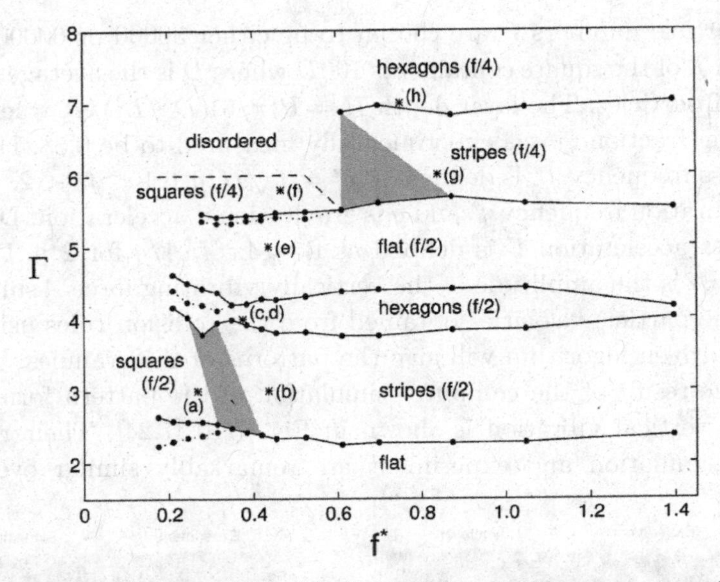

Fig. 7.11 Phase diagram obtained in the experiments. The parameter values for the comparison in Fig. 7.10 of patterns from the simulations and experiments are indicated by (a) through (h). The transitions from a flat layer to square patterns are hysteretic: solid lines denote the transition for increasing Γ, while dotted lines denote it for decreasing Γ. Shaded areas show transitional regions between stripes and squares [7.24].

entire range of (f^*, Γ) examined with patterns obtained at seven values of (f^*, Γ). The simulation shows the mechanism and condition of pattern formation. It also predicts what types of patterns would be expected. They found that the patterns become striped, squared and hexagonal, according to the different vibration acceleration and frequency. The intrinsic frequencies of the patterns are usually 1/2 or 1/4 of the vibration frequency.

7.3 Granular Flow

7.3.1 *Jamming of granular flow*

(1) Jamming in 2D or 3D hoppers

Jamming is often seen in traffic flows, emptying from auditoria after shows, or in escaping of large groups of people from fire locales.

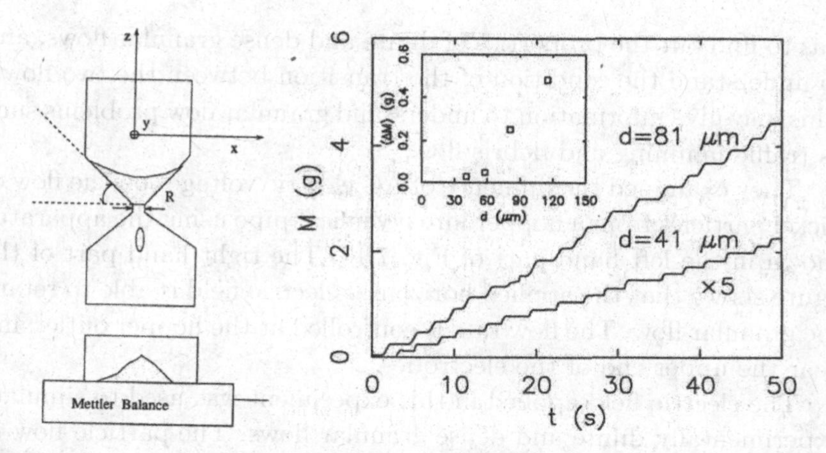

Fig. 7.12 Dynamic phenomena of flow granular materials in a hopper [7.29].

Jamming of granular flows can be used to simulate such jamming to find rules that govern jamming, and eventually, to reduce such jamming in real life.

Wu *et al.* studied dynamic phenomena of flowing granular materials in a hopper [7.29]. The change of the outgoing mass with time is presented in Fig. 7.12, showing a series of steps. The flow is not very smooth — horizontal steps imply that, as time goes on, few to no grains drop onto the scale, which means that air bubbles appear when sands flow through the bottleneck. Bubbling means periodic jamming of sand flow.

Two-dimensional granular flow [7.30] and oscillation pipe flow [7.31] were studied with hopper flow experiments. Hopper flow mass rate $Q \sim cvA$, where v is sand flow speed and A is the area of an opening. In a two-dimensional case, if d is the size of opening, the sand flow speed $v \sim d^{1/2}$, and area of the opening $A \sim d$, so $Q \sim d^{3/2}$; while in a three-dimensional case, $v \sim d^{1/2}$, area of the opening $A \sim d^2$, so $Q \sim d^{5/2}$, which leads to a mass flow rate being proportional to $d^{D-1/2}$.

(2) Dilute and dense flow under electric field

A transition between dilute and dense granular flow was studied experimentally by Chen Wei *et al.* [7.32]. The purpose of the study was to find out the properties of dilute and dense granular flows, and to understand the condition of the transition between the two flows. This may give information to understand granular flow problems such as traffic jamming and debris flow.

They examined the influence of AC electric voltages on the flow of nickel particles from a hopper into a vertical pipe using the apparatus shown in the left hand part of Fig. 7.13. The right hand part of the figure shows that the applied horizontal electric field is able to retard the granular flow. The flow rate is controlled at the hopper outlet and near the upper end of the electrodes.

The electric field applied in this experiment was used to simulate experimentally dilute and dense granular flows. The particle flow in the steady-state rate Q_B is a dilute flow, while the particle flow in the flowing rate Q_{A2} is a dense flow in the sense that the flow comes from a dense column of particles from above the electrodes.

At relatively weak electric fields when $V < V_c$ (=2.0 kV), a transition of the granular flow in the pipe from flow rate regime A to

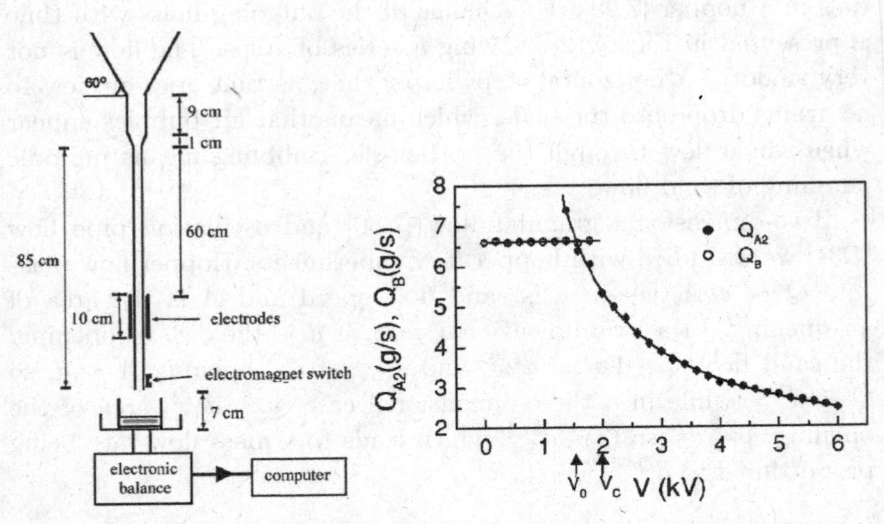

Fig. 7.13 The left side is a sketch of the apparatus. The right side shows flow rates Q_{A2} and Q_B vs. V. Q_B is essentially constant; Q_{A2} decreases monotonically with increasing V. Regime B disappears at a voltage $Vc = 2.0$ kV. The two curves meet at V_0 [7.32].

flow rate regime B is observed. When $V > V_c$, regime B of the flow disappears and the flow stays in regime A_2. The flow rate at regime B, Q_B, is practically unaffected by V. The flow rate of regime A_2, Q_{A2}, decreases with V monotonically with a power law, $Q_{A2} \sim V^{-0.8}$, which shows that the applied electric field can indeed retard the nickel granular flow. At voltage V_0 slightly below V_c, Q_{A2} equals Q_B.

The retardation effect of the electric field may be attributed to particles forming clusters at the two sidewalls near the electrodes, when V is large enough. Due to the inhomogeneity of the electric field, the polarized particles are pulled towards the two sidewalls; the friction between the particles and the walls is enhanced.

The above experiment is explained with a calculation [7.33]. The equations of motion of N particles with natural boundary condition were solved numerically for a given initial condition to find particle coordinates and momenta at each time. To establish the equation of motion, the forces on particles in a tube include the gravitational force, collision between particles (sliding, not rotating), collision between particle and wall, force on particles induced by inhomogeneous electric field and interaction between particles due to electric field. The initial condition reflects the fact that the particles of average size d drop from the top with a constant interval at a random position, and then fall freely with gravitational acceleration, g.

Supposing that all forces act on centers of mass (no rotation), the forces between the two particles are as follows: the normal force (elastic, dissipative) is

$$F_{j\to i}{}^{n} = K_h{}^{*}(R_i + R_j - |\vec{r}_{ij}|)^{3/2} - K_d{}^{*} m_{eff}{}^{*}(\vec{v}_{ij} \cdot \hat{r}_{ij}), \qquad (7.28)$$

the tangential force (dissipative, friction)

$$F_{j\to i}{}^{s} = -K_s \cdot m_{eff} \cdot (\vec{v}_{ij} \cdot \hat{s}_{ij}) - sgn(\vec{l})\mu|\vec{F}^{n}_{j\to i}|, \qquad (7.29)$$

where K_h, K_d, K_s, μ are parameters. The interaction between polarized particles is

$$\vec{f}_{ij} = \frac{3}{4\pi\varepsilon_0} \times \frac{\vec{p}_i \cdot \vec{p}_j \cdot (1 - 3 \cdot \cos\theta \cdot \cos\theta)}{|r_{ij}|^4} \cdot \hat{r}_{ij}. \qquad (7.30)$$

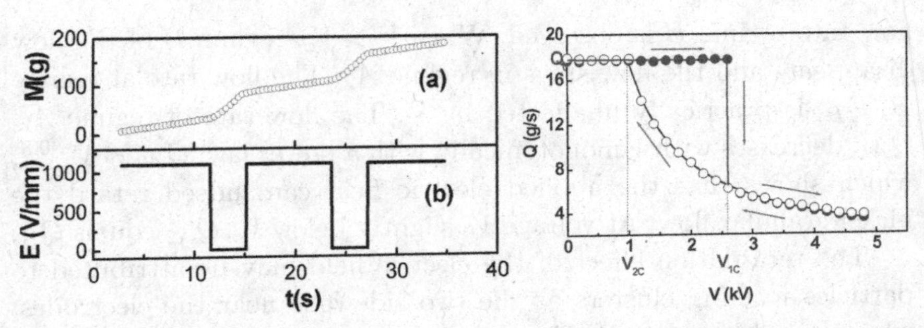

Fig. 7.14 The change of the average flux of flowing mass as function of an external electric field [7.33].

According to the dipole approximation, the polarization can be expressed as

$$\vec{p} = 4\pi\varepsilon_0 R^3 \vec{E}. \tag{7.31}$$

The force on polarized particles in an inhomogeneous field is

$$\vec{f} = -\nabla(\vec{p} \cdot \vec{E}) = 4\pi\varepsilon_0 R^3 \nabla E^2. \tag{7.32}$$

The result of the simulation is depicted in Fig. 7.14. It can be seen that the grain flow flux can be controlled by an electric field. If the initial state is a dilute flow, voltage applied to the tube V is less than a critical value V_c, grain flux Q remains a certain value Q_0, or the flux is V-independent. When $V = V_c$, the flux Q suddenly decreases, and grain flow experiences a transition from dilute to dense flow. When $V > V_c$, Q decreases monotonically with V. If the initial state is a dense state, Q decreases monotonically with V. The authors found that one-dimensional and two-dimensional simulations are consistent with the experimental results.

7.3.2 *Self organization criticality*

(1) Sand pile formation

Bak, Tang and Wiesenfeld (BTW) [7.34, 7.35] carefully studied the process of sand pile formation in the 1980s and proposed a concept of self organization criticality (SOC).

When one drops sand, one grain at a time, at random position onto a tabletop [7.36], during the formation of a pile, if the slope of the pile becomes too shallow, the addition of grains will tend to increase it; if the local slope of the pile exceeds a threshold, avalanches are induced by adding a single grain. When the sands are wet, the slope may become very steep; when the sands are getting dry, avalanches may occur, and the slope becomes a proper one accordingly. Therefore, the pile "self organizes" into a steady state in which its slope fluctuates about a constant angle of repose. Statistically, there are many small avalanches, fewer medium ones, and very few large ones; or in other words, the response of the pile is no longer restricted to being small and localized. In this sense, the pile is said to be "critical".

(2) Cell automation

Using the methods of cell automation, Bak, Tang and Wiesenfeld proposed a "sand pile" model when they discovered an example of a dynamical system displaying self-organized criticality. The model is a cellular automaton. At each site on the lattice there is a value that corresponds to the slope of the pile. This slope builds up as grains of sand are randomly placed onto the pile, until the slope exceeds a specific threshold value at which time that site collapses, transferring sand into the adjacent sites, increasing their slope. This random placement of sand at a particular site may have no effect and it may cause a cascading reaction that will affect every site on the lattice.

The iteration rules for the 1D model of a sand pile formation are as follows [7.37].

Figure 7.15 shows the right half of a one-dimensional sand pile. Z_n represents the height difference of two adjacent seats n and $n+1$: $Z_n = h_n - h_{n+1}$. When the height difference is above a certain critical value Z_c, $Z_n > Z_c$, the grain at seat n drops to the next lower seat $(n+1)$. The upper-seat height h_n reduces by one, while the lower-seat h_{n+1} increases by one, so the height difference between the two seats reduces by two, or

$$Z_n \to Z_n - 2; \tag{7.33}$$

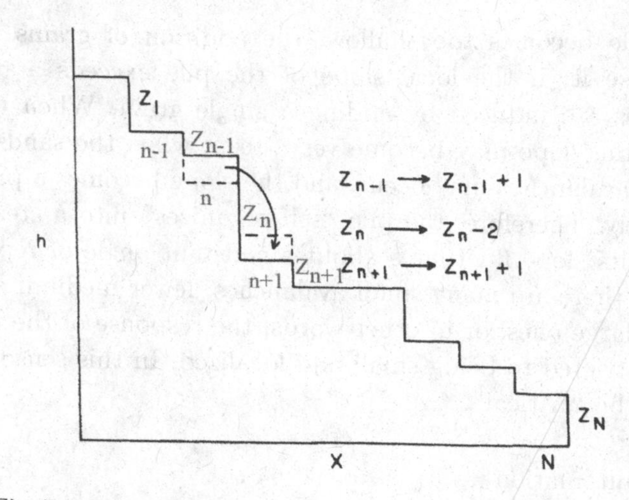

Fig. 7.15 Sketch diagram of a 1-dimentional sand pile [7.38].

At the same time, since the height of the upper seat reduces by one, the height difference between seat n and the seat above the upper seat, seat $(n-1)$, increases by one:

$$Z_{n-1} \rightarrow Z_{n-1}+1; \tag{7.34}$$

since the height of the lower seat reduces by one, the height difference between seat $(n+1)$ and the seat below the lower seat also increases by one:

$$Z_{n+1} \rightarrow Z_{n+1}+1. \tag{7.35}$$

The above three equations are a nonlinear discrete diffusion equation set. The nonlinearity comes from the threshold condition $Z_n > Z_c$. The state of variable Z_n at time $t+1$ depends on the grain's own state and the state of its nearest neighbors at time t. When $Z_n < Z_c$ at seat n, dropping grains pile up on this seat until a critical state comes, namely when a critical condition is satisfied:

$$Z_n = Z_c(n = 1, 2, 3, \ldots, m). \tag{7.36}$$

Similarly, one may discuss sand pile formation using a 2D cellular automation model which describes the interaction of an integer variable Z with its nearest neighbors [7.35]. The iteration rules for the 2D model are as follows: starting with a flat surface $z(x, y) = 0$

for all x and y, add a grain of sand, and height difference at point (x, y) increases by one:

$$z(x, y) \rightarrow z(x, y) + 1. \tag{7.37}$$

An avalanche happens if $z(x, y) > z_c$, and the height differences at point (x, y) and its four nearest neighbors are updated synchronously as follows:

$$\begin{aligned} z(x, y) &\rightarrow z(x, y) - 4, \\ z(x \pm 1, y) &\rightarrow z(x \pm 1, y) + 1, \\ z(x, y \pm 1) &\rightarrow z(x, y \pm 1) + 1. \end{aligned} \tag{7.38}$$

Figure 7.16 shows a process of local avalanches after a sand grain drops onto the center of this 5 × 5 2D "sand pile" under the critical condition Z_c =3. In this process, sand grains on eight seats (marked in black) are conducted at least one avalanche before the system reaches a stable state.

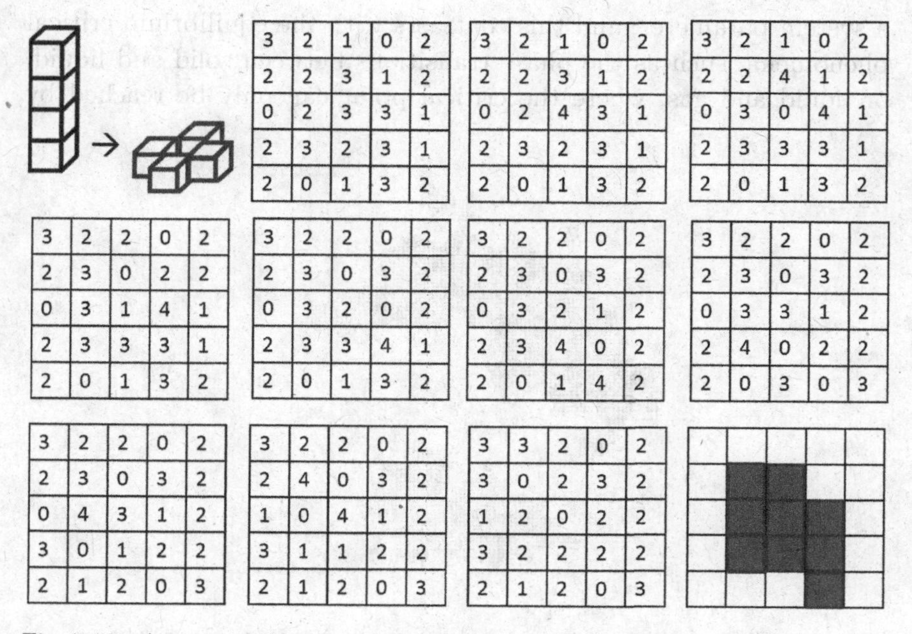

Fig. 7.16 A process of local avalanches after a sand grain drops onto the center of this 5 × 5 2D "sand pile".

(3) SOC is the origin of power law

STW also showed that the SOC is the origin of power law. Similar processes have been conducted with 2D and 3D sand pile systems of a different array of sizes [7.35]. Figure 7.17 shows a structure obtained for a 100 × 100 2D array. Each cluster marked in black is reached through avalanches originated by dropping a grain in a single site.

Figure 7.18 shows a log-log plot of the distribution $D(s)$ of cluster size s for a 2D system. The curve is consistent with a straight line, indicating a power law $D(s) \sim s^{-\tau}$, critical exponent $\tau \sim 0.98$. The fact that the curve is linear over two decades indicates that the system is at a critical point with a scaling distribution of clusters.

(4) Differences between SOC and equilibrium critical state

In the sand pile model, the correlation length of the system and the correlation time of the system at critical states go to infinity — this is the same as equilibrium critical phenomena. However, in the sand pile model, the critical state is realized without any adjustment of a system parameter, and this contrasts with the equilibrium critical phenomena, such as the phase transitions between solid and liquid, or liquid and gas, where the critical point can only be reached by

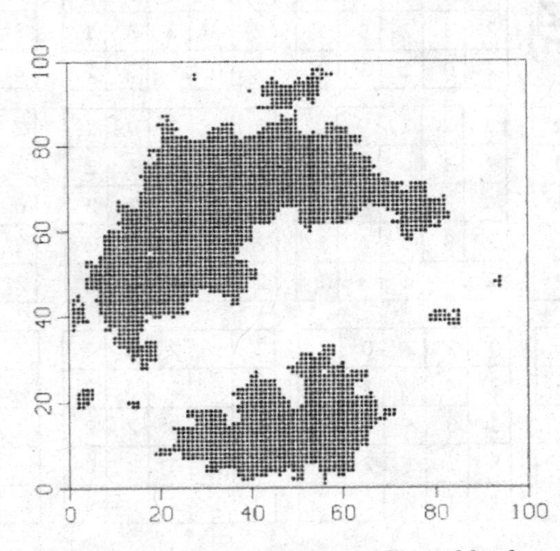

Fig. 7.17 Self-organized critical state of minimally stable clusters for a 100 × 100 array [7.35].

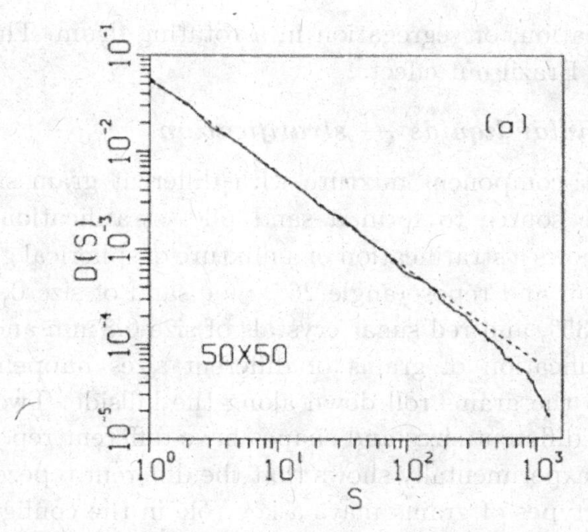

Fig. 7.18 Distribution of cluster sizes at criticality in a 2D system of 50×50 array, averaged over 200 samples [7.35].

precise parameter (usually temperature) tuning. Hence, in the sand pile model we can say that the criticality is self-organized.

Once the sand pile model reaches its critical state there is no correlation between the system's response to a perturbation and the details of a perturbation. Generally, this means that dropping another grain of sand onto the pile may cause nothing to happen, or cause the entire pile to collapse in a massive slide. [Wikipedia]

Some studies have shown that not all sand piles are of SOC. It was found that small sand piles in water, and drum and rice piles formed of short and round rice do not show SOC, while large, non-uniform sand piles in water and rice piles formed of long rice show SOC [7.39].

7.4 Grain Segregation

Two granular materials differing in density, size, or friction coefficient, coefficient of elastic restitution, exhibit a distinct segregation. The segregation can be distinguished into at least two different modes. One is segregation by shearing, which is often observed in grain stratification, or segregation in a rotating drum. The other is observed in a Brazil nut effect.

7.4.1 *Granular liquids — stratification*

When a multi-component mixture with different grain sizes drops from a single source to form a sand pile, stratification appears. Figure 7.19 shows a stratification of a mixture of spherical glass beads of size 0.15 mm and repose angle 26°, blue sand of size 0.4 mm and repose angle 35°, and red sugar crystals of size 0.8 mm and 39°.

The stratification of grains of different sizes happens spontaneously when the grains roll down along the hillside. Two granular systems with different sizes and shapes have different repose angles. Makes [7.40] experimentally shows that the different repose angles of two different types of grains plays a key role in the configuration of particle segregation.

A possible stratification mechanism in granular mixtures has been proposed as follows. The repose angle depends on the local composition on the surface. $\theta_{\alpha\beta}$ is the repose angle of a rolling grain of type α on surface of β. Large grains roll more easily on top of small grains (surface "looks" smoother) than small grains roll on top of large grains, or $\theta_{21} <$ pure species $\theta_{\alpha\alpha} < \theta_{12}$, as depicted in Figs. 7.20 and 7.21.

The equations of motion are

$$\frac{\partial R_\alpha}{\partial t} = -v_\alpha \frac{\partial R_\alpha}{\partial x} + \Gamma_\alpha, \tag{7.39}$$

$$\frac{\partial h}{\partial t} = -\sum_\alpha \Gamma_\alpha, \tag{7.40}$$

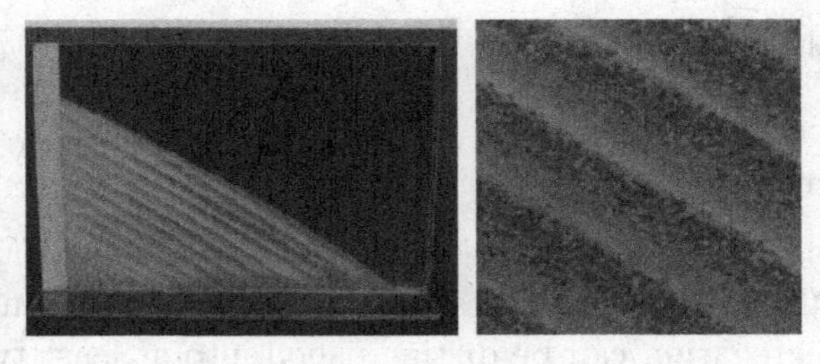

Fig. 7.19 Stratification. Courtesy of [7.40].

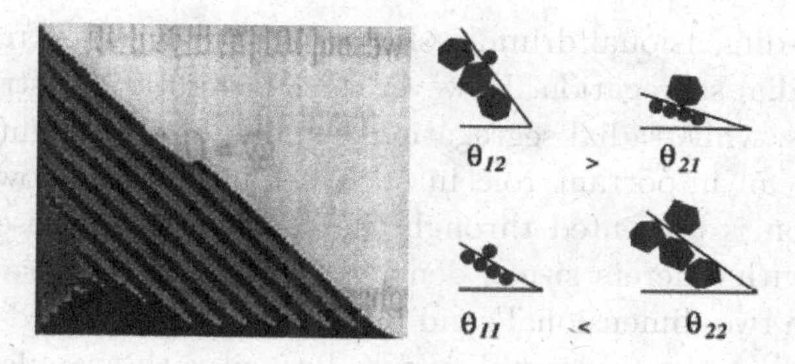

Fig. 7.20 Explanation of repose angle for different cases [7.40].

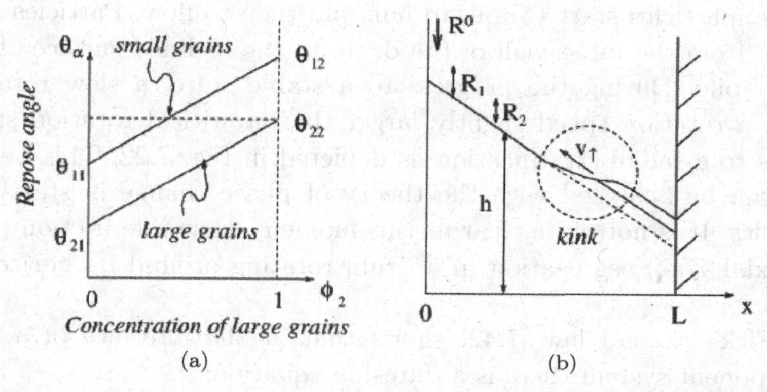

Fig. 7.21 (a) Dependence of the repose angle on the two types of rolling grains on the concentration of the surface of large grains ϕ_2. An essential ingredient to obtain stratification is that $\theta_{22} > \theta_{11}$. For the numerical integration, we use the linear interpolation between $\phi_2 = 0$ and $\phi_2 = 1$ as plotted here. (b) Picture with the different quantities appearing in the text. The dashed circle is the kink [7.40].

where $R_\alpha(x,t)$ is the thicknesses of the rolling grains $\alpha = 1, 2$, for small and large grains, respectively, $h(x,t)$ the height of the sand pile, $\phi_\beta(x,t)$, the volume fraction of grains of b type in the static phase, v_α, the downhill convection velocity of species α, Γ_α describes the interaction of the rolling grains with the surface and concentrations $\phi_1 + \phi_2 = 1$ and $\phi_a = -\Gamma_\alpha/(\partial h/\partial \mathrm{t})$.

7.4.2 *Rotation drum*

(1) Radial segregation

The configuration of particle mixing and segregation in a horizontal drum rotating around its axis depends on the rotation speed.

In a three-dimensional drum, a slow rotation brings a stratification-type of radial segregation. However, in a fast rotation, stratification disappears, while radial segregation remains. It is thought that friction plays an important role in such behavior. In a slow rotation, this friction is presented through the difference of repose angles of particles with different size or density — just as in the case of stratification in a two-dimensional sand pile formation described in the last section. When the horizontal surface formed with granular particles becomes inclined, and the inclined angle is larger than the repose angle, particles start to roll, to fall, and then to flow. Particles thus move from the inner wall of the drum to the inclined surface of the grain pile. During the progress to a stable state, a slow rotation with a rotation speed slightly larger than a critical rotation speed leads to a radial stratification as depicted in Fig. 7.22. This behavior can be analyzed with the theory of phase change in statistical physics. It is not quite clear at this moment what role friction plays in axial grain segregation in a drum rotating around its horizontal axis.

Fick's second law [7.42] shows that at the interface of a two-component system there is a diffusion equation:

$$\frac{\partial C(z,t)}{\partial t} = \frac{\partial}{\partial z}\left(D\frac{\partial C(z,t)}{\partial z}\right) \tag{7.41}$$

where z axis is the mixer axis, $C(z,t)$, relative concentration by the volume of the smaller particles, D, a diffusion coefficient, and L, the drum length. The initial condition is

$$C(z,0) = \begin{cases} 0, \; -L/2 \le z < 0 \\ 1, \; 0 < z \le L/2. \end{cases} \tag{7.42}$$

The boundary condition is

$$\left.\frac{\partial C}{\partial z}\right|_{z=-\mathrm{L}/2} = \left.\frac{\partial C}{\partial z}\right|_{z=\mathrm{L}/2} = 0. \tag{7.43}$$

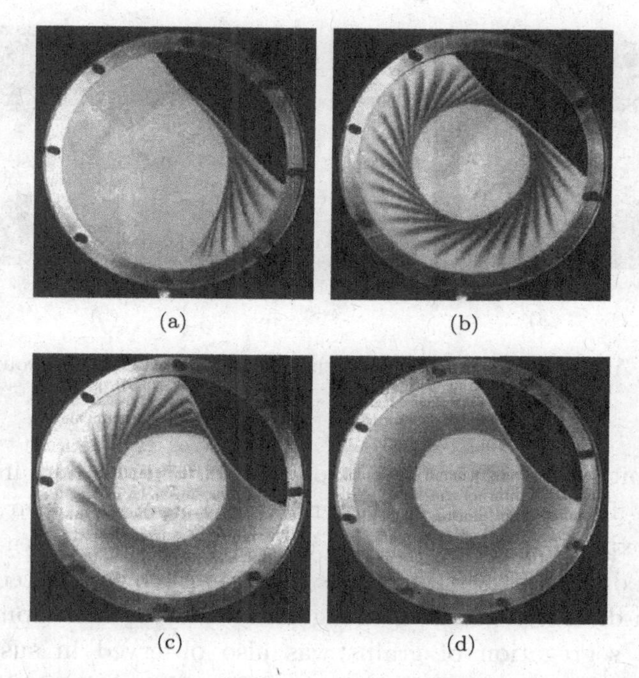

(a) (b) (c) (d)

Fig. 7.22 A cross section of segregation of a mixture of sugar (white) and iron power (black) in a horizontal drum with 25 cm diameter. The top two photos show a radial segregation when the drum is under a slow rotation of 0.54 rpm when stratification can be observed. The bottom two photos show how stratification disappears at a fast rotation (>3 rpm) while radial segregation remains [7.41].

If $D =$ constant

$$C(z,t) = \frac{1}{2} + \frac{2}{\pi}\sum_{k=1}^{\infty}\frac{1}{2k-1}\exp\left\{-\frac{(2k-1)^2\pi^2 Dt}{L^2}\right\}\sin\left\{\frac{(2k-1)\pi z}{L}\right\}. \tag{7.44}$$

This solution is consistent with the experiment [7.42].

(2) Axial segregation

As Oyama reported in 1939 [7.43, 7.44], a mixture of two granules of the same types but of different sizes axially segregate in an elongated cylindrical container which is rotated horizontally around its axis. The process starts with just a few slow rotations of the drum. After

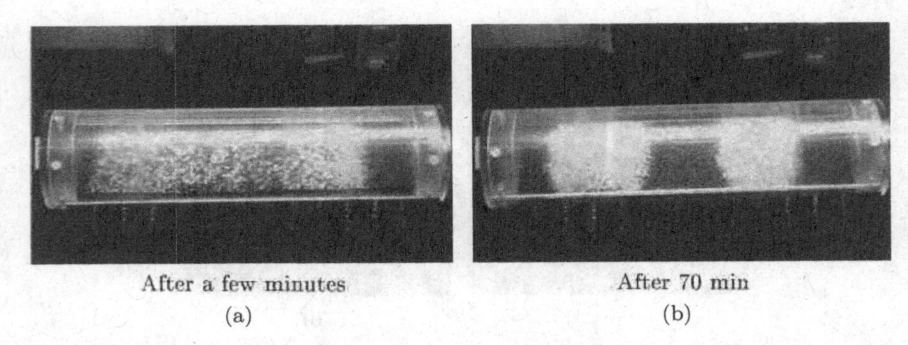

After a few minutes (a) After 70 min (b)

Fig. 7.23 An axial segregation in a drum rotating horizontally around its axis [7.45].

a sufficiently long time, the segregation continues, first into three, and then five zones typically observed after about an hour.

The experiment described in Fig. 7.23 is an axial segregation of a 100% filled mixture of poppy seeds (white) and mustard seeds (black) in a 7 cm-diameter and 25 cm-long drum rotating at 15 rpm [7.45].

Axial segregation of grains was also observed in suspensions, reported by a group in City College of City University of New York. The mixture is a combination of Triton X-100, $ZnCl_2$, and water. The axial segregation depicted in Figs. 7.24–7.27 show that a higher rotation speed corresponds to a longer period of the aggregation bands; the period does not change with rotation time, and this is quite different from what one sees in segregation of dry grains, where the longer the rotation time, the shorter the period is.

The axial segregation (Fig. 7.28) of dry grains was explained with Savage's phenomenological model [7.47]. Suppose there are two populations of spheres A and B, with θ_A and θ_B being the kinetic angles of repose of the corresponding spheres at a given rotation speed ω, and $C_A(x)$ the local concentration of A at x. The kinetic angle of the mixture of spheres A and B,

$$\begin{aligned}\theta(x) &= \text{weighted average of } \theta_A \text{ and } \theta_B \\ &= \theta_A C_A(x) + \theta_B(1 - C_A(x)) \\ &= \theta_B + \Delta\theta C_A(x),\end{aligned} \tag{7.45}$$

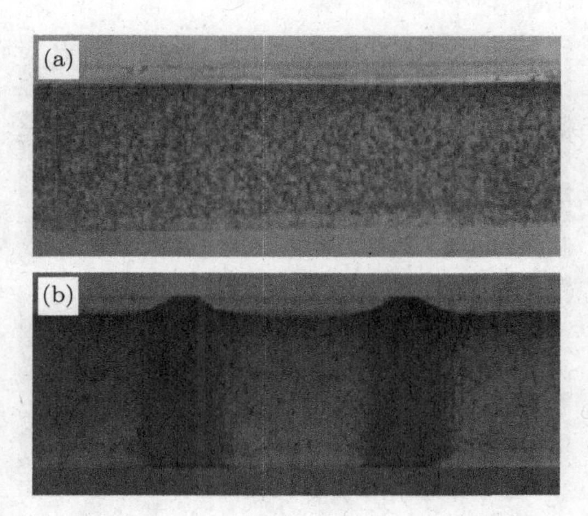

Fig. 7.24 10% suspension (a) before and (b) after shearing, with the inner cylinder rotating at 9 rpm when the Couette is filled up to 95% of the available gap volume [7.46].

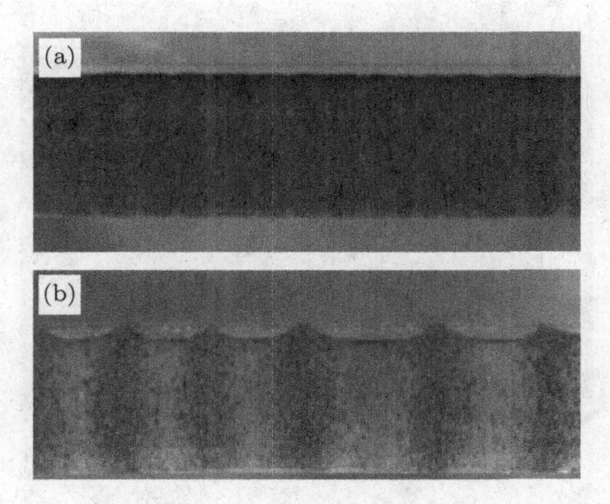

Fig. 7.25 A 15% suspension at the 95% fill level is sheared at (a) 2.5 rpm (b) 9 rpm.

where $\Delta\theta = \theta_{\mathrm{A}} - \theta_{\mathrm{B}}$ is the difference of kinetic angles of the two spheres. If it is focusing on population B, there exist two effects. On the one hand, there is a flux $\Phi_{\mathrm{B}x}(\Delta\theta)$ due to difference in kinetic angle, which tends to drive particles along the direction of the x-axis. On the other hand, there is an opposing flux $\Phi_{\mathrm{BD}}(\Delta\theta)$

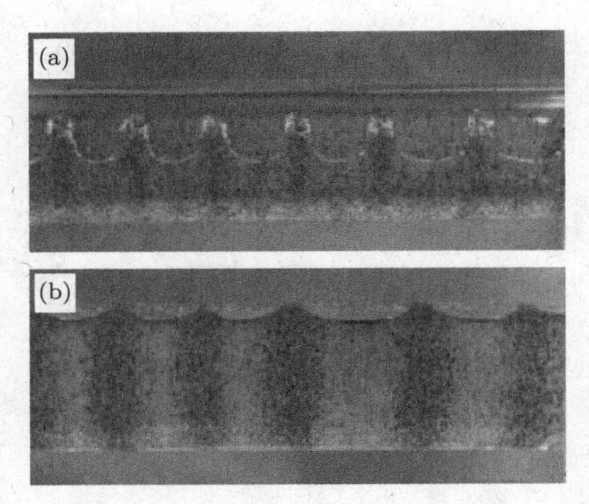

Fig. 7.26 The mixture of 15% suspension (a) 50% fill level (b) 95% fill level, sheared at 9 rpm [7.46].

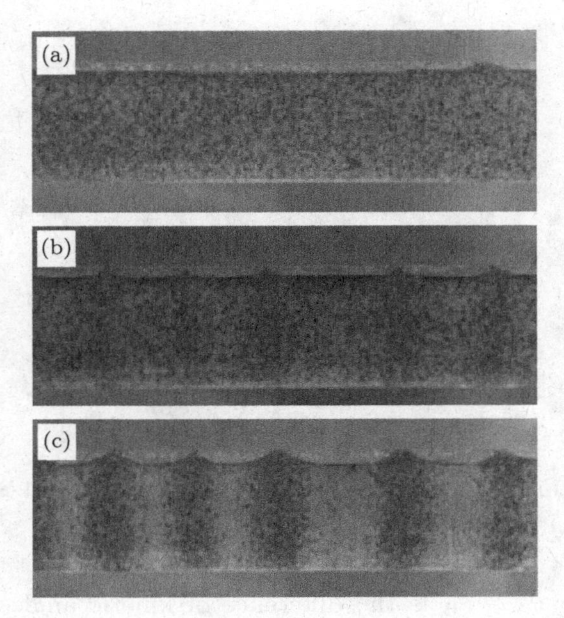

Fig. 7.27 Time-dependent band formation: 15% suspension, 95% fill level sheared at 9 rpm after (a) 1 min.; (b) 20 min; (c) 3 h [7.46].

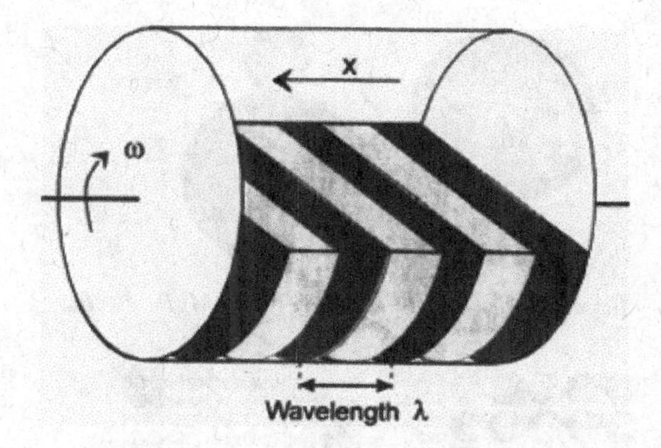

Fig. 7.28 The axial segregation [7.47].

due to diffusion described by Fick's law involving a diffusion coefficient D. The total flux of B-type spheres along the x-axis is then given by

$$\Phi_{\mathrm{B}x} = -\Delta\theta \frac{\partial C_{\mathrm{B}}}{\partial x} - D \frac{\partial C_{\mathrm{B}}}{\partial x}. \tag{7.46}$$

The horizontal flux is created by differences in the kinetic angles of repose but is opposed by a diffusive component that tends to equalize the concentrations. The formation of bands results from concentration between the two effects.

7.4.3 *Segregation by vertical vibration — Brazil nut problem*

A mixture of particles of different sizes experiencing vertical vibration leads to particle segregation, with larger particles moving up while smaller ones sink down. This is called the Brazil nut problem (BNP) [7.48–7.51].

The Brazil nut problem was explained with the following analysis [7.52]. Suppose that the large spheres have mass and radius of M and R, respectively, and small spheres have m and r. In Fig. 7.29, the top two graphs show two different configurations of sphere packing. On the left side, the large sphere is on top of the small one, and the energy is $E_u = mgr + Mg(R+2r)$, while on the right side, the small sphere is on top of the small one, and the energy is $E_l = mg(r+2R)+MgR$.

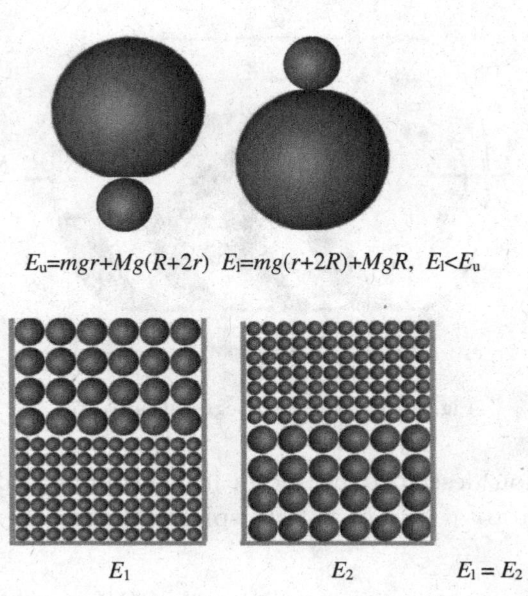

E_u=mgr+Mg(R+2r) E_l=mg(r+2R)+MgR, E_l<E_u
E_1
E_2
$E_1 = E_2$

Fig. 7.29 While the upper packing shows an energy difference, the lower sphere packing shows that, as far as energy is concerned, neither configuration is more favorable than the other [7.52].

It is clear that $E_l < E_u$. The lower two graphs of Fig. 7.29 show two different configurations of multiple sphere packing. The energy of the left side is denoted as E_1, $E_1 \propto v^2/2 + (v + V/2)V \propto (v + V)^2$, and $E_1 = E_2$. As far as energy is concerned, neither configuration is more favorable than the other.

However, in the two-dimensional case, the energy of the planar square lattice (left graph) is larger than that of triangular lattice (right graph), and the triangular lattice is energetically more favorable than planar square lattice. It is clear that the triangular lattice is denser than the planar square lattice, and in general, nature favors denser packing, which has lower internal energy, and is more stable.

This rule is also true in the configuration shown in Figs. 7.30, where the triangular symmetry of the upper part is greatly disturbed by the introduction of the large disk [7.53]. The upper part is less dense, and the less dense portion moves toward the top, dragging the intruder along with it to a more stable state.

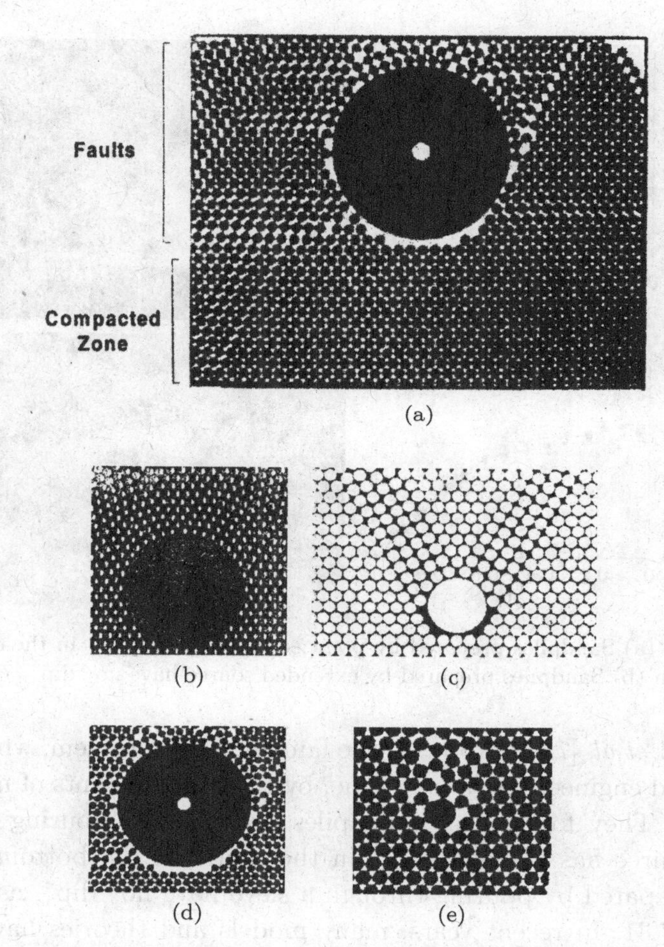

Fig. 7.30 The triangular symmetry of the upper part is greatly disturbed [7.53].

7.5 Granular Solid

7.5.1 *Counterintuitive phenomenon: construction history [7.54]*

(1) Experiments

Sand piles prepared by pouring from a point source (localized) have the stress dip in the center at the bottom, while those prepared by pouring through a sieve (extended) have none.

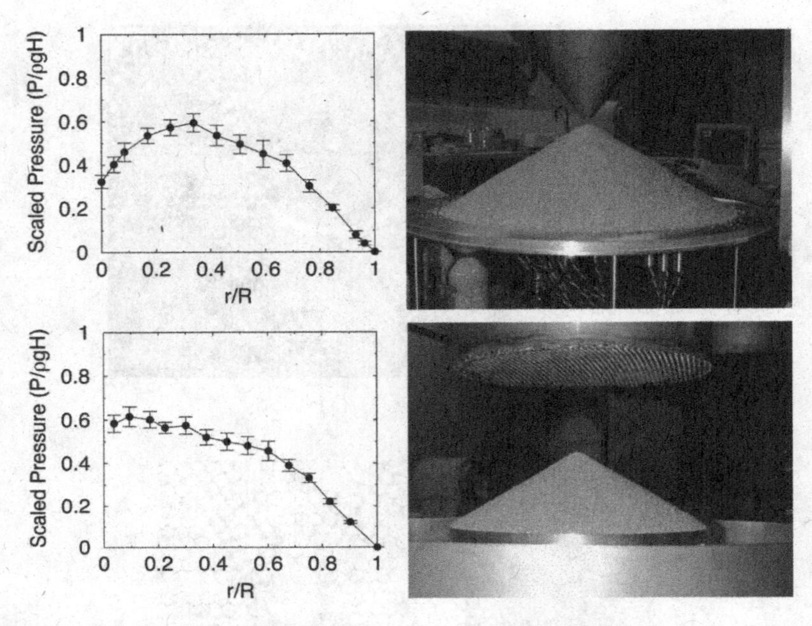

Fig. 7.31 (a) Sandpiles prepared by point source have a "dip" in the center of the bottom (b) Sandpiles prepared by extended source have no "dip" [7.55].

Vanel *et al.* [7.55] proposed the famous "dip" problem, which has perplexed engineers for a long time, by their experiments of memory in sand. They found that sand piles prepared by pouring from a point source has a stress "dip" in the center of the bottom, while those prepared by pouring through a sieve have no "dip", as shown in Fig. 7.31. In recent years, many models and theories have been given to interpret this problem. I will detail these models in this section.

(2) Force chains

Grains only interact at the contact points, which form a random network inside the granular material. A static sand pile looks like a conical solid, but its internal forces are never homogenous in contrast to those inside a continuous solid [7.2, 7.3] because the discrete particles only interact at contact points, which tend to form a random network inside the granular material. Figure 7.32 gives a visualization of the forces in a granular medium as viewed between two

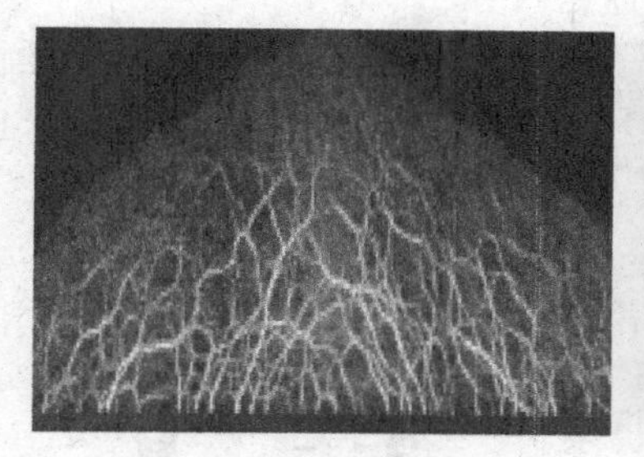
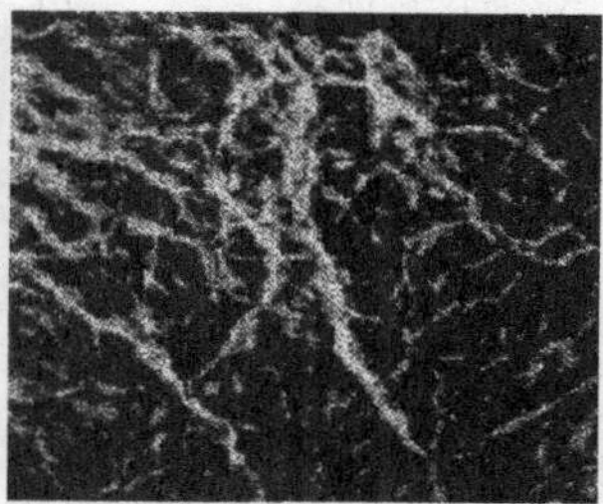

Fig. 7.32 The force chains in a 2D (left) [7.57] and a 3D sand-pile (right) [7.58]. Right: 3 mm diameter Pyrex glass beads in 70 × 70 × 40 mm^3 box saturated by glycerol and water mixture under 200 N compression are viewed using the photoelastic method.

crossed-circular polarizers [7.56]. Using photoelastic measurements, scientists have found the privileged force paths (force chains) for the force propagation in granular materials [7.56–7.58]. Simulations also show that force transmission in sand piles is inhomogeneous.

The following experiment gives another proof of force chains.

Granular material is packed in a container, leaving one hole open to allow the vibration wave of a loudspeaker to be transmitted into the grains inside container through a bar. If one initiates a sound wave at one point in the material, the transmitted signal at a second position is sensitive to the exact arrangement of all particles in the container. The inset of Fig. 7.33 shows a schematic view of the apparatus. Figure 7.33 shows the transmitted rms (root mean square) magnitude of the acceleration A_d (in units of g) versus time showing the effect of a temperature pulse ($\Delta T = 1$ K) on the sound propagation in a granular medium. If heater H and the second detector D_2 are not in the same force chain, the pulses of A_d are much fewer.

(3) Ordered versus disorder packing

Geng *et al.* [7.60] pointed out that for ordered packing (using mono-diameter-disks to form 2D HCP packing) the stress response has a strong propagative feature and has a bottom "dip". As shown in Fig. 7.34, with the increase of disorder (using mixing of disks with two different diameters) the "dip" decreases. For greatly disordered packing (using pentagons), the response profile always has one peak with width broadening linearly as pile depth increases, and the bottom

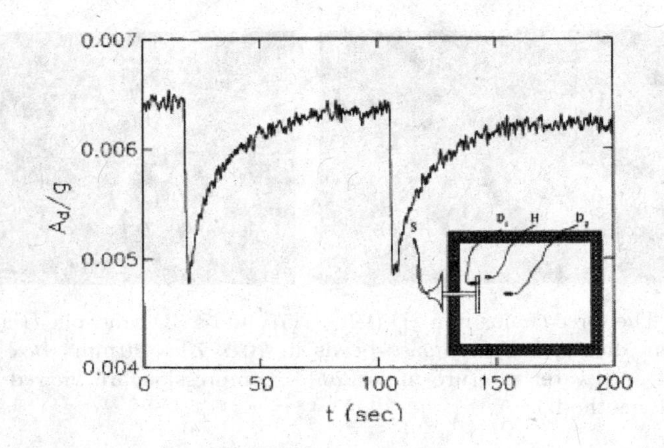

Fig. 7.33 An experiment that shows a force chain [7.4, 7.58].

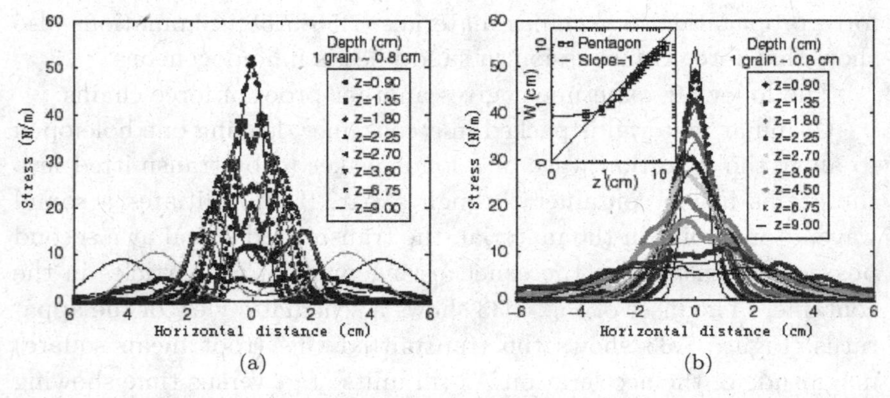

Fig. 7.34 (a) The stress profile at different depths for ordered packing. Near the bottom, there are two diffusive peaks; (b) Stress profile at different depths for disordered packing; the inset shows the linear dependence of peak width on depth [7.59].

stress forms a "plateau," which verifies that the disordered packing is ruled by elastic theory [7.60].

In the 2D heap memory experiment [7.57], it was found that for heaps prepared by fixed height point source, the mean nearest neighbor contact angle preferred six-fold directions, which means a highly anisotropic heap. There is a "dip" at the bottom of this heap, as shown in Fig. 7.35. For heaps deposited by an extended source,

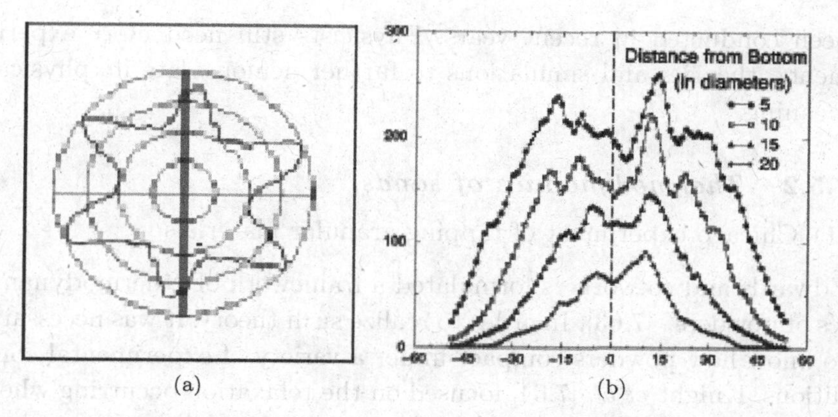

Fig. 7.35 (a) The 6-fold preferred contact angle direction. (b) The stress profile (point source) with the distance from bottom and a "dip" at the bottom center [7.62].

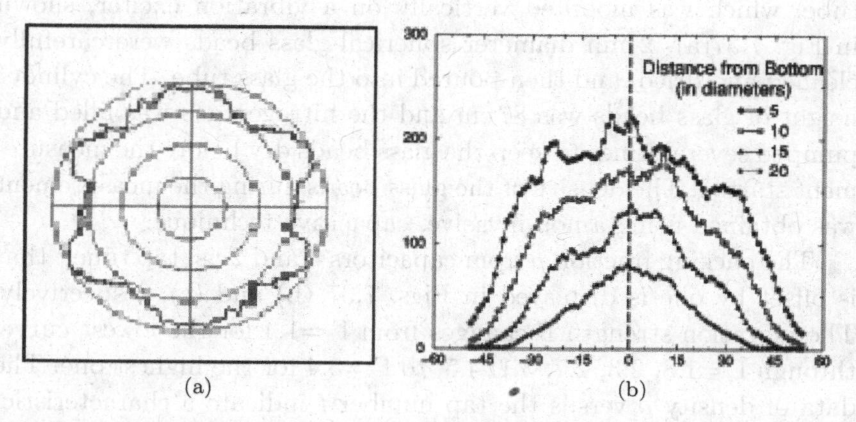

Fig. 7.36 (a) The randomly distributed contact angle direction. (b) The stress profile (extended source) with the distance from bottom and no "dip" at the bottom center [7.62].

the contact angle is almost randomly distributed, which means an isotropic heap and there is no convincing "dip" at the bottom of such a heap, as shown in Fig. 7.36.

Finally, it should be pointed out that a common shortcoming [7.63] shared by all recent models for the analysis of stress distribution is that the exact physical origin or justification of constitutive relation is not clearly known. Many computer simulations have been conducted in recent years. Physicists still need more experiments, theories and simulations to further demonstrate its physical meaning.

7.5.2 *Thermodynamics of sands*

(1) Chicago experiment of tapping granular materials

Edwards and co-workers formulated a framework of "thermodynamics of powders" [7.63]. In order to realize such theory, it was necessary to know how powders compact under a variety of experimental conditions. Knight *et al.* [7.64] focused on the relaxation occurring when the system is prepared in a low density state and then vibrated to induce decay toward the steady state density.

Their experiment was conducted with a 1.88 cm diameter glass tube, which was mounted vertically on a vibration exciter, shown in Fig. 7.37(a). 2 mm diameter spherical glass beads were carefully cleaned and dried, and then poured into the glass tube. The cylinder height of glass beads was 87 cm and the nitrogen gas was filled and pumped several times to keep the glass beads dry before the measurement started. The density of the glass beads during the measurement was obtained using a non-invasive, capacitive technique.

The packing fraction ρ from capacitors 4 and 2 vs. tap time t that is offset by one is displayed in Figs. 7.37 (b) and (c), respectively. The vibration strength Γ changes from Γ =1.4 for the lowest curve, through Γ =1.8, 2.3, 2.7, 3.1, 4.5, to Γ =5.4 for the highest one. The data of density ρ versus the tap number t indicate a characteristic, non-exponential relaxation process that depends on both vibration strength and the depth of upper surface into the cylinder. The vibration strength, Γ, is defined by the ratio of the peak acceleration of a vertical tap, a, to the gravitational acceleration, g, namely $\Gamma = a/g$.

The curves in Figs. 7.37 (b) and (c) are fitted to the whole data range with the following expression:

$$\rho(t) = \rho_f - \frac{\Delta\rho_\infty}{1 + B\ln\left(1 + \frac{t}{\tau}\right)}, \tag{7.47}$$

where the parameters ρ_f, the final steady state density that is approximately equal to $\rho(t=\infty)$, $\Delta\rho_\infty = \rho(t=\infty) - \rho_0$, τ, relaxation time,

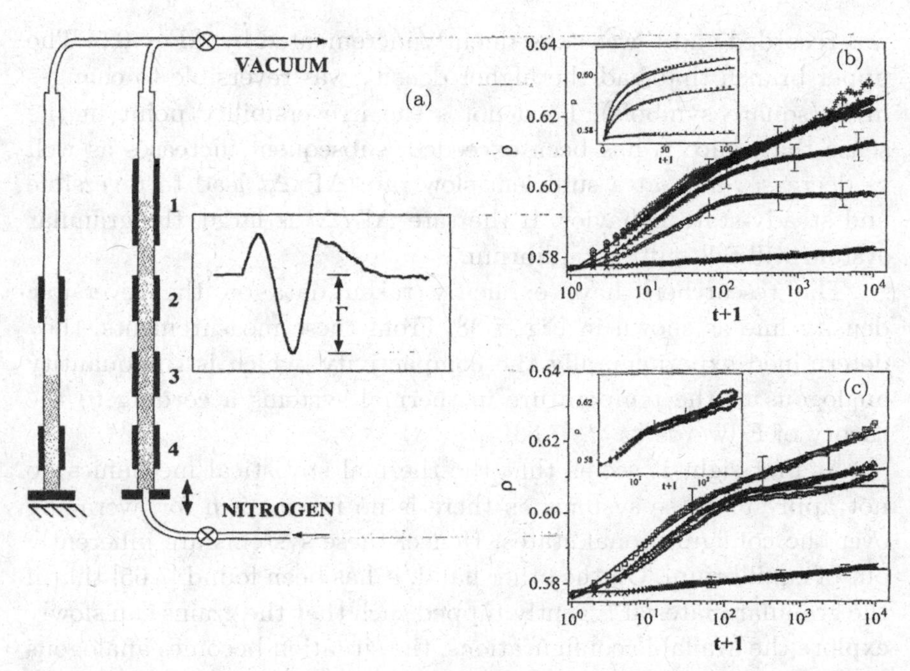

Fig. 7.37 (a) A schematic drawing of the apparatus. (b) and (c) The packing fraction ρ from capacitors 4 and 2, respectively, vs. tap time that is offset by one. The curves are parameterized by vibration strength Γ [7.64].

and B, a fitting parameter, are constants that depend only on the vibration strength Γ.

(2) Small density fluctuation about the reversible steady state

To examine the volume fluctuations about a steady-state density for a granular system, it is essential that the fluctuations be measured in reversible steady-state conditions. For this reason, the above experiment was analyzed further to find out the condition for reversible steady state [7.65].

The beads were prepared in a low density initial configuration by inserting high pressure nitrogen gas upward from the bottom of the tube. Then the vibration strength Γ was slowly first increased (solid symbols in Fig. 7.38) and then decreased (open symbols). At each value of Γ the system was tapped 10^5 times after which the density was recorded and Γ was subsequently incremented by $\Delta\Gamma \sim 0.5$. The upper branch that had the higher density was reversible to changes in Γ (square symbols). Γ^* denotes the irreversibility point in the sense that, once it has been exceeded, subsequent increases as well as decreases in Γ at a sufficient slow rate $\Delta\Gamma/\Delta t$ lead to reversible and steady-state behavior. If the rate $\Delta\Gamma/\Delta t$ is large, the granular system will fall out of equilibrium.

The researchers have explicitly taken data on the reversible density line as shown in Fig. 7.38. From these measurements, they determined experimentally the compactivity, which is the quantity analogous to the temperature in thermal systems according to the theory of Edwards *et al.* [7.63]

At first sight it seems that the thermal statistical mechanics do not apply to these systems as there is no mechanism for averaging over the configurational states. Hence, these systems are inherently out of equilibrium. On the other hand, it has been found [7.65] that if the granular material is gently tapped such that the grains can slowly explore the available configurations, the situation becomes analogous to the equilibrium case scenario. It has been shown that the volume of the system is dependent on the applied tapping regime, and this dependence is reversible, implying ergodicity.

In ordinary statistical mechanics, the fluctuation-dissipation theorem allows the determination of the response of a system to a

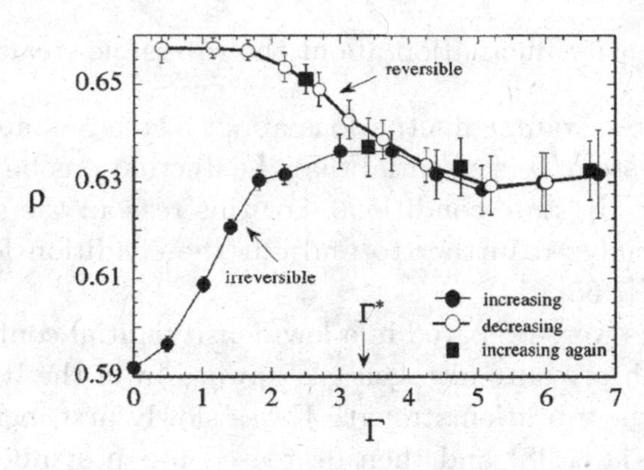

Fig. 7.38 The dependence of ρ on the vibration history [7.65].

small perturbation from its thermal fluctuations about equilibrium. In Ref. [7.65], the authors derived similar information about the granular system from its fluctuations about its steady-state density. Figure 7.39 shows the fluctuations in the density measured at the bottom capacitor as a function of acceleration Γ.

Figure 7.40 plots the average values of the variance, $\langle \delta V^2 \rangle = \langle (V - V_{ss})^2 \rangle$, for several steady-state volumes, V_{ss}, along the reversible

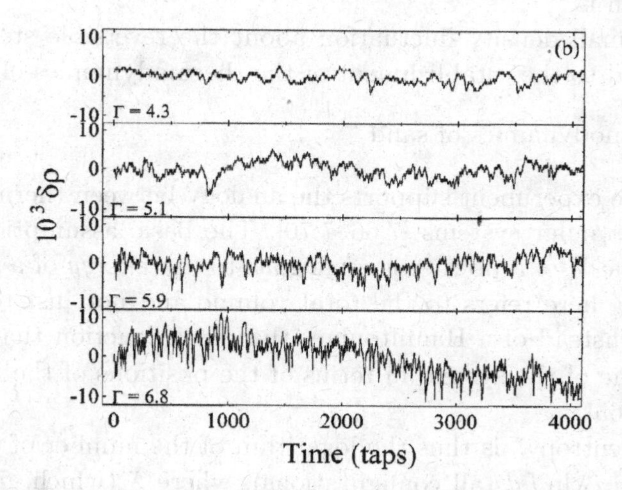

Fig. 7.39 Fluctuations in the volume density $\delta\rho(t) = \rho(t) - \rho_{ss}$ after the system has had sufficient time to relax to a steady-state density ρ_{ss} [7.64].

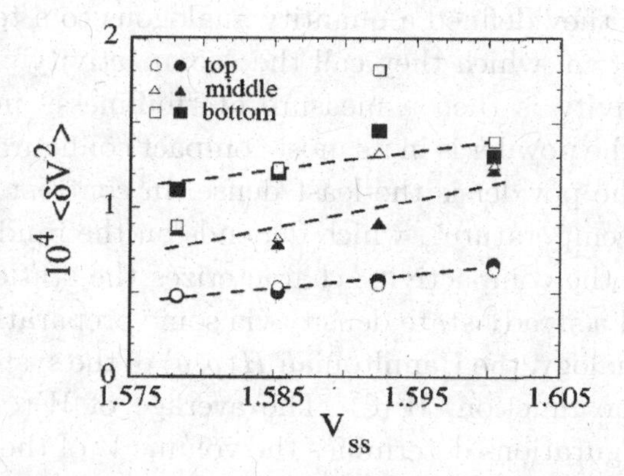

Fig. 7.40 The average variance of the experimental volume fluctuations (open symbols) as a function of the steady-state volume [7.64].

branch of Fig. 7.38 over the range $4 < \Gamma < 7$. The dashed line through the data for the top of the pile represents a linear fit to the function $\langle (V - V_{ss})^2 \rangle = a + bV_{ss}$, where $a = 7.2 \times 10^{-4}$ and $b = 4.9 \times 10^{-4}$. This implies that the magnitude of the fluctuations goes to zero at $\rho \approx 0.68$, which means that near the close-packed density, the density fluctuation is very small. Figure 7.40 indicates $\langle (V - V_{ss})^2 / V_{ss}^2 \approx 1/40000$, which also tells us how small the size of fluctuation is.

The small density fluctuation about the reversible steady state would lead to the establishment of the thermodynamics of sands.

(3) Thermodynamics of sand

The above experiment supports the analogy between thermodynamics and granular systems [7.66–7.70]. The basic assumption is that the volume V of a powder is analogous to the *energy* of a statistical system. V here refers to the total volume and not just to the free volume. Instead of a Hamiltonian, there is a function that specifies the volume of the system in terms of the positions of the individual grains [7.65].

The "entropy" is thus the logarithm of the number of configurations: $S = \lambda \ln \int d$ (all configurations), where λ (which gives S the dimension of volume) is the analog of Boltzmann's constant k (which gives S the dimensions of energy).

Using this they defined a quantity analogous to a temperature in a thermal system, which they call the "compactivity": $X = \partial V/\partial S$. The compactivity is then a measure of "fluffiness" in the powder: when $X = 0$, the powder is in its most compact configuration, whereas for $X = \infty$ the powder is the least dense. In contrast to the notion of "granular temperature", which depends on the random motion of the particles, the compactivity characterizes the static system after it has reached a steady-state density via some preparation algorithm.

In their analogy, the Hamiltonian $H(p, q)$ of the system is replaced by the volume function, $W(\zeta)$. The average of $W(\zeta)$ over all the jammed configurations determines the volume V of the system in the same way as the average of the Hamiltonian determines the average energy E of the system.

S. Edwards and co-workers also derived the Boltzmann equation, the Gibbs distribution, entropy, free energy for granular materials in analogy with the classical ones for thermal systems. Readers can refer to Refs. [7.69, 7.70] for the details.

7.6 Granular Gas

7.6.1 *Experiment of sand as Maxwell's demon*

The second law of thermodynamics can be expressed as the assertion that in an isolated system, entropy never decreases. Maxwell realized that, in an open system, there should be an energy-controlling mechanism that can decrease the entropy. J.C. Maxwell humorously imagined in 1871 that there could be a "demon" which can identify the velocity of each molecule that flies near a hole between two identical compartments, and allow molecules with higher speed to be cumulated in one of the compartments. Eventually, the temperature of one compartment would be higher than the other.

This "demon" does not exist in molecular systems, but in biological entities. According to the Prigogin theory of dissipation structure, open systems, such as biological entities, can absorb energy and matters, and release their entropy to the environment. In this sense, Maxwell's demon is the prototype of the dissipation structure.

A granular system is a typical dissipation system. One of the amazing properties of the granular gas is that sand performs as Maxwell's demon [7.71].

As shown in Fig. 7.41, a box with bottom area 12 cm^2 and height 20 cm is divided into two identical parts by a wall with a hole at height 2.3 cm. Sand of 1 mm in diameter is initially put in one chamber. When the box mounted on a shaker is vertically vibrated at a frequency 50 Hz and vibration amplitude 0.3 cm, the sand is eventually distributed homogeneously in the two chambers. However, the symmetry is spontaneously broken when the vibration frequency is reduced to below 30 Hz. That is to say, during the vibration, there is a net flux flow from one chamber to another. Eventually, small but "hot" sands (with high speed) are in one chamber, leaving the rest, "cold" sands, in the other [7.70].

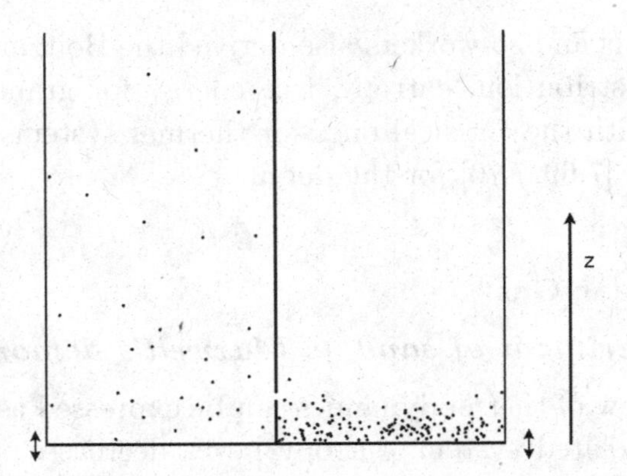

Fig. 7.41 Sand as Maxwell's demon [7.72].

7.6.2 *Model of flux function*

To understand this process, J. Eggers [7.71] proposed a flux function model which is clearly given below [7.72, 7.73]. Imagine the container described above is filled with small grains. The grains are fluidized by vertically vibrating the set-up with a sawtooth-shaped signal with frequency $f$and amplitude a. The container has a small, rectangular aperture of size $S = w\Delta h$ at a height h. In the following derivation of the flux function, the grains are assumed to be frictionless spheres, such that only normal restitution contributes to the dissipation. Furthermore, any collisions with the walls (extending all the way up to infinity) and bottom are assumed to be perfectly elastic.

It is also assumed that the dissipative granular gas in a compartment is always in a steady state and that macroscopic flow is absent, the hydrodynamic equations [7.71] reduce to

$$\nabla p = -\rho m g e_z, \quad \nabla \cdot \mathrm{J} = I. \tag{7.48}$$

In the first equation, which expresses momentum conservation, p is the pressure and ρ is the number density of particles. The other parameters are the particle mass, m, acceleration of gravity, g, and the unit vector in the vertical direction, $\mathbf{e}_z$.

The second equation, where $\mathbf{J}$ and I are the heat flux and the energy dissipation rate per unit volume, respectively, represents the energy balance in the system. When integrated over the container volume V, it gives (with use of Gauss' theorem)

$$\int_{\partial v} \mathrm{J} \cdot \mathrm{dA} = \int_{v} I dV \quad \text{or} \quad Q_{\text{in}} = Q_{\text{diss}}, \tag{7.49}$$

where ∂V stands for the boundaries of the container, i.e., the walls and the bottom. Since the walls (and bottom) are assumed to be non-dissipative, the only contribution to the left integral stems from the energy input rate (Q_{in}) at the bottom. This should be balanced by the right integral, which can be interpreted as the total dissipation rate in the container due to the inelastic particle collisions.

Using the ideal gas law for dilute granular gases, $p = m\rho T$, where T is the granular temperature, defined as one third of the velocity fluctuations in the system, the first equation of Eq. (7.48) can be integrated, and it gives

$$\rho(z) = \frac{gN_k}{\Omega T(z)} \exp\left\{-\int_0^z \frac{g}{T(\varsigma)} d\varsigma\right\}. \tag{7.50}$$

Here N_k is the total number of particles in the compartment, and Ω is its ground area. The key assumption in this derivation of the flux function is that the temperature is constant within the gas. For dilute gases this is in fair agreement with molecular dynamics simulations, besides a small region just above the bottom, where the energy input leads to a local increase of the temperature. With this assumption, the evaluation of the integral in Eq. (7.50) results in the barometric height distribution

$$\rho(z) = \frac{gN_k}{\Omega T} \exp\left\{-\frac{gz}{T}\right\}. \tag{7.51}$$

To write the temperature as a function of N_k, we first need to express Q_{in} and Q_{diss} in terms of N_k and T.

The standard way is to assume a local Maxwell–Boltzmann distribution for the velocity, calculate the probability for a collision between two particles at positions $\boldsymbol{r}_1$ and $\boldsymbol{r}_2$, with velocities $\boldsymbol{v}_1$ and $\boldsymbol{v}_2$ respectively, average over the ensemble, and integrate over the container volume. This gives

$$Q_{\text{diss}} = 4\frac{\sqrt{\pi}}{\Omega}mgr^2N_k^2(1-e^2)\sqrt{T}. \tag{7.52}$$

A similar procedure for the energy input at the bottom gives

$$Q_{in} = 2mgN_k af\left(1+\frac{2af}{\sqrt{\pi T}}\right) \approx 2mgN_k af, \tag{7.53}$$

since the typical velocity of a particle in a dilute granular gas ($\sqrt{T}$) must be larger than the velocity of the bottom ($2af$). By equating Eqs. (7.52) and (7.53), one can finally express the temperature in terms of the number of particles in the container:

$$T = \frac{\Omega^2 a^2 f^2}{4\pi(1-e^2)^2 r^4 N_k^2}. \tag{7.54}$$

Denoting the particle fraction n_k in that compartment with respect to the uniform distribution, one has $n_k = N_k/N_{av}$, where N_k is the number of particles in compartment k and $N_{av} = N_{tot}/K$, with N_{tot} the total number of particles in the K-compartment system. The flux (in particles per second) from the compartment is now found to be the density of particles moving towards the wall ($\rho(z)/2$) at height h multiplied by the average velocity normal to the aperture ($\sqrt{(2T/\pi)}$) and the area of the aperture (S), $F(n_k) = (\rho(z)/2)\sqrt{(2T/\pi)}S$ or

$$F(n_k) = An_k^2 \exp(-Bn_k^2), \tag{7.55}$$

where

$$A = \sqrt{2}(1-e^2)gr^2S/(\Omega^2 af), \tag{7.56}$$

and the dimensionless driving parameter

$$B = \frac{4}{\pi}\frac{gh}{(af)^2}(1-e^2)^2(\pi r^2 N_{av}/\Omega)^2. \tag{7.57}$$

The time rate of change $\mathrm{d}n_k/\mathrm{d}t$ of the particle fraction in the k-th compartment is given by the inflow from its two neighbors minus the outflow from the compartment itself:

$$\frac{dn_k}{dn_t} = F(n_{k-1}) - 2F(n_k) + F(n_{k+1}), \qquad (7.58)$$

with $k = 1, \ldots, K$. For simplicity, we will use cyclic boundary conditions $n_{K+1} = n_1$. Numerical calculation of Eq. (7.58) gives the bifurcation of n_k as shown in Fig. 7.42.

This shows that for $B < 1$ (at higher frequencies as $B \sim 1/f^2$) there is only one stable equilibrium, namely the uniform state, for the driving factor $B > 1$ (at frequencies lower than a certain critical value). The symmetry is broken, and there would be an asymmetric and dynamic equilibrium; a clusterization occurs. In this way, the entropy of the dissipative open system is decreased.

A dynamical equilibrium is established in which the particle flux from the densely populated cluster is balanced by the flux from the diluted compartment. This is possible because, in a unit time, particles in the dense compartment experience many energy-dissipating collisions and are therefore slow. Likewise, particles in the diluted compartment are relatively fast, and are thus able to generate the same flux through the slit as the slower particles, but more particles in the densely populated compartment [7.72].

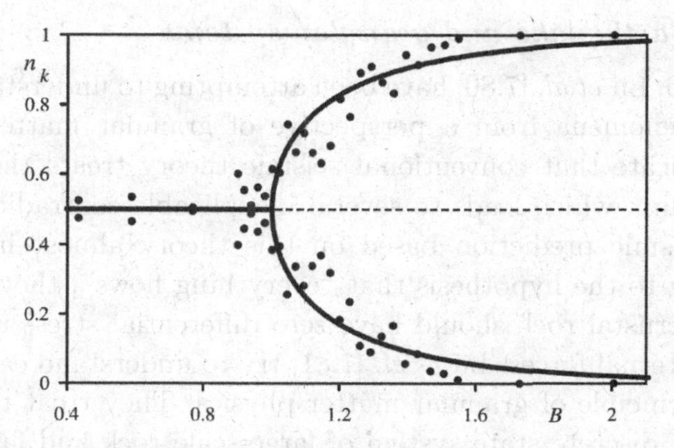

Fig. 7.42 Bifurcation diagram for the 2-box system ($k = 1, 2$). The solid line represents stable and the dashed line unstable equilibria of the flux model. The dots are experimental measurements. In both cases, the transition to the clustering state is seen to take place via a pitchfork bifurcation [7.73].

Many other properties of granular gas are also explored. They include clusterization in multi-compartments [7.72, 7.75], compartmentalization of bidisperse granular gas [7.75, 7.76], and granular fountain [7.77].

Granular systems are far from well understood. Some important aspects have not even been touched in this context, such as wet sand [7.77] and granular flow [7.78]. The applications of the knowledge of granular materials are not extensively introduced here either: a thorough understanding of granular materials is extremely important in reducing jamming problems in traffic/transportation and in evacuation from stadiums or theaters during a fire, as well as in reducing disasters brought on by some granule-related natural phenomena, such as sand storms or debris flows.

To the question: "Can a granular material be described by hydrodynamic equations, most specifically those equations which apply to an ordinary fluid?" L. Kadanoff [7.79] said "it seems to me that the answer is 'no!'", and granular materials form a particularly rich example of a non-equilibrium system, with behaviors which are, at this moment, not fully understood.

7.7 Applications on Granular Systems

7.7.1 *Earthquake and granular systems*

Since 2010, Lu *et al.* [7.80] have been attempting to understand earthquake phenomena from a perspective of granular matter physics. They indicate that conventional seismic theory treats the crust as a continuum, which leads to several inexplicable contradictions and makes seismic prediction based on this theory almost impossible. According to the hypothesis that "everything flows", they proposed that the crustal rock should have zero differential stress under prolonged external forces. Lu *et al.* [7.81] try to understand earthquakes via the principle of granular matter physics. They treat the earth's crust as a discrete state system of large-scale rock and fault gouge, in which the scale of large-scale rock masses ranges from several kilometers to several hundred kilometers. During crustal movement, the forces acting between the rock masses form force chains, and an earthquake occurs along the force chains where the rock structure is weak. Strain sensors buried in sandpits can be used to measure seismic precursor signals [7.82].

(1) Hyposes of the crust as a discrete state

The crust is usually thought of as a continuum. In 1910, Reid [7.83] proposed the elastic rebound hypothesis. This hypothesis argues that the strain accumulated continuously before the earthquake, and the earthquakes occur when the tectonic force exceeds the limit of rock elasticity, because the rocks in the crust are fault-slipping and the rocks themselves are elastic. Rocks that are deformed elastically at the time of fracture would bounce back in the opposite direction after the force disappears, and the state before the deformation is recovered. This bounce can produce tremendous speed and strength, and the energy stored for long-time would be released in an instant when the earthquake occurs. The elastic bounce has been regarded as the main cause of earthquakes, which is one of the basic concepts of traditional seismology. However, there are some unexplained points in this hypothesis. This leads to the unexplained "heat flow paradox". If the earthquake is caused by a brittle fracture of the rock, due to the slip of the rock at the fault, and the experimental results show that the rock friction coefficient is large [7.84], most of the energy released in an earthquake must be converted to heat by friction. However, the measurement of the temperature change near the hypocenter shows that there is little heat released during an earthquake. The seismic wave energy actually released is about two orders of magnitude lower than the calculation value of the work done by brittle fracture of rocks. On other words, if earthquakes are caused by brittle fracture of rocks, the energy of the earthquake would be much greater. This major contradiction is called "heat flow paradox".

Lu *et al.* [7.80] treat the earth's crust as a discrete system of massive rock and fault gouge. To them, one tectonic plate of the crust consists of hundreds of rock masses, each measuring several or even thousands of kilometers. The stress distribution in the fault between the rock masses is different from that in the rock masses. The fault gouge filled in the fault boundaries is composed of granular materials such as gravel and sediment, and the fault width can be $10^{-1} - 10^{3}$ m. The volume fraction of fault gouge increases with depth, and its average density is about 2/3 of that of rock, i.e. the vertical stress caused by its own weight is about 2/3 of that of rock. At the same time, considering that the crust and mantle systems have basic characteristics of slow dynamic behaviors, they propose a new perspective of the physical mechanism of earthquake occurrence based on the observed facts.

According to the knowledge of granular matter physics, the geotectonic stress and the self-weighting stress of the original crust superpose to form the real-time stress in the crustal rock. According to the discrete state characteristics of the crustal rock, the geotectonic forces are spread in the earth's crust in the form of rock stick-sliding and force chains. The tectonic force is transferred to the next rock by pressing the fault gouge, and the fault gouge is composed of granular matters. The relation between the effective elastic modulus and depth change h is $E_{eff} \propto h^{5/9}$, causing the shear stress due to the tectonic force on rock mass increases with depth. The deeper the rock is in plastic state, the lower the failure strength. Therefore, shallow-focus earthquakes generally occur in the deep crust, mainly as a plastic slide, while the slip velocity is generally in the range of $1 \sim 10\,\mathrm{m/s}$.

(2) Assumptions of "everything flows"

There exists large differential stress in rocks in the crustal elastic zone. The pulling stress on the solid is positive while the pressing one is negative. When the shear stress is zero, the corresponding normal stress is the principal stress. The maximum principal stress is σ_1 and the minimum one is σ_3, and yield strength is generally defined as the strength of the solid. In geology, the rock strength is often characterized by the differential stress $\sigma_D = \sigma_1 - \sigma_3$. According to Fig. 7.43, in the case of confined compression, the rock strength increases with increasing stress. The confining pressure is equivalent to the external block or support, which can make the rock withstand more pressure without being damaged. It is observed in some confining experiments

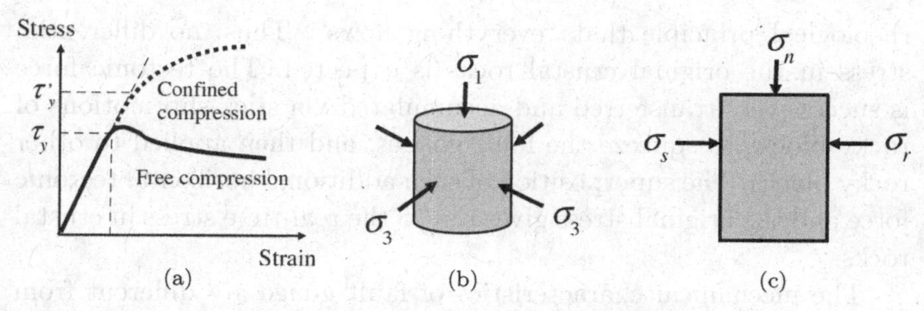

Fig. 7.43 (a) Stress-strain relations for free and confined compressions, and τ_y and τ_y' denote the yield stresses of a material under free and confined compressions, respectively; (b) the differential stress $\sigma_1 - \sigma_3$ under a confined compression; (c) a scratch of confined shear action, where σ_s, σ_n and σ_r are external stress, normal pressure and resistive blocking stress, respectively [7.80].

that the differential stress $\sigma_1 - \sigma_3$ can be continuously increased even under pressure as high as 10 GPa.

The stress-strain relationship and stress distribution have been well described in the elastic range for solids under free compressions, i.e., in the absence of confining pressure. The relationship between compressive stress σn and shear stress σs is $\sigma\text{s}/\sigma\text{n} = \nu/(1-\nu)$, where ν is the Poisson's ratio, the ratio of the tangential strain to the compressive strain. The value of the Poisson's ratio is dependent on the specific material and $\sigma\text{s} = \sigma\text{n}/3$ for $\nu = 0.25$. The conditions for the above relationship are: deformation is small and in the elastic range; no external stress in the shear direction, that is, no confining pressure; and the compression time is not too long.

If the above conditions are not established, such as rocks are under pressure for a long time, the above inequality $\sigma\text{s} \neq \sigma\text{n}$ will not exist. Heraclitus, an ancient Greek philosopher, used the phrase "everything flows" to expound his philosophical thoughts. Barnes summarized a variety of views on the laws of yield and rheology in his review article entitled *Everything Flows* [7.85].

Speaking of the stress distribution in earth's crust, one would notice: without taking the tectonic force into account, the vertical and the horizontal stresses must be in balance due to the effect of gravitational pressure over a long period of time according to the rheological principle that "everything flows". Thus, no differential stress in the original crustal rocks is expected. The tectonic force is successively transferred and accumulated via stick-slip motions of rocky blocks to squeeze the fault gouges, and then applied to other rocky blocks. The superposition of such additional horizontal tectonic force and the original stress gives rise to the real-time stress in crustal rocks.

The mechanical characteristics of fault gouge are different from rocks as it consists of granular matters. Thus the elastic modulus of the fault gouge is much lower than that of rocks, and will become larger with increasing pressure. This character of the fault gouge leads to a tectonic force that increases with depth in a nonlinear fashion. The distribution and variation of tectonic stress in the crust are then specified.

Under the long-term effect of billion years due to the gravitational pressure of the crust itself, the concept of "everything is flowing" implies that, when the stress exceeds the elastic limit or creep occurs under compression, the deformation of the crust would make its Poisson's ratio exceed the elastic range. If $\nu \to 0.5$, according to $\sigma_s/\sigma_n = \nu/(1-\nu)$, one has $\sigma_s/\sigma_n \to 1$, that is, shear stress tends to be equal to the normal stress, i.e., the rock is fluid-like as long as the rock masses are sufficiently ancient as a result of long-term variation. Thus, in the locations where ρh is the same, the shear stress tends to be the same as the gravitational stress, namely $\sigma_s(h) = \sigma_n(h) = \rho g h$, the vertical stress and horizontal stress in the crustal original rocks are the same, and there is no differential stress.

(3) Crustal elasticity-plasticity change

The self-weight of the crustal rock makes its elastic-plastic transition. Weisskopf [7.86] and Zhao [7.87] found that, for high enough mountains, the pressure due to gravity would cause the mountain rock to change from an elastomer to a plastomer and flow, causing the mountain to sink.

By comparing the energy required for rock to melt and the work done by gravitational potential energy, they estimated as follows. Suppose a rock with cross section A, thickness δ and density ρ falls from height H to the ground, the work done by the gravity is $A\delta\rho gH$. The heat absorbed by the melting rock would be $A\delta\rho\Lambda_m/m_{\text{mol}}$, where Λ_m is the molar melting heat of the rock and m_{mol} is its molar mass. Let the gravitational work equal the endothermic melting, then $H = \Lambda_m/gm_{\text{mol}}$. If the major component of the rock is SiO_2, with $m_{\text{mol}} = 60$ g/mol and $\Lambda_m = 8.52$ kJ/mol, the calculated maximum height of a mountain is about 14 km.

Similarly, crustal rocks undergo elastic-plastic transition under their own weight. According to the above calculation, the maximum depth of elastic-plastic transition of quartz in crustal rocks is no more than 14 km. Factors such as water content, porosity and temperature, as well as inhomogeneities of the rock, would decrease its yield strength, resulting in the elastic-plastic transition starting from shallow depth. There are three crustal rock layers along with the rock depth: elastic, partial plastic and completely plastic layers. In the partial plastic layer, when the proportion of plastic rock reaches about 10%, there would appear topological connection of plastic regions. Below this depth, the strength variation of rock shows plastic characteristics. Since the crustal lithosphere has a long history of evolution, the concept of "everything flows" is well established. One should also notice that the rock in the crust is very uneven, and its viscosity coefficient is closely related to the temperature and stress state. With the depth of 10–20 km in the crust, the temperature is about 300–600°C, and the self-weight pressure is between 270–540 MPa.

In the partial plastic layer, the shear strength of rock is determined by the plastic character. The equivalent friction coefficient of plastic sliding is smaller than that of brittle fracture by more than an order of magnitude, and the shear strength of rock is much smaller than that of brittle fracture.

When the solid is in a plastic state, it is a non-Newtonian fluid and its shear strength can be approximated by the Bingham fluid relationship $\tau_s = \tau_y + \eta\dot{\gamma}$, where τ_y is the shear yield strength, η is the viscosity and $\dot{\gamma}$ is the shear rate. The change in shear strength τ_s is related to η, and the viscosity η is determined by the friction between atoms or molecules, much less than the macroscopic friction coefficient. Since it is difficult to obtain the exact value of viscosity η, in the usual range of shear rate $\dot{\gamma}$, $\tau_s \approx \tau_y + \mu \mathrm{p} \sigma_n$ can be used to approximate the change of shear strength of plastic state with positive pressure, where μ_p is the equivalent friction coefficient of plastic state and can be measured experimentally.

Under the restricted compression for a long time, the role of confining pressure is equivalent to the application of a lateral reaction force, limiting deformation, which would have an impact in the longitudinal direction. When the compression time is long enough, according to the principle of force balance and the concept of "everything flows", the pressures in all directions of the solid in the whole container tends to be the same. The stress in the horizontal direction σ_s is equal to the stress in the vertical direction $\sigma_n, \sigma_s = \sigma_n$, which is achieved by adjusting the microstructure of the solid.

Based on some observations and analysis [7.88–7.90] parts of rocks begin to undergo plastic transformation at depths of 6–7 km; in the depth range between 7–20 km, it is a mixed layer of elastic-plastic state, called the partially plastic layer; whereas in areas deeper than 20 km, most or all of the rocks are converted to plastomers. The geological situation of the crust is different from place to place, and the depth of elastic-plastic transition would be quite different in different locations.

(4) The conditions of earthquakes

Earthquakes are bound to produce volume expansion and the expansion only occurs when the barrier is broken. The conditions for earthquakes are as follows: the tectonic force of the crust exceeds the sum of the rock breaking strength, frictional force at the boundary of the fault and the total stopping force. Thus, shallow-focus earthquakes are in fact plastic slips when rocks break through barriers.

The necessary and sufficient condition of earthquake occurrence is

$$F \geqq F_s + F_r. \tag{7.59}$$

As shown in Fig. 7.44, F is the shear force generated by the tectonic force which propagates and accumulates through the stick-sliding movement of the rock block pressing onto the fault gouge; F_r

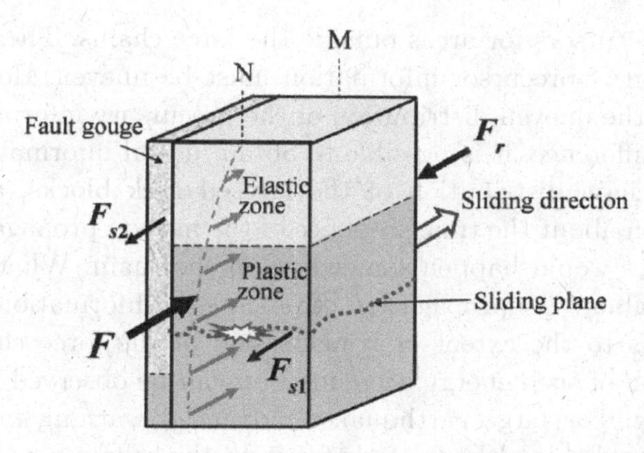

Fig. 7.44 Sketch of various forces applied on part of the crust. Arrows starting from the dash curve represent tectonic forces; M and N respectively are the frontier and posterior of the rock block [7.80].

is the resistance force opposite to the rock slip direction; F_s is the resistance force of the rock itself to rupture and slip. F_s is divided into two parts, namely $F_s = F_{s1} + F_{s2}$, where $F_{s1} = \tau_b(h) \times S$, the product of the strength of the rock burst shear stress $\tau_b(h)$ and the shear area S, and F_{s2} is the resistance of the rock mass exerted by the surrounding fault gouge and other friction.

(5) Pro-earthquake precursory information

Considering the granular matter characteristics of the rock block-fault gouge system that generates earthquakes, the precursor information can propagate very far away but not homogeneously. It is possible to obtain useful precursory information by observing the effects of the process of the accumulation of tectonic forces that trigger earthquakes, the geological conditions in the local area, and the effects of micro-cracks and micro-slips of rocks. From the observation of the sand pit, groundwater level, rock strain, and terrestrial electricity, it can be seen that some earthquakes with precursory information can propagate as far as about 10^3 km. The inevitable result is force chains in granular matter systems. A rock block can be considered as a granule. One or two rock blocks constitute the granular system with length scale of force chains up to 10^3 km. There are no obvious precursors for areas outside the force chains. Therefore, the distribution of precursor information must be uneven. However, by observing the uneven distribution of the precursory information and the time difference, it is possible to obtain useful information about the force-chain distribution of the layered rock blocks, as well as information about the time sequence of the motion propagation. The earthquakes would happen somewhere in the chain. What Lu *et al.* have said about the propagation of precursory information being far away refers to the extent of transmission of the force chains. The information of such energy accumulation can be observed before the earthquake. For large earthquakes, advanced warning information would be available although it is far from the source.

In the process of earthquake incubation, the energy accumulates in the form of the stick-slipping movement of rock masses and the extended propagation of force chains. Therefore, most of the earthquake precursor information is not generated from the epicenter. When and where a specific earthquake occurs depends on the energy accumulation and geologic conditions in the force chains. It is often observed that "before the earthquake the anomalies migrate and converge from the periphery to the epicenter, while after the earthquake the anomalies spread from the epicenter to periphery" [7.91]. This phenomenon is consistent with the principle of granular matter physics.

If one can detect the stick-slip movement of each rock mass and of force chains, recognize the displacement, force distribution and variation in force chains, and distinguish the types of stress steps within the rock block, one should be able to determine the areas where earthquakes may or may not occur. Earthquakes may occur in the immediate vicinity if fast lock-in and stress steps are observed in rock layers and are accumulated to very high values. The degree of the earthquake is determined by the nearby formation resistance and rupture strength, which is closely related to the geological conditions. Based on the above analysis and available observational data, it is possible to determine the location of the earthquake, the magnitude and approximate time of the earthquake.

It is even difficult to determine the exact time of an earthquake. There are two reasons for this difficulty. First, the mantle driving force is weakened or has disappeared in some cases, such as energy release of the regional mantle via a small plug-flow transition. The second is the force chain relaxation of rock layers. The relaxation of non-uniform force chains of granules (rock masses) or other interference may weaken the force chains or cause them to disappear, which is also a characteristic of granular matters.

7.7.2 *Creep and sand soilization*

More than one fifth of the total continental area of this planet is occupied by deserts. Preventing the deserts from expanding is an eager desire of many. Besides, there are also efforts to convert sands into soil (soilization of sand). In this section, we will first introduce the concept of creep of granular matter and soil. We will examine how the sandy land of a desert is transformed to soil, and try to find out the characteristics of creep during the process of sand soilization.

(1) The concepts of consolidation and creep [7.92]

The consolidation of soil is a water drainage process which reduces voids in the soil, as well as its total volume, leading to soil settlement. There are two phases of soil consolidation — primary and secondary. Creep mainly describes deformation of materials under certain stress. Soil creep happens after the primary consolidation, and it corresponds to the secondary consolidation settlement under an action of long-term loads. It is possible to divide creep behavior into volumetric and deviatoric (or shear) creeps according to the acting stress. The volumetric creep is caused by a constant volumetric stress, and the deviatoric creep is caused by a constant deviatoric stress.

In physics of granular systems, creep is the tendency of granular materials to move or deform slowly under external stresses. Everyone knows that a skyscraper cannot be built on a sand beach without laying a solid foundation at the very beginning. To reduce the creep effect on building settlement and/or inclination, the pile-raft foundation of the main building of Dubai Towers Doha is composed of 192 cast-on-site spun piles. The pile diameter is 1.5 meters and its length is about 50 meters. Concrete consumption in the foundation is 44,726 m^3 [7.93]. On the other hand, the Tower of Pisa, Italy, has inclined for about 5.5° due to creep of soil under the tower foundation. Under the same amount of external stresses, the less the creep strain ε is in the same period of time, the stronger the ability against the creep would be. Generally speaking, soil has less creep than sand. This suggests that creep characteristics could be adapted as a criterion to judge the degree of sand soilization in a certain sense.

An alternative way of dividing the creep behavior is based on the type of strain-time behavior shown in Fig. 7.45. According to the shape of the strain-time curve, one can divide creep into primary, secondary and tertiary phases. The primary phase can be defined as a creep deformation during which the strain rate decreases continuously with time. Deformation at a constant rate (material flow) is denoted as the secondary phase. In the case of the tertiary or the accelerated phase the strain rate is continuously increasing and this leads to creep rupture. Generally, volumetric creep consists only of the primary phase of the creep deformation, i.e. it tends to stabilize. If the deviatoric stress is low, only the primary creep phase will appear; however, when the deviatoric stress is above a level of shear mobilization, the primary phase will be followed by the secondary

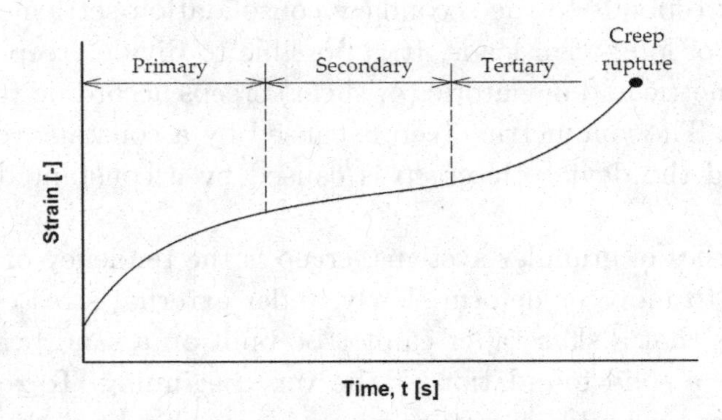

Fig. 7.45 Illustration of the primary, secondary and tertiary phases of the creep [7.92].

phase which can lead to the tertiary phase and consequently, creep rupture.

(2) Creep models

Creep behaviors can be approximately described equivalently either by the Kelvin–Voigt exponential function or Buisman's logarithmic function.

(i) Kelvin–Voigt model [7.92]

Kelvin and Voigt [fo:kt] rheological model is a viscoelastic model consisting of Hook's elastic element (spring) connected in parallel with Newton's viscous element (dashpot), see Fig. 7.46. The strain ε is characterized by elastic deformation delayed by time effects. In this model, elastic and viscous elements can only be deformed together and to the same extent. Hence instantaneous deformation of this rheological model is zero. Stress σ will be distributed between both model components

$$\varepsilon = \varepsilon_s = \varepsilon_{vs},$$
$$\sigma = \sigma_e + \sigma_{vs}, \tag{7.60}$$

where subscripts e and vs stand for elastic and viscous, respectively. After the combination of equations, the differential equation according to Kelvin and Voigt is

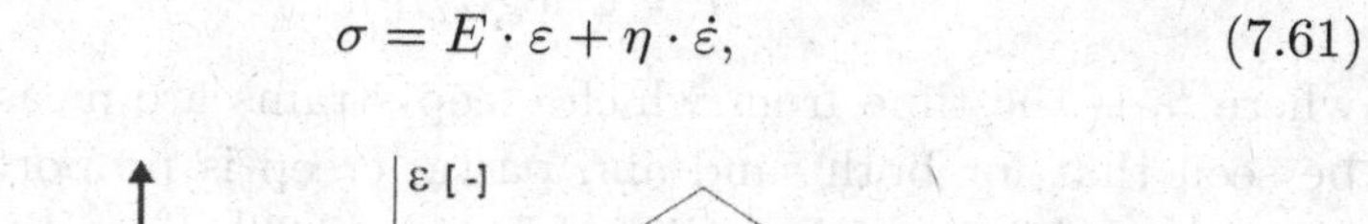

$$\sigma = E \cdot \varepsilon + \eta \cdot \dot{\varepsilon}, \tag{7.61}$$

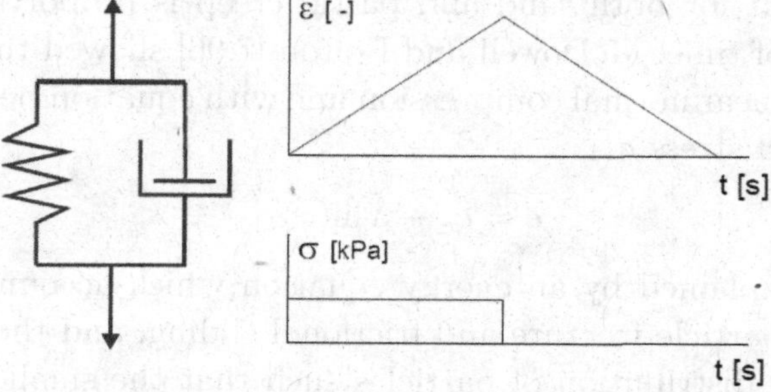

Fig. 7.46 Kelvin and Voigt rheological model with demonstration of the stress σ and strain ε behavior [7.92].

where E is the elastic or Young's modulus, and η is the viscosity. Solving this equation for a constant stress, one arrives at the relation, also called the creep function,

$$\varepsilon(t) = \frac{\sigma}{E}\left[1 - \exp\left(\frac{E}{\eta}.t\right)\right] = \frac{\sigma}{E}\left[1 - \exp\left(-\frac{t}{\lambda_{\mathrm{K}}}\right)\right], \tag{7.62}$$

where λ_{K} is a constant called the time of retardation which determines the time-dependent deformation behavior of the parallel-connected components of the Kelvin–Voigt model. Solving this equation at time t= λ_K leads to the result

$$\varepsilon(\lambda_{\mathrm{K}}) = (\sigma/E)(1 - 1/e) = 63.2\%(\sigma/E) = 63.2\%\varepsilon_{\max}. \tag{7.63}$$

This means that in the creep phase at the time corresponding to the retardation time the value of the creep strain has increased to 63.2% of the maximum strain $\varepsilon_{\max}$, which will be finally reached at the end of the load interval. Similarly like relaxation time λ_{M} for relaxation tests, the retardation time λ_{K} is relevant in creep tests and is unique for all materials. The creep behavior in the Kelvin–Voigt rheological model for the linear viscosity is illustrated in Fig. 7.46.

(ii) Buisman's logarithmic function [7.94]

It is well known that granular materials creep under constant effective stress [7.95] such that creep strain e is proportional to log time:

$$e = C\log(t/t_0), \tag{7.64}$$

where t_0 is the time from which creep strains are measured. It can be seen that for both sand and pasta, creep is proportional to the logarithm of time. McDowell and Bolton [7.96] showed that the existence of a linear normal compression line with equation between creep strain e and stress σ :

$$e = e_c - \lambda \ln(\sigma/\sigma_c) \tag{7.65}$$

could be explained by an energy equation which accounts for dissipation by particle fracture and frictional sliding, and the generation of a fractal distribution of particles such that the smallest particles continue to fracture under increasing stress, becoming statistically stronger and filling voids. A micro-mechanical analysis was extended to account for creep and showed that Eq. (7.65) is consistent with Eq. (7.64) for a granular material subjected to creep at constant stress under one-dimensional conditions, if the time dependence of particle strength is accounted for. Equation (7.65) has been written in a dimensionally consistent manner. The stress σ_c is simply a stress on the normal compression line, and voids ratio e_c is the voids ratio at that applied stress. According to Eq. (7.65), an aggregate should be in equilibrium with a voids ratio e_c under an applied stress σ_c, where σ_c is proportional to the average strength of the current smallest particles σ_s, so that

$$\sigma_c = k\sigma_s. \tag{7.66}$$

The constant k is independent of particle size, because as the smallest particle size reduces, the smallest particles are in self-similar geometrical configurations, according to the micro-mechanical argument proposed by McDowell and Bolton [7.96]. Further compression can only occur if the stress level increases above σ_c or if the average strength of the smallest particles σ_s decreases. It is the fracture of the smallest particles which gives the reduction in voids ratio, because even when a wide distribution of particle sizes has formed, if some large particles break, there are no available voids for the large fragments to fill. Substituting Eq. (7.66) into Eq. (7.65) gives:

$$e = e_c - \lambda \ln(\sigma/k\sigma_s). \tag{7.67}$$

The purpose for writing Eq. (7.67) in this way is to examine the effect of time-dependent particle strength. It is well established in materials literature [7.97, 7.98], that ceramics exhibit time-dependent strength. This is caused by slow crack growth as moisture in the environment interacts with flaws in the material. The end result is that for a tensile test on a ceramic specimen, if the standard test used to measure the tensile strength σ_{TS} takes a time t_{TS}, then the stress which the sample will support safely for a time t is given by the equation

$$\left(\frac{\sigma}{\sigma_{TS}}\right)^n = \frac{t_{TS}}{t}, \tag{7.68}$$

where n is the slow-crack growth exponent. Data for n are very limited, but n is $10 \sim 20$ for oxides at room temperature; for carbides and nitrides, n can be as large as 100 [7.97]. It is now possible to examine the effect of the dependence of σ_s on time at constant stress level in Eq. (7.67). If σ_{so} is the average particle strength which could be measured at time $t = t_o$, then the average strength σ_s after a time t, according to Eq. (7.68), would be

$$\sigma_s = \sigma_{so} \left(t_o/t\right)^{1/n}. \tag{7.69}$$

Substituting Eq. (7.69) into Eq. (7.67) gives:

$$e = e_c - \lambda \ln(\sigma/k\sigma_{so}) - \frac{\lambda}{n} \ln(t/t_o). \tag{7.70}$$

Hence the reduction in voids ratio Δe as a function of time after time t_o is simply:

$$\Delta e = \frac{\lambda}{n} \ln(t/t_o) = \frac{2.3\lambda}{n} \log_{10}(t/t_o) \tag{7.71}$$

so the log time effect is observed.

McDowell also justifies a log e–log σ normal compression law using particle strength data [7.99, 7.100].

(3) Creep measurement of granular matters [7.92]

(i) Measurement of volumetric creep of granular matter

The main problem of the oedometer apparatus is friction between the specimen and the wall of the metal ring. Frictional forces reduce the transmission of the external load to the soil specimen, which can lead to an inhomogenous state of stress and physical nonlinearity in the soil. Lower soil density from the top downwards and from the center to side of the specimen is then observed. This defect can be reduced by an increase in the D/H ratio (where D is the diameter of the specimen and H is the height of the specimen) and using industrial Vaseline or a Teflon film in the internal surface of the ring. There are also other methods, like using rings with a special design. One example can be a floating ring with double-ended compression, i.e. both porous discs serve as a piston (see Fig. 7.47).

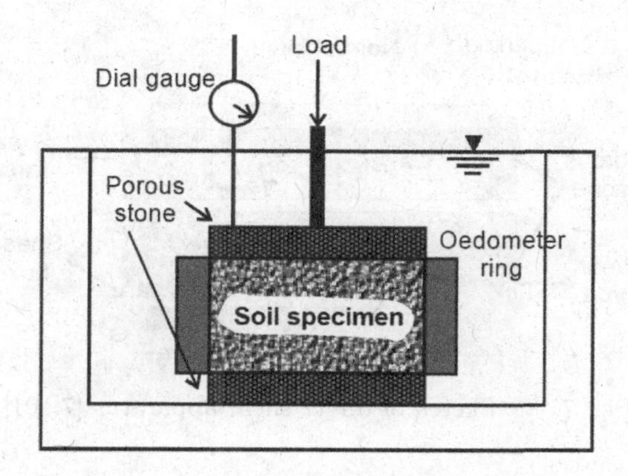

Fig. 7.47 Basic scheme of the oedometer [7.101].

(ii) Measurement of deviatoric creep of granular matter

Tests under simple shear are widely used in geotechnical laboratories all over the world for the evaluation of the shear strength and investigation of the post-peak behaviour. Generally the simple shear test is called direct shear test and used laboratory equipment is denoted as a direct shear apparatus. A sketch of the simplest direct shear test apparatus is given in Fig. 7.48. The general procedure of this test in the case of soil material can be divided into two parts. The first part is the application of the normal stress for the consolidation of the tested material. After consolidation (in the case of the direct shear test this means mainly primary consolidation, i.e. dissipation of the pore water pressure) the shearing is gradually applied. Depending on the equipment, the direct shear test can be either strain-controlled or stress-controlled. The main advantage of the direct shear test is its extreme simplicity. It was noted by Maslov [7.102]: "Tests under simple shear best reflect the actual conditions of the possible shear of the soil structure, as they provide visible and convincing evidence of shear failure. If the drainage conditions are properly maintained, their results coincide with those obtained from compressive tests on triaxial apparatus". Nevertheless, there are several disadvantages with the direct shear test, where the main ones are that drainage conditions cannot be controlled and shear stress on the failure plane

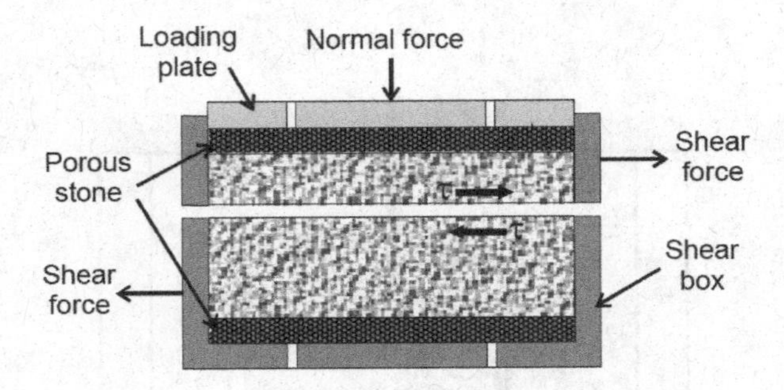

Fig. 7.48 Sketch of direct shear apparatus [7.101].

is not uniform. A detailed description of the direct shear test together with an example of the soil behavior under this test can be found in [7.101, 7.103].

(4) Sand soilization

(i) Efforts in recovering sand degradation

To modify sand body, Yi *et al.* [7.104, 7.105] first mixed sand with an organic solution (containing 2% modified sodium carboxymethyl cellulose (CMC) and 5% compound fertilizer) at a weight ratio of 1:0.15. The mixture was then placed on top of a $15 \sim 25$ cm thick plain sand layer underlain by a $20 \sim 30$ cm thick gravel layer on the ground mimicking desert landform conditions. After the water was evaporated, the water-soluble solid state organic substance could bond sand particles, and dissolved in water again to form the solution with the cohesiveness and the adhesion to bond the sand particles. The modified sand body allowed a common sand body to be in pedogenesis, changed the interaction relation between sand body particles, and formed a particle pore structure with binding force (cohesive force). A field experiment of plants growing on a 2 km^2 desert land of Inner Mongolia, China, was conducted.

Another sand modification method was also proposed which was especially suitable for high-wind-erosion and high-water-erosion areas [7.106].

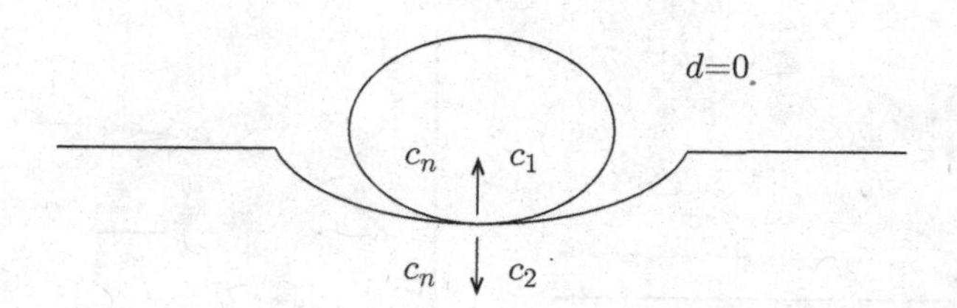

Fig. 7.49 Contact constraint is exactly satisfied [7.108].

(ii) Mechanical model of sand soilization

To explain why soilized sand can maintain its eco-cycle to sustain the growth of plants, Yi *et al.* [7.107] suggested to concentrate on the interactions between granular particles. The constraint of sand in its discrete state is a contact constraint. Contact configuration is described by the (generalized) distance function $d = \Phi(q)$ with $\Phi(q) \geq 0$. Contact forces are compressive, $c_n \geq 0$, and they act only when the contact constraint is exactly satisfied, or $\Phi(q)$ is complementary to c_n or $\Phi(q)c_n = 0$, or $\Phi(q) \perp c_n$ [7.109].

However, according to Yi *et al.* [7.107], when soil is in a rheological state (especially an ideal viscoelastic state, in which the water content of soil is between its liquid limit ω_{L} and plastic limit ω_{P}), the constraint among its granules is an omni-directional integrative (ODI) constraint, which provides tension and moment in addition to pressure and friction while allowing for some displacement in the constrained tensile, rotational and tangential directions. ODI constraint has the properties of omni-directionality and restorability. When this constraint exists among soil granules, any granule can become connected with others along any direction of contact (omni-directionality), and even after separating, they can reconnect when contact occurs again (restorability).

When the water content of soil is below its plastic limit ω_P, with the evaporation of water, the constraint among the granules gradually transitions from the ODI constraint to a fixed one. When soil is in an ideal solid state (the water content approaches 0), a fixed constraint acts among its granules, and can almost fully restrain the degrees of freedom of the soil granules. Thus, the granular arrangement is fixed.

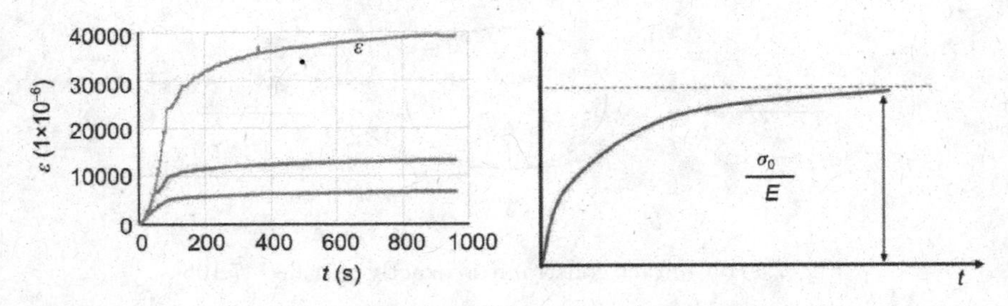

Fig. 7.50 Left: Creep strains of soilized sand in its ODI state with different water content. Right: Typical curve of the Kelvin model in viscoelasticity theory [7.107].

Therefore, the transformation of soil between the rheological state and the solid state is essentially the transformation between the ODI constraint and the fixed constraint.

(iii) Creep of soilized sand

Yi *et al.* [7.107] measured the creep strains of soilized sand in its ODI state with different water content. The left panel in Fig. 7.50 shows the creep curves of a sand sample with three different water contents (between ω_L and ω_P) after acquisition of the ODI constraint. These curves fit well with the Kelvin model in viscoelasticity theory shown in the right panel. The creep curves of the "soilized" sand satisfy the Kelvin creep equation $\varepsilon(t) = \sigma_0/E(1 - e^{-Et/\eta})$, where σ_0 is the constant applied stress, η is the viscosity, and E is the elastic modulus. When the water content in the "soilized" sand decreases below ω_P, the ODI constraint among the granules gradually transitions to the "fixed constraint" as the water evaporates. As the water content approaches 0, η also approaches 0 and E reaches a constant value; thus, the above equation becomes $\varepsilon = \sigma_0/E$, (a linear elastic stress-strain relation). Hence, the ODI constraint is transformed into the fixed constraint.

7.7.3 *Final words: granular system is a complex fluid*

The study of granular systems is still in its early developing stage. Some of the studies that are included in this introduction may not be mature, even while they are worthy of spreading word on. An open and critical mind is encouraged in order to learn and know the progress of the subject.

To understand the nature of granular systems, as some final words of the chapter and of the book, it is interesting to mention a work by Yujie Wang's group [7.109]. They found that granular materials flow more like complex fluids than liquid or solid systems. They used X-ray tomography to determine the microscale relaxation dynamics of hard granular ellipsoids subject to an oscillatory shear.

Kou *et al.* [7.109] measured and plotted the probability density functions of particle displacements of the granular ellipsoids in the shear (y) direction (d_y) as a function of $d_y/\langle d_y^2\rangle^{1/2}$ for different values of γ. The finding that the data are well described by a Gumbel law [7.110] (Eq. (7.72), with $\lambda = 0.605$ and $A(\lambda) = 1.313$) is also shown for each γ. The same quantitative behavior as shown here for the y direction is found for the x and z directions.

$$f(d_y) = A(\lambda)\exp\left[-\frac{|d_y|}{\lambda} - \exp\left(-\frac{|d_y|}{\lambda}\right)\right]. \tag{7.72}$$

Here λ, strain amplitude, is a length scale, d_y is the displacement of the particle in the y direction and $A(\lambda)$ is a normalization constant. The Gumbel law is similar to a Gaussian distribution for small displacements but has a heavier tail for larger displacements, and possesses a shape parameter that is independent of the amplitude of the shear strain and of the time.

Binder and Kob [7.111] discussed the dynamical observables of simple liquids or network forming liquids. One of the simplest time correlation functions is the mean squared displacement (MSD) $\langle \Delta r^2(t)\rangle = \langle |\boldsymbol{r}_i(t) - \boldsymbol{r}_i(0)|^2\rangle$ of a tagged particle. In the case of low temperatures, the time correlation function MSD is complex, and shows three regimes: apart from the ballistic regime at very short times and diffusive regime at long times, there is a "cage regime" at intermediate times at which the MSD is basically constant, namely it shows a crossover to a plateau, a behavior found in a crystal for which the long time motion of the particles is of vibrational nature, i.e. the particles oscillate around their equilibrium position and there is no relaxation. The granular system does not exhibit the "cage effect", and this proves that it is not a simple liquid.

This Gumbel law can be interpreted as the consequence of the interplay between two relaxation mechanisms: diffusion on small length scales induced by the roughness of the particles and by friction, and irreversible relaxation events on larger scales [7.109].

References

[7.1] S. Luding, Theory and simulation of hard granular media, Workshop of Simulation and Theory of Granular Media, (2003), Beijing.

[7.2] P. G. de Gennes, *Rev. Mod. Phys.* **71** (1999) S374.

[7.3] J. Duran, *Sands, Powders, and Grains,* Springer (1997).

[7.4] H. M. Jaeger, S. R. Nagel, R. P. Behringer, *Rev. Mod. Phys.* **68** (1996), 1259.

[7.5] H. A. Janssen, Z. Vereins, *Dtsch. Eng.* **39** (1895), 1045

[7.6] L. Vanel, E. Clement, *Eur. Phys J. B.* **11** (1999), 525.

[7.7] P. G. Gennes, Physica A **261** (1998), 267

[7.8] X. Jia, C. Caroli, Phy. Rev. Lett **82** (1999), 1863

[7.9] P. Evesque and J. Rajchenbach, *Phys. Rev. Lett.* **62** (1989), 44.

[7.10] H. K. Pak, E. Van Doorn, and R. P. Behringer, *Phys. Rev. Lett.* **74** (1995), 4643.

[7.11] J. B. Knight, *Phys. Rev. E* **55** (1997), 6016.

[7.12] E. L. Grossman, *Phys. Rev. E* **56** (1997), 3290.

[7.13] K. M. Aoki *et al*, *Phys. Rev. E* **54** (1996), 874.

[7.14] K. Liffman, G. Metcalfe, and P. Cleary, *Phys. Rev. Lett.* **79** (1997), 4574.

[7.15] F. Melo, P. B. Umbanhowar and H. L. Swinney, *Phys. Rev. Lett.* **75** (1995), 3838; *Nature* **382** (1996), 793.

[7.16] K. Kim and H. K. Pak, *Phys. Rev. Lett.* **88** (2002), 204303.

[7.17] E. Serre, P. Bontoux and R. Kotarba, *Int. J. Fluid Dyn.* **5** (2001) Article 2, 17.

[7.18] H. W. Mueller, *Phys. Rev. E* **49** (1994), 1273.

[7.19] J. B. Swift and P. C. Hohenberg, *Phys. Rev. A* **15** (1977), 319.

[7.20] C. J. Chapman and M. R. E. Proctor, *J. Fluid Mech.* **101** (1980), 759.

[7.21] M. C. Cross, P. C. Hohenberg, *Rev. Mod. Phys.* **65** (1993), 851.

[7.22] S. K. Song, Pattern formation review, Soft Matter Physics Lecture paper, Fudan University, (2011).

[7.23] H. N. Zhu, 2D Pattern formation of granular material, Soft Matter Physics Lecture paper, Fudan University, (2011).

[7.24] C. Bizon, M. D. Shattuck, J. B. Swift *et al.*, *Phys. Rev. Lett.* **80** (1998), 57.

[7.25] D. C. Rapaport, *J. Comput. Phys.* **34** (1980), 184.

[7.26] D. C. Rapaport, *The Art of Molecular Dynamics Simulation,* Cambridge University Press, Cambridge (1995).

[7.27] S. McNamara and W. R. Young, *Phys. Rev. E* **53**, 5089 (1996).

[7.28] S. Miller, S. Luding, Event-Driven Algorithm for Granular Media, Diploma Thesis, http://www.icp.uni-stuttgart.de/Jahresberichte/01/node32.html

[7.29] X.-L. Wu et. al, *Phys. Rev. Lett.* **71** (1996), 1363; T. L. Pennec, K. J. Måløy et. al, *Phys. Rev. E* **53** (1996), 2257.

[7.30] S. Hørlück, P. Dimon, *Phys. Rev. E* **63** (2001), 31301.

[7.31] L. Bocquet, W. Losert *et al.*, *Phys. Rev. E* **65** (2001), 11307.

[7.32] W. Chen *et al.*, *Phys. Rev. E* **64** (2001), 61305.

[7.33] W. Chen, A Lecture on the National Conference of Soft-matters.

[7.34] P. Bak, *How Nature Works: The Science of Self-organized Criticality.* New York: Copernicus (1996).

[7.35] P. Bak, C. Tang and K. Wiesenfeld, *Phys. Rev. Lett.* 59 (1987), 381.

[7.36] K. Christensen, N. R. Moloney, Complexity and Criticality, Fudan University Press, Shanghai, November 2006, 250.

[7.37] Per Bak, Chao Tang, Kurk Wiesenfeld, *Phys. Rew. A* **38**, 364 (1988). Chen Zhong, Sheng Yihua, *The Science of Modern Systematic Science*, Shanghai Science and Technology Literature Press, Shanghai (2005), 444 (in Chinese).

[7.38] P. Bak, C. Tang, K. Wiesenfeld, *Phys. Rew. A* **38** (1988), 364.

[7.39] Shi-xiong Li, Ling-kan Yao, Liang-wei Jiang, J. Southwest Jiaotong Univ. **39** (2004), 366 (in Chinese).

[7.40] H. A. Makes, *Phys. Rev. Lett.* **78** (1997), 3298.

[7.41] J. M. N. T. Gray and K. Hutter, *Continuum Mech. Thermodyn.* **9** (1997), 341.

[7.42] G. H. Ristow, *Pattern Formation in Granular Materials*, Springer-Verlag, Berlin, (2000), 115.

[7.43] Y. Oyama, *Bull Inst. Phys. Chem. Res. (Tokyo), Rep.* **18** (1939), 6001.

[7.44] J. Duran, *Sand, Powder, and Gains, An Introduction to the Physics of Granular Materials*, Springer-Verlag, New York, (2000).

[7.45] G. H. Ristow, *Pattern Formations in Granular Materials,* Ch. 7, Springer-Verlag, New York, (2000).

[7.46] M. Tirumkudulu, A. Tripathi, A. Acrivos, *Phys. Fluids,* **11** (1999), 507; *Phys. Fluids,* **11** (1999), 1962.

[7.47] J. Duran, *Sand, Powder, and Gains, An Introduction to the Physics of Granular Materials*, Springer-Verlag, New York, (2000), 182–183.

[7.48] T. Rosato *et al., Phys. Rev. Lett.* **58** (1987), 1038.

[7.49] W. Reisner and E. Rothe, in *Bins and Bunkers for Handling Bulk Materials, Rock and Soil Mechanics* (Trans. Tech., Clausthal- Zellerfeld, West Germany (1971).

[7.50] Z. T. Chowhan, *Pharm. Technol.* **19** (1995), 56.

[7.51] S. Luding*et al., Pharm. Technol.* **20** (1996), 42.

[7.52] J. Duran, *Sand, Powder, and Gains, An Introduction to the Physics of Granular Materials*, Springer-Verlag, New York (2000), 156.

[7.53] J. Duran *et al.*, *Phys. Rev. Lett.* **70** (1993), 2431.

[7.54] Chengliang Qian's BS thesis: A brief theoretical review of granular materials, Fudan University, June (2003).

[7.55] L. Vanel, D. Howell *et al.*, *Phys. Rev. E* **60** (1999), R5040.
[7.56] J. F. Geng and P. R. Behringer, *Phys. Rev. E* **64** (2001), 060301.
[7.57] C. H. Liu, S. R. Nagel, D. A. Schecter, S. N. Coppersmith, S. Majumdar, O. Narayan and T. A. Witten, *Science* **269** (1995), 513.
[7.58] Chu-heng Liu and Sidney R. Nagel, Sound in a granular material: Disorder and nonlinearity, *Phys. Rev. B* 48 (1993), 15646; Sound in sand, *Phys. Rev. Lett.* **68** (1992), 2301.
[7.59] J. F. Geng, D. Howell, E. Longhi, E. Longhi, G. Reydellet, L. Vanel, E. Clément and S. Luding, *Phys. Rev. Lett.* **87** (2001), 035506.
[7.60] G. Reydellet and E. Clement, *Phys. Rev. Lett.* **86** (2001), 3308.
[7.61] V. M. Kenkre, J. E. Scott, E. A. Pease and A. J. Hurd, *Phys. Rev. E* **57**(1998), 5841.
[7.62] From a lecture of Workshop of Simulation and Theory of Granular Media, (2003), Beijing.
[7.63] S.F. Edwards, in *Granular Matter, An Interdisciplinary Approach*, edited by Anita Mehta (Springer-Verlag, New York, 1994), pp. 121–140.
[7.64] J. B. Knight, C. G. Fandrich, C. N. Lau, H. M. Jaeger and S. R. Nagel, *Phys. Rev. E* **51** (1995), 3957.
[7.65] E. R. Nowak, J. B. Knight, E. Ben-Naim, H. M. Jaeger and S. R. Nagel, *Phys. Rev. E* **57** (1998), 1971.
[7.66] S. F. Edwards and R. B. S. Oakeshott, *Physica A* **157** (1989), 1080.
[7.67] H. A. Makse, J. Brujic, and S. F. Edwards, Statistical mechanics of jammed matter, in *The Physics of Granular Media*, eds. H. Hinrichsen and D. E. Wolf, Wiley-VCH (2004), pp. 45–85.
[7.68] S. F. Edwards and D. V. Grinev, *Phys. Rev. E* **58**, 4758 (1998).
[7.69] S. F. Edwards, J. Brujic, H. A. Makse, arXiv:cond-mat/0503057 v**12** Mar 2005.
[7.70] H. Hinrichsen , D. E. Wolf , *The physics of granular media*, Wiley-VCH, Weinheim, Great Britain, (2004).
[7.71] J. Eggers, *Phys. Rev. Lett.* **83** (1999), 5322.
[7.72] D. van der Meer, K. van der Weele, P. Reimann and D. Lohse, *J. Stat. Mech.* **7** (2007), 07021.
[7.73] K. van der Weele, D. van der Meer, M. Versluis and D. Lohse, *Europhys. Lett.* **53** (2001), 328.
[7.74] J. J. Brey, M. J. Ruiz-Montero and F. Moreno, *Phys. Rev. E* **63** (2001) 061305, and the references cited in.
[7.75] K.-C. Chen, C.-C. Li, C.-H. Lin, L.-M. Ju, and C.-S. Yeh, *J. Phys. Soc. Jpn.* **77**, 084403 (2008).
[7.76] M. Y. Hou, H.E. Tu, R. Liu, Y. C. Li, and K. Q. Lu, *Phys. Rev. Lett.* **100** (2008), 068001; M. Y. Hou, P. Evesque, *Res. Signpost Phys. Sci.* **3** (2008), 1.
[7.77] D. van der Meer, K. van der Weele and P. Reimann, *Phys. Rev. E* **73** (2006), 061304.
[7.78] Z. Fournier, D. Geromichalos, S. Herminghaus, M. M. Kohonen, F. Mugele, M. Scheel, M. Schulz, B. Schulz, Ch. Schier, R. Seemann and A. Skudelny, *J. Phys.: Cond. Matter* **17** (2005), S477.

[7.79] L. P. Kadanoff, *Rew. Mod. Phys.* **71** (1999), 436.
[7.80] Kunquan Lu, Zexian Cao, Meiying Hou, Zehui Jiang, Rong Shen, Qiang Wang, Gang Sun and Jixing Liu, The mechanism of earthquake, *Inter. J. Mod. Phys. B* **32** (2018), 1850080. (Refs. [7.83]–[7.91] are re-cited from this paper.)
[7.81] Kunquan Lu, Meiying Hou, Zehui Jiang, Qiang Wang, Gang Sun and Jixing Liu, Understanding earthquake from the granular physics point of view – causes of earthquake, earthquake precursors and predictions, *Inter. J. Mod. Phys. B* **32** (2018), 1850081.
[7.82] Lu KunQuan, Hou MeiYing, Wang Qiang, Peng Zheng, Sun Wei, Sun XiaoMing, Wang YuYing, Tong XiaoHui, Jiang ZeHui & Liu JiXing, *Chin. Sci. Bull.* **56** (2011), 1071.
[7.83] H. F. Reid, The Mechanics of the Earthquake, The California Earthquake of April 18, 1906, *Report of the State Investigation Commission*, Vol. 2, Carnegie Institution of Washington, Washington, D. C. (1910).
[7.84] F. V. De Blasio, *Introduction to the Physics of Landslides: Lecture Notes on the Dynamics of Mass Wasting*, Chap. 2, Springer Science+Business Media B. V. (2011).
[7.85] H. A. Barnes, *J. Non-Newton. Fluid Mech.* **81** (1999), 133.
[7.86] V. F. Weisskopf, *Am. J. Phys.* **54** (1986), 110.
[7.87] Kaihua Zhao, *Qualitative and Half-quantitative Physics*, High Education Publishing, Beijing, (2008).
[7.88] D. L. Kohlstedt, B. Evans and S. J. Mackwell, *J. Geophys. Res.* **100** (1995), 17587.
[7.89] R. P. Wintsch and M. W. Yeh, *Tectonophys.* **587** (2013), 46.
[7.90] W. M. Behr and J. P. Platt, *Earth Planeta. Sci. Lett.* **303** (2011), 181.
[7.91] G. M. Zhang, Z. X. Fu and X. T. Gui, *Introduction to Earthquake Prediction* (Science Press, Beijing, (2001) (in Chinese).
[7.92] F. Havel, Dissertation for doktor ingeniør, *Creep in soft soils*, Norwegian University of Science and Technology, Trondheim, 2004. https://brage.bibsys.no/xmlui/bitstream/handle/11250/231200/124915_FULLTEXT01.pdf?sequence=1.
[7.93] https://zhidao.baidu.com/question/462646505.html (in Chinese).
[7.94] G. R. McDowell, J. J. Khan, *Gran. Matt.* **5**, 115 (2003). (Refs. 7.95–7.98 are cited from this paper.)
[7.95] P. V. Lade and C.-T. Liu, *J. Engin. Mech.* (1998) 912.
[7.96] G.R. McDowell & M.D. Bolton, *Géotehnique* **48** (1998), 667. Note there is a discussion on the above paper: G.R. McDowell & M.D. Bolton, *Géotechnique* **50** (2000), 315.
[7.97] M. F. Ashby & D. R. H. Jones, *Engineering Materials 2 — An Introduction to Microstructures, Processing and Design.* Oxford: Pergamon Press, (1986).
[7.98] R. W. Davidge, *Mechanical Behaviour of Ceramics.* Cambridge: Cambridge University Press (1979).
[7.99] G. R. McDowell, (2005). *Géotechnique* **55** (2005), 697.
[7.100] J. de Bono, G. R. McDowell, *Géotechnique* **68** (2018)., No. 5, 451

[7.101] B. M. Das, *Fundamentals of Geotechnical Engineering*, Brooks/Cole, Thomson Learning$_{\rm TM}$, printed in USA, 2000, cited from [7.92]. Recent edition of this book by B. M. Das and N. Sivakugan is published by Cengage Learning, Boston, (2016).

[7.102] N. N. Maslov, *Problems in Geotechnical Investigations*, Svirstroy Publications, Leningrad, Vol. IV, (1935). p. 177 (cited from [7.92]).

[7.103] R. F. Craig, *Soils Mechanics*, 5th ed. Chapman & Hall, Printed in Great Britain by T. J. Press, ISBN 0 412 39590 8, pp. 427 (cited from [7.92]).

[7.104] Zhijian Yi, Chaohua Zhao, *Engineering* **2** (2016), 270.

[7.105] Zhijian Yi, Chinese Patent CN103348831A, *Modified sand body*, applied on 06.06.2013, disclosed on 16.10.2013.

[7.106] Kaihan Chen, Xiaowen Li, Chinese Patent CN103650692A, *Sand modification method suitable for high-wind-erosion and high-water-erosion areas*, applied on 29.11.2013, disclosed on 26.03.2014.

[7.107] Z. J. Yi, C. H. Zhao, J. Y. Gu *et al.*, *Sci. China-Phys. Mech. Astron.* **59** (2016), 104621.

[7.108] M. Anitescu, Hard Constraint Methods for Multi Rigid Body Dynamics with Contact and Friction, Argonne National Laboratory, Banff, June 8, 2005. http://www.mcs.anl.gov/~anitescu/Presentations/2005/mihaiBanffJune05.pdf, p. 5.

[7.109] B. Q. Kou, Y. X. Cao, J. D. Li, C. J. Xia, Z. F. Li, H. P. Dong, A. Zhang, J. Zhang, W. Kob, Y. J. Wang, Granular materials flow like complex fluids, *Nature* **551** (2017), 360–372.

[7.110] S. T. Bramwell, *Nature Phys.* **5** (2009), 444.

[7.111] K. Binder and W. Kob, *Glassy Materials and Disordered Solids, An Introduction to Their Statistical Mechanics*, World Scientific, Singapore, (2005), pp. 385–386, ISBN 981-256-510-8.